Desertification and Development:
Dryland Ecology in Social Perspective

UNESCO in the framework of its Man and the Biosphere (MAB) Programme has assisted in the distribution of this book.

Desertification and Development:
Dryland Ecology in Social Perspective

Edited by

Brian Spooner
Department of Anthropology
University of Pennsylvania
Philadelphia, U.S.A.

H. S. Mann
Central Arid Zone Research
Institute, Jodhpur,
India

1982

ACADEMIC PRESS
A Subsidiary of Harcourt Brace Jovanovich, Publisher
London   New York
Paris   San Diego   San Francisco   São Paulo
Sydney   Tokyo   Toronto

ACADEMIC PRESS INC. (LONDON) LTD.
24–28 Oval Road
London NW1 7DX

U.S. Edition published by
ACADEMIC PRESS INC.
111 Fifth Avenue
New York, New York 10003

British Library Cataloguing in Publication Data

Desertification and development.
1. Desertification   2. Ecology
I. Spooner, B.   II. Mann, H. S.
304.2′5 (expanded)   GB611

ISBN 0-12-658050-2

LCCCN 82-71004

Filmset in 'Monophoto' Baskerville by Eta Services (Typesetters) Ltd., Beccles, Suffolk
Printed in Great Britain by Galliard (Printers) Ltd, Great Yarmouth

Contributors

Rasik Bhadresa: University of London King's College, 68 Half Moon Lane, London SE24 9JF, U.K.

L. P. Bharara: Central Arid Zone Research Institute, Jodhpur (Raj.), India 342003.

Siegmar-W. Breckle: University of Bielefeld, Fakultaet fuer Biologie, Postfach 8640, D-4800 Bielefeld, F.R.G.

Francis P. Conant: Hunter College of the City University of New York, 695 Park Avenue, New York, NY 10021, U.S.A.

R. W. Dennell: University of Sheffield, Department of Prehistory and Archaeology, Sheffield S10 2TN, U.K.

R. P. Dhir: Central Arid Zone Research Institute, Jodhpur (Raj.), India 342003.

Lee Horne: University of Pennsylvania, Department of Anthropology, Philadelphia, PA 19104, U.S.A.

N. S. Jodha: International Crop Research Institute for the Semi-Arid Tropics, Patancheru P.O., Andhra Pradesh 502324, India.

H. S. Mann: Central Arid Zone Research Institute, Jodhpur (Raj.), India 342003.

S. P. Malhotra: Central Arid Zone Research Institute, Jodhpur (Raj.), India 342003.

Mary A. Martin: Washington University, St. Louis, MO 63130, U.S.A.

Douglas J. Merrey: University of Pennsylvania, Department of Anthropology, Philadelphia, PA 19104, U.S.A.

P. D. Moore: University of London King's College, 68 Half Moon Lane, London SE24 9JF, U.K.

A. Endre Nyerges: University of Pennsylvania, Department of Anthropology, Philadelphia, PA 19104, U.S.A.

Stephen Sandford: Overseas Development Institute, 10–11 Percy Street, London W1P OJB, U.K.

Ann Schulz: Department of Government, Clark University, Worcester, Massachusetts, 01610, U.S.A.

Brian Spooner: University of Pennsylvania, Department of Anthropology, Philadelphia, PA 19104, U.S.A.

A. C. Stevenson: University of London King's College, 68 Half Moon Lane, London SE24 9JF, U.K.

M. S. Swaminathan: The International Rice Research Institute, Los Baños, Laguna, Philipines.

Helmi R. Tadros: American University of Cairo, Social Research Center, 113 Sharia Kasrg el Aini, Cairo, Egypt.

Foreword

Dr Brian Spooner and Dr H. S. Mann are to be congratulated on bringing together this useful volume on "Desertification and Development: Dryland Ecology in Social Perspective". This book attempts something which has not really been done so far by choosing to focus on the human factor as the pivotal force in all aspects relating to desertification. We know less about human behaviour and perception than about the natural processes causing either the destruction or diminution of the biological potential of land. Consequently, we are less capable of organizing human activity towards achieving planned objectives than we are of engineering physical and biological processes. The book is, therefore, a valuable companion to the recommendations of the U.N. Conference on Desertification (UNCOD) held in Nairobi in 1977.

At UNCOD a World Plan of Action to Combat Desertification (PACD) was developed based on the following premises and programme thrusts:

(a) The level of existing knowledge allows for immediate action; hence, priority should be given to applying the available scientific information and technologies.

(b) Combat of desertification needs to be an integral part of national plans for the development of land resources in areas prone to desertification; and

(c) Urgent short-term relief measures, although important, should not be considered as a substitute for long-term programmes with preventive measures.

While setting as its goal the worldwide halt of desertification by the year 2000, PACD identified a seven-year (1978–1984) programme for immediate implementation. It was stressed that the implementation of the Plan was to be carried out by national governments with support, where necessary, of international and/or bilateral assistance.

Dr M. K. Tolba, Executive Director of the U.N. Environment Programme has recently identified the following constraints which have retarded the effective implementation of PACD:

(a) National governments in many desertification-prone developing countries have not included desertification control programmes within their development plan nor within their international aid schemes.

(b) As desertification hazards do not recognize political boundaries, their control requires cooperative and joint projects at a regional level. Political constraints have impeded progress in implementing regional projects, including those covered by the six studies considered and endorsed by UNCOD.

vii

(c) Programmes of the U.N. agencies and bodies have yet to be fully coordinated and interrelated for maximum impact and effectiveness.

(d) There is an urgent need to develop local scientific and technical capabilities at the national and regional levels. This would be an effective means of assimilating and adapting available scientific knowledge and technological means and applying them in the field.

(e) There remain gaps in scientific knowledge that need to be filled.

(f) Financial resources allotted to desertification control activities within international aid schemes (grant, assistance, loans, etc.) are too little and the total available resources are far short of what is required.

The resources available to combat the menace of desertification are in no way commensurate with the extent and complexity of the problems faced. The prospects for concerted and coordinated action are dim and there does not seem to be any indication that the governments of the world are willing to take the problems seriously. They are discouraged by the figure of additional resources needed (U.S. $1.8 billion per year for 20 years), while they ignore the cost of productivity lost every year due to desertification (close on $25 billion).

The present compilation by Spooner and Mann assumes significance in the context of both what has happened and what has not happened since 1977. The book provides remedies to some of the major maladies identified by Dr Tolba. It calls attention to the global need for protecting the productivity of one third of its surface prone to desertification and to generating an appropriate blend of political will, professional skill and people's action. The dimensions of eco-catastrophes awaiting many countries characterized by a relentless growth in demographic pressures are not widely appreciated. For example, the additional population of 15 million now being added each year in India alone is equivalent to the estimated total global population about 12 000 years ago when agriculture or settled cultivation had its birth. The papers contained in Chapters 15 and 16 of this book show how India provides examples of the greatest potential and the greatest vulnerability.

The book is appropriately divided into two major parts, one dealing with a global perspective and the other with regional programmes. The importance of integrated studies has been brought out clearly in every chapter. Also, the need for a careful action-reaction analysis is evident from the several examples given. In Chapter 6, a case is cited where removal of the cause of grazing led to other disadvantageous changes in vegetation. This is why a systems approach to the study of such complex problems dealing with interaction between man and his environment is essential.

I am glad that in Chapter 11 it has been stressed that planning to improve pastoral productivity and range quality should start from an understanding not only of grazing pressures but also of patterns of co-adaptation and interdependence. Thus, animal grazing behaviour and plant toxicity balance each other in such a way that the disturbance of either is likely to have an adverse effect on animal productivity or plant palatability. This is a good example of the underlying refrain of the book, namely the need for positive approaches that seek to build on existing potential rather than negative orientations that assume the need for a fresh start.

In addition to the various chapters which are all written in a scholarly and readable manner, the Preface and Introduction greatly add to the value of the book. The introductory chapter written by Spooner argues that human activities immediately or directly related to desertification processes can be understood and therefore can be realistically treated only in terms of a much larger economic and political context which determines the direction of change. This theme is carried through the entire volume and related to each section and each chapter in brief editorial introductions.

I am glad the book is being published in 1982 which marks the 10th anniversary of the U.N. Conference on the Human Environment held in Stockholm on June 1972. The intention of the Stockholm Conference was "to safeguard and enhance the human environment for present and future generations of man". Unfortunately, however, most human activities fall under the category "Every new source from which man has increased his power on earth has been used to diminish the prospects of his successors. All his progress has been made at the expense of damage to the environment which he cannot repair and could not foresee".

Dr Spooner and Dr Mann as well as the various authors of this book deserve our gratitude for their excellent contribution to the cause of protecting the quality of life on our Spaceship Earth. The Academic Press is to be congratulated for sponsoring the publication of this outstanding contribution to current knowledge on the linkages and feed-back relationships existing between desertification and development.

M. S. Swaminathan
Director General
International Rice Research Institute
Independent Chairman, FAO Council
Los Baños, Laguna, Philippines

Preface

The Need for a Social Perspective

The human tragedy that followed the recent Sahelian drought drew attention to a more general phenomenon which has become known as desertification. An exact and generally acceptable definition of desertification is elusive, but it is recognized in the appearance or intensification of desert conditions in arid, semi-arid and sub-humid regions—the world's drylands—where the long-term increase in human activity has suddenly accelerated in recent history as a result of population growth and economic development.

In December 1974 the United Nations General Assembly called for the organization of a United Nations Conference on Desertification (UNCOD), and established a secretariat within the United Nations Environment Programme (UNEP) for that purpose. The Secretariat was given the mandate to prepare for the Conference a synthesis of existing knowledge on dryland ecology—with special attention to the human factor—and a plan of action for a world-wide campaign to combat desertification. The Conference was duly convened in Nairobi from August 29th to September 9th, 1977. The Executive Director of UNEP, Dr Moustafa Tolba, presided as Secretary-General, and Dr M. S. Swaminathan, the Director-General of the Indian Council of Agricultural Research (ICAR) and Secretary to the Government of India, chaired the Committee of the Whole which prepared the final draft of the Plan of Action. As approved on the final day of the Conference this Plan of Action was intended to harness the financial and administrative resources of the world at the international and regional levels, and to encourage national governments to rethink socio-economic policies and development strategies, in order to protect and reclaim the potential productivity of the 30% of the land surface of the world (45 million km² distributed through two-thirds of the world's 150 nations) that is dry and therefore judged to be already suffering from, threatened by, or vulnerable to, desertification.

Desertification was diagnosed and defined by ecologists, but it involves human factors that are beyond the limits of ecological explanation. Although in their analysis of desertification processes some scientists have focused on geomorphological and climatological factors, most have at least implied the primacy of human activities in their explanations. For a while desertization and desertification were by some distinguished in order to avoid confusion of

xi

what was in some cases assumed to be a natural process with what was obviously the result of human activities. By the time of UNCOD, however, all attention was focused on desertification as an ecological process resulting from human activity. But despite the conscious role allotted to human activity in the process, in the campaign to combat it no role was allowed to scientists specializing in the explanation of human behaviour—social scientists.

Despite the relative scarcity of social scientists interested in ecological problems, in the academic and professional investigations of the conditions and processes of desertification so far a consistent theme has been generated by the difference in orientation between the social and natural sciences. In both the social and the natural sciences each discipline tends to give priority to those factors that constitute its own special interest, and then develop them in an independent explanatory scheme. This difference between disciplines is at its greatest between the social and the natural sciences. At the same time demand is steadily growing on all sides for a greater contribution from the social sciences. But fulfilment of this demand is hindered by difference of opinion about the role of that contribution. Anthropology, since it comprehends a range of disciplinary interests that span the boundaries between the natural and social sciences, and since of all the social science disciplines it has the most experience with the type of population most commonly affected by desertification, has a special part to play in this arena.

When it was decided to convene the tenth quinquennial Congress of the International Union of Anthropological and Ethnological Sciences (Xth ICAES) in New Delhi in December, 1978, on the theme of Anthropology and the Challenges of Development, the ideal opportunity was created for developing the social perspective on desertification by investigating both the social dimension of desertification and the social science role in combating it. India, and particularly ICAR, and its Central Arid Zone Research Institute (CAZRI), were already playing an important role in the anti-desertification campaign. India had hosted the ESCAP Regional Meeting on Desertification in April 1977. One of the six case studies developed for UNCOD in order to demonstrate the dangers of desertification, and the benefits of specialist experience and existing technologies, had been commissioned from CAZRI. CAZRI became the executing agency for the Government of India in the Transnational Monitoring Project for Southwest Asia, one of six transnational projects designed and developed to spearhead the international campaign. Dr H. S. Mann, Director of CAZRI, who was Project Director for the case study and Indian National Representative for the Transnational Project, had contributed to many of the projects that have laid the groundwork for the international effort. Especially important among those projects have been the meetings and studies in this subject area which were sponsored by the International Geographical Union's Working Group on Desertification, the

International Science Associations' Nairobi Seminar on Desertification, the United Nations Environment Programme, UNESCO's Programme on Man and the Biosphere, and the United Nations University. During the year following UNCOD, CAZRI has added a stimulus to the campaign by celebrating its Silver Jubilee with an international conference on the state of arid lands research (February 1978). Most recently, in October 1981, CAZRI was the host institution for the first follow-up conference for Asia and Australia (organized by ESCAP, the Economic and Social Commission for Asia and the Pacific).

India shares with her neighbours to the north west her experience of the conditions that predispose to desertification. In the Transnational Project (which has since lapsed in consequence of political upheaval in two of the participating countries) India worked with Afghanistan, Iran and Pakistan to capitalize on that shared experience. Each of these three neighbouring countries of Southwest Asia has also contributed to the international effort. Among them Iran was especially conscious of desertification as a national problem, and had instituted a dryland ecology research programme under the Department of the Environment (Tehran) in 1975 (continued under the Ministry of Agriculture 1977–1979), in an area known as Turan on the northeastern margin of the central deserts. Brian Spooner was principal investigator of the Turan Programme, and in that capacity became involved in a number of related international projects.

Between 1976 and 1978 the Turan Programme and CAZRI were closely associated in the Transnational Project for Southwest Asia. CAZRI is a large research institution with a comprehensive programme in which all the scientific disciplines relevant to the use of drylands are represented and Human Factor Studies have been an important component from the beginning. The Turan Programme was relatively small and short-lived but designed from its inception with social science (specifically, anthropology) in a coordinating role. Given the common interests, approaches and aims of these two research teams, the combination of their experience and strengths might be expected to produce useful advances in our understanding of desertification processes and in our ability to apply existing knowledge to combat them.

The Xth ICAES provided the framework for this combination. The research associates of the Turan Programme and the research staff of CAZRI put together a post-plenary session on desertification. Thanks to the personal interest of Dr Swaminathan, the backing of ICAR and the financial support of the Smithsonian Institution, the editors of this volume were able to put together a symposium programme that represented fairly accurately the complementarity of their interests and illustrated some of the rewards of the international and interdisciplinary dialogue in which they had been involved.

As conveners of the symposium, the editors gratefully acknowledge the recognition of the President of the Xth ICAES, Dr L. P. Vidyarthi, who granted post-plenary status to our meeting. The attachment of the symposium to the programme of the International Congress in Delhi enabled us to attract additional participants with relevant field experience in other parts of the world. We thank Mr S. P. Malhotra, Head of the Division of Economics and Sociology at CAZRI, not only for organizing the ICAES relationship, but also, by tireless attention to innumerable details in his capacity as Local Organizing Secretary, for making the meeting such an enjoyable experience for all participants. Finally, it would be remiss of us not to remember here the inaugural address of Dr Ranbir Singh, Vice Chancellor of the University of Udaipur, which set the tone for the meeting by introducing its social, historical and administrative themes.

The resulting papers and discussions, which were summarized in a report (Spooner and Mann, 1979), while not representative of the total range of work on the subject constituted a significant step towards the solution of the problem of integrating the various disciplinary views and harnessing the potential interdisciplinary contribution to the international campaign to combat desertification. We hope this volume will be seen as a further step. It has been developed from a selection of the original papers, all of which have been substantially reworked on the basis of interaction in the symposium and the later scientific dialogue that it generated. Our aim has been to present a unitary statement on the problem of desertification that would redress the balance of the existing literature by directing attention to some of the most important social and sociological aspects. To non-social scientists it may appear that we have overweighted the social side of the scales. In reply we would claim that the usual limitations of space prevent us from restating in any detail what is already known and accepted. The few chapters that have little or no explicit social dimension are included because they are based on research that was developed collaboratively with social scientists and deal with aspects of the desertification process that have received relatively little attention elsewhere. We should like to avoid or counteract any polarization of social science and natural science approaches and emphasize rather a range of approaches from positivist or materialist to relativist and idealist; we have sought to emphasize approaches that we believe have been and remain under-represented—with unfortunate consequences—in this campaign so far. The social arguments that characterize the greater part of the book are intended not to contradict existing interpretations of natural processes, but to complement them by illustrating some of the parts played in human activities. Finally, we should like to draw attention to the fact that human activities are explored here at two levels: the record of behaviour and the realm of perception and ideas.

The book is organized in sections, and each section and each chapter has editorial introductions. The theme moves from the general to the particular. After Chapter 1 which introduces the social dimension of desertification with special attention to the historical and political context, Part I looks at the problem globally: the three chapters in Part I A each give examples of particular social science disciplinary perspectives which are rarely worked out but demonstrably facilitate important insights. Chapter 2 is written by a political scientist, Chapter 3 by an economist and Chapter 4 by an archaeologist. Part I B contains three case studies of different human situations which show why particular conventional corrective measures do not work. Part II presents integrated samples of two research programmes—from India and Iran—whose association provided the rationale for this enterprise.

August, 1982   *Brian Spooner*
                                                               H. S. Mann

UNESCO in the framework of its Man and Biosphere (MAB) Programme has assisted in the distribution of this book.

Contents

"Society, as much as nature, resists men's plans: it is not wax at the hands of the scientist, the planner, the legislator."

Passmore 1974:100

Introduction

1

Rethinking Desertification: the Social Dimension

Brian Spooner

University of Pennsylvania
Department of Anthropology
Philadelphia, PA 19104, U.S.A.

By investigating the social dimension of desertification we not only deepen our understanding of it—we enlarge the number of factors that we consider in relation to it. In this way we expand the limits of the context in which it is conceptualized—which means that we finish up re-evaluating our conception of it. In making this re-evaluation we become aware of deficiencies in the conceptualization so far, and are able to outline ways in which it should be modified and broadened, in order to comprehend the type of perspectives which are instanced in the body of the volume. —Ed.

Desertification constitutes a serious threat to world food production and to the quality of life and the environment—but in a rather more complex way than is generally represented. The circumstances of the discovery of desertification led to a particular structuring of the campaign to combat it. From the beginning the campaign combined the seeds of conflict in the form of political imbalance. The data that were gathered to further the campaign have served to fuel the conflict. No resolution of the conflict is in sight: in fact it has received little open discussion. Meanwhile, the campaign languishes.

This introductory chapter[1] seeks to disentangle some of the complexity—which is the social dimension—in the desertification debate. It falls into four sections. The first reviews the background to the campaign. The second discusses the organization of related activities and of relevant information, giving special attention to the inherent conflicts of interest that seem to have been responsible for the difficulties encountered in formulating and implementing practical measures to combat desertification. The third looks at the

1

concept of desertification, as it has developed and continues to develop, as a rationalization of the ideas generated in the campaign. The final section outlines the prospect for dealing with desertification in so far as it may affect production and environment at the global level, and formulates an approach to it that may be not only more acceptable politically but more comprehensive scientifically.

I. BACKGROUND

The modern consciousness of ecological degradation can be traced back at least to the nineteenth century; the conviction that degradation cannot be allowed to continue began to appear only in the 1950s when the international development effort got under way and the limits of the world's capacity to produce enough food to satisfy anticipated levels of demand seemed to come within sight. That development effort was based on the assumption, very simply, that increased production and improved health and well-being (and reduced population growth) could be induced by the injection of investment and the transfer of technology into existing low-growth systems of production. In the 1960s, as it became clear that results were falling considerably short of expectations, this assumption began to be questioned. At the same time ecological degradation was diagnosed by specialists as endangering the overall success of the effort. Implicit in this diagnosis was the related assumption that such degradation was not only increasing and intensifying but was caused by the traditional systems that development was designed to transform. The situation was particularly grave in areas subject to either continuous or regular annual water deficit—the world's dry lands.

The Sahelian drought, which lasted from 1968 to 1973 and had particularly severe effects in many areas that had recently been the scene of intensive development efforts, can now be seen to have dealt the deciding blow to the paradigm of development that was based on those assumptions—although at the level of implementation the change has been slow, largely because of the difficulties of operationalizing and mobilizing the concepts that have begun to take their place. The effects of the drought also led to the formal establishment, at the official level, of the term "desertification," when at the end of 1974 the United Nations General Assembly passed a resolution calling for an international conference on desertification (UNCOD), to be held in 1977.

Since the case for a campaign against desertification is *prima facie* ecological and is argued in ecological terms, while the case for development is economic and social and is argued in economic, social and political terms, it is not usually recognized that both spring from different orientations towards the same historical process—the struggle of increasing numbers of people for an optimum distribution of available resources.

The difference in orientation reflects a difference in social background and social identity. Failure to see the intimate relationship between desertification and development is not surprising since the two concepts derive from different orientations towards various aspects of the global problem of production and environment. To put it simply, desertification implies that people must change their ways to fit nature; development implies that the use of resources must be reorganized with the aid of newly available technologies to fit people's expectations. Neither gives sufficient attention to the intractable nature of social or cultural factors. The effects of the Sahelian drought were so startling that for a time the ecological emphasis implied in the concept "desertification" eclipsed the development orientation. A chronological survey of the literature since the early 1970s shows that enormous progress has been made towards resynthesizing the two orientations, but we have still not brought the problem of desertification fully into focus.

The United Nations Organization has played the major organizing role in the desertification–development debate. But it has probably been able to do this only because of the pervasive, but mistaken, sense—especially early on— that desertification is an ecological and not a political problem. As the political dimensions of the problem have become clearer the coherence and cohesion of the United Nations organizing role have dissipated.

The resolution of the United Nations General Assembly specified that "a world map should be developed showing areas vulnerable to desertification, all available information on desertification and its consequences for development should be gathered and assessed, and a plan of action to combat desertification should be prepared with emphasis on the development of indigenous science and technology" (UNCOD, 1977a:2). Between early 1975 and the summer of 1977 the UNCOD Secretariat aided by various United Nations agencies and a number of interested governments, accumulated and synthesized an impressive amount of information on the nature of desertification, the rate of its advance and the seriousness of its effects (although they were not able to do much on indigenous science and technology, since the apparatus for gathering and organizing this information was not available to them, and in fact barely exists), and compiled a draft Plan of Action which if passed by the Conference would provide the basis of corrective action at the national, regional, and international levels. This draft was discussed and modified at five regional preparatory meetings, and unanimous approval of it (in a form significantly modified to accord with political considerations) constituted the culmination of the Conference. So far efforts to implement this Plan of Action have been disappointing.

At the present time it would seem that everyone—all governments—

acknowledge that desertification is a serious global problem, but similar unanimity is lacking on the analysis of its component processes and on how it is related to other problems that are also considered serious. As a result there is significant disagreement about how the problem should be dealt with. The Plan of Action states *what* should be done, but not *how* it should be done. More exactly it states what should be done politically and economically to combat desertification, but it does not (and could not, even if the compilers and the delegates were conscious of them) attack the political and economic conditions that are an integral component of desertification processes. It is symptomatic of this situation that neither the Desertification Unit, set up in Nairobi to co-ordinate the Plan, nor the intergovernmental Consultative Group, which includes representatives of both prospective donor and recipient countries, and which met twice in May 1978 and March 1980 to facilitate the organization and especially the funding of the measures agreed on in the Plan, has produced tangible results. It appears that governments generally do not perceive desertification (as it is presently defined) to be among the most pressing of their problems, whether it is viewed in local or in global perspective.

In the context of this volume these governmental perceptions demand a careful reconsideration of the campaign to combat desertification. The hypothesis that underlies the reconsideration that follows is that desertification must be reconceptualized, if anything is to be gained from its discovery. The original conceptualization—as far as it was ever expressed in a generally accepted definition—was partial and inadequate as a result of stunned reaction to the efficient cause, which was the natural disaster in the Sahel. With the lapse of time a more balanced definition should now be possible, that would take into account the intimate interrelation of desertification and development.

II. THE ORGANIZATION

The desertification debate has been dominated by UNCOD and the bureaucratic processes leading up to it and deriving from it. The nature of the problem can be seen in some of the conflicts that developed at the Conference itself. The delegates were presented with a comprehensive synthesis of existing knowledge (in the form of an Overview and four Component Reviews), examples from twelve countries of experience in specific projects (Case Studies), and six feasibility studies demonstrating practical ways to achieve transnational co-operation (that is, co-operation at the level of ecological and regional rather than national and political units) to combat desertification.[2] On the basis of this material they were told that they should accept existing knowledge as adequate for the immediate purpose of establishing an international Plan to combat desertification. They should, therefore, devote their attention during the Conference to the problem of organizing the successful application of this knowledge. They were told that their task lay in the organization of programmes and resources in order to make possible (in the words of the Plan of Action) "the immediate adaptation and application of existing knowledge". For, "Desertification can be halted and ravaged land reclaimed in terms of what is known now. All that remains is the political will and determination to do it" (UNCOD, 1977a:61). Like all U.N. conferences, therefore, UNCOD was—implicitly at least—a political conference, in that it was concerned primarily with organization and mobilization.

Organization on this scale transcends the province of ecology where desertification was diagnosed. Answers to the problems of ecological management inevitably beg questions of management of the political economy. As often happens in such international forums, the discussions were conducted on two levels. While ostensibly the delegates were discussing means and guidelines for the organization of programmes in which they would co-operate to mobilize resources and combat desertification, many were using the discussion to bargain about relations between the parties to the Conference. Most delegates saw the solution to desertification in the mobilization of resources, but many also blamed the incentives for exploitation of people and resources that they considered to be inherent in the present system, and saw the solution in the reorganization of the world economic order. While all the delegates accepted the ecological explanations of desertification and the technical solutions, many were more concerned with causation at another level: that of the economic and political conditions that generate land use decisions. The organizers of the Conference pursued a strategy designed to keep deliberations at the former level, but the "political will and determination" that they sought to stimulate were more abundant at the latter level—though more difficult to harness.

These two levels of discussion are evident in other arenas of the anti-desertification debate. They are inherent in the political process, and it is unrealistic to hope to keep them entirely separate. The campaign to organize for the purpose of conserving resources can never entirely free itself from the campaign to reorganize the distribution of resources. The consequent dialectic between overt discussion of how to organize in the existing system and the underlying theme of how to reorganize the system is particularly noticeable in two other arenas. Most obviously, it arises in the relations between populations which are at risk or suffering from desertification and the planners and implementers of management programmes designed to combat desertification. But perhaps most significantly, it characterizes the relations between natural scientists concerned with the viability of physical and biological systems and social scientists concerned with the viability of social and cultural systems.

For example, management programmes designed by range scientists to address the long term ecological balance in the relationship between animals and carrying capacity in the arid and semi-arid rangelands of the world are based on values and perceptions different from those of pastoralists. Coming from a different cultural environment and a different social class and trained in different land use systems, the ecologists are led to define the universe of the problem differently and to place a different emphasis in the aims that they pursue in relation to it. The ecologist is primarily concerned with the long term productivity of the resource; the pastoralist is primarily concerned with survival—first in the short term and then in the long term, and survival is for him not only his own personal survival but survival of his socio-cultural unit, which relies upon the productivity of the herds. It is obvious that these concerns overlap—but they are centred on different priorities, which are in turn based on different values. In the interaction between the ecologist and the pastoralist over the implementation of a management programme that would redress the balance in the ecological system of which the pastoral population is a component, the explicit bargaining concerns specific elements of the management programme; implicitly the values of the ecologist are pitted against the values of the pastoralist—a conflict that will be resolved eventually in the larger political process.

In the study of ecosystems where the productivity of natural resources is reduced as the consequence of activities in the human use system that incorporated them, the ecologist's reaction has commonly been to focus on the degraded resource, and attribute the degradation to the immediate cause in the form of the social group exploiting it, as in many cases of traditional pastoralists and degraded rangelands. Desertification then, is caused by social factors, but solutions are generally designed by focusing on the natural symptoms of the problem (for example, reduction in the quantity and quality of vegetation) and by attempting to rearrange the social factors in relation to them. Judging by the record, this approach commonly fails to lead to a satisfactory solution, and besides often brings about new adverse social factors which may accelerate desertification. Cause is translated easily into fault, and central authorities with large urban constituencies are indulged in their prejudices against marginal rural populations. Cultural discrimination of urban against rural increases; the population concerned suffers further reduction in its range of economic options and tends to become an increased burden on its immediate natural resources. The ecologist focuses on natural processes and sees the fault in the behaviour of the human population which failed to reorganize its activities in the way prescribed. The social scientist is invited to devise ways to encourage the people to confine their activities within boundaries prescribed by the ecologist. As with the UNCOD delegates some social scientists are happy to accept these terms of reference and seek to apply their expertise as a service in the ecologists' programme; others seek to redefine the terms of reference: they seek to redefine the situation in terms of the interests of the human population and to develop an ecologically feasible strategy that will serve those interests. In dealing with desertification the social scientist tends to look for the ultimate social cause, which is likely to be outside the affected area.

Where an ecologist is studying an ecosystem without a human component, or where his research interests allow him to ignore the role in the ecosystem of other members of his own species, he implicitly determines relative values for the various species in his study. For example, in an open steppe ecosystem the survival interests of grasses and forbs, shrubs, herbivores and predators are obviously in conflict. The ecologist stands outside the system but bases his research design implicitly on certain inter-related assumptions such as 1) the system should not run down, 2) the number of species should not decrease, and so on. However objective his research design, the ecologist is led by his assumptions to discriminate in favour of the survival of the system. The survival of the system may, of course, be in the long term best interests of all the component species. It is not, however, in the best interests of all living individuals in the system, some of whom will, for example, fall prey to predators. A reduction in the number of predators would, therefore, be in the best interests of some at least of the living herbivores and a reduction in the number of herbivores would be in the best interests of many living plants. If a gazelle or a shrub could produce a study of the same ecosystem, therefore, we might expect their results to differ from those of the ecologist inasmuch as they would, as a matter of course, be based on different assumptions. The ecologist can argue in terms of the survival of species and of the system, because survival on that level suits his own social values best. A member of the system, such as the gazelle, whose personal interests are at stake, would argue for his own survival first. Both arguments may be equally objective and scientific, but differ on grounds of morals and personal interests, which are socially relative. The conflict between them is resolved politically as a function of the difference in power of the populations in question. This reduction of the relationship between the ecologist and his subject matter to questions of morals and politics is exaggerated, and in the case of plants may seem absurd, but by bringing out the fact that the ecologist's argument is based on the attribution of a value to the survival of a species or a system which might conflict with the values of individuals, it serves to focus attention on the moral and political aspects of the problems that develop between scientists and local populations in the treatment of declining ecosystems, especially in the case of desertification. It is worth noting that this problem arises in all the applied sciences—social, biological, and physical. In the physical and biological sciences it has generally been possible to ignore it, but it has effectively paralysed the theoretical development of applied social science.

When the ecologist includes a human population in the system he is studying, his recommendations for treatment and management are likely to conflict with the perceptions and values of that population. These conflicts can be presented as differences between scientific understanding and uneducated superstition and self interest, but it may be more realistic to minimize the difference between science and lore and regard them as differences between socially-derived perceptions based on an interest in the long term survival of a total ecosystem and its usefulness to humanity in general on the one hand and perceptions based on less long term interests in individual economic and group cultural survival on the other.

Pragmatically, there has to be a compromise between such differing perceptions. Such compromise is the essence of any national political process. The ecologist's case is commonly reinforced by the central authority, which often shares his assumptions. However, the government depends on the political process and, ultimately therefore, on the various interest groups which achieve participation in it. The population involved in a process of desertification may or may not have a voice in the political process. In any case, to the degree to which the populations using the resources that are at risk of desertification are represented in the larger political process, they are divided into different interest groups. Any proposal to manage the threatened resources according to the scientist's values unavoidably benefits some interest groups at the expense of others. Since the final arbiter is the political process and relative power (in which the claim to be on the side of absolute values is an important but not a determining factor), it may not in the long term be a realistic ecological management policy to allow the terms of reference for the solution of desertification problems to be defined exclusively according to those supposedly absolute values.

The campaign to combat desertification, as it has been organized—logistically and intellectually—so far, has implied an adversary relationship between those who (consciously) fought it and those who (perhaps unconsciously) caused it. The adversary relationship has often also had the nature of a class conflict. But the question of who stands to gain or lose in particular situations has seldom been carefully investigated. Without such investigation the social dimensions of the concept cannot be understood. Throughout the preparations for the Conference, and since, there has been no general agreement on a definition of desertification—even without the social dimensions (cf. UNCOD, 1977b:108). The concept has evolved and continues to evolve as the progressive rationalization of the logistical structuring of the campaign. The following section discusses this continuing process of rationalization.

III. THE CONCEPT

Desertification does indubitably occur. In popular usage its meaning seems obvious. But determination of the details that should be included or excluded from even a working definition for use in conservation or planning outside particular disciplinary forums has proved difficult.

Desertification is detrimental to settlement, health, and recreation, but most importantly to food production. But what are the chances that it will significantly affect the quantity or quality of world food production in the foreseeable future? Are its effects most likely to be direct, in reduction in acreage or productivity? or indirect, in problems caused by the resulting redistribution of population and increased pressure on distribution networks, in the organization of labour, of investment, or of society generally? Will it be a serious problem at the international level? Finally, what might be done to counteract or, failing that, to reduce its severity. Agreement on answers to these questions depends on more careful investigation of the conceptualization implicit in the campaign.

Despite the enormous outpouring of literature on the subject of desertification that occurred between 1975 and 1978 as direct and indirect results of the Conference, there is still room for another appraisal of the problem— partly because now that four years have elapsed since the Conference some of the dust raised in the logistical, emotional and political flurry of organizing the Conference has settled, and partly because as time has passed more information has arrived and some earlier perspectives on the phenomenon will now bear modification or revision.

Despite considerable efforts on the part of the Secretariat to cost out all the implications of desertification, there has been a tendency all along simply to define desertification as a type of ecological decline which is *per se* evil. This tendency alone may explain the relative political failure of the global campaign to combat it. In general it has not been made entirely clear just who is suffering or stands to lose by specific cases of desertification. The disadvantages of desertification are commonly left as self evident. It is well, therefore, at the outset to make it clear that—at least at the political level— they are self evident only given the assumption that the resources which are thereby degraded are irreplaceable for the population that currently depends on them. Outside certain intellectual circles the argument that future generations will suffer carries little weight. Most governments, because of the exigencies of the individual politician's career and the recurrence of elections, have perspectives of less than five years. Here we are not limited by such a perspective, but in order to understand the concept of desertification it is necessary to understand the variation in perspectives on it and the basis of that variation. This aspect of the problem was only implicit at the Conference and has received little open discussion. The following discussion cannot be comprehensive but attempts to include at least some indication of all the significant questions.

In developing the concept of desertification it is necessary to ask, first: what factors are responsible for productivity and for environmental quality? They are socio-demographic (sc. numbers of people, social structure, spatial distribution and organization of labour), basic needs and their cultural elaboration (tastes, values and expectations), technological repertoire, climate (temperature, wind and precipitation), and the condition or productivity of natural resources (especially soil and vegetation), to which must be added national and international factors of political economy that affect choice of crop, investment patterns, terms of trade and distribution networks. Prediction or planning for environment and development involves predicting or planning for each of these variables. Desertification, as it is usually defined, comprehends only one of these variables—resources (and then only in certain environments)—though it is of course intimately interrelated with all the others. It is possible to see, therefore, that however dangerous the threat of desertification may be other threats may with some justification attract more political attention.

Secondly, what is the case for giving desertification high priority? It is now conventional wisdom that desertification is:

> "the impoverishment of arid, semi-arid and subhumid ecosystems by the combined impact of man's activities and drought. It is the process of change in these ecosystems that can be measured by reduced productivity of desirable plants, alteration in the biomass and the diversity of the micro and macro fauna and flora, accelerated soil deterioration, and increased hazards for human occupancy" (Dregne, 1977:324).

Defined in this way, desertification is generally understood to be affecting large areas of every continent, especially Africa, and to be increasing. Potentially productive but threatened lands have been estimated at 45 million square kilometres or 30% of the land surface of the world, distributed through two thirds of the world's 150 nations. The rate of desertification is said to be increasing and "some experts have suggested that it has reached at least 50 000 km² per year" (UNCOD, 1977a:6–7).

The accompanying map (map 1) shows four classes of desertification: slight, moderate, severe, and very severe. The last category is defined as "economically irreversible" (Dregne, 1977:328). However, "there just are not many large areas where economically irreversible desertification has occurred" (ibid.:329). This judgement was "shared by all persons contacted—thus far—who have helped improve the first draft" (ibid. note) of his article. The late Professor Mikhail Petrov suggested to Dregne that "very severe desertification might be limited to small areas up to 20 km wide around oases" (ibid.:330). If this is true, it suggests not only (as Dregne concluded)

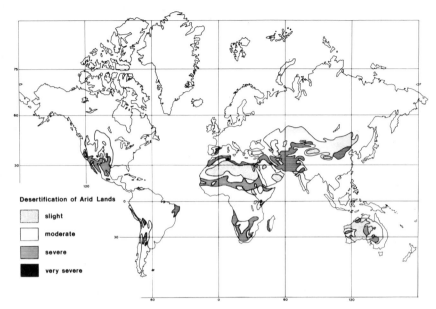

Map 1. Status of desertification in arid regions of the world.

that "therefore the true extent of the very severely affected areas will not be shown on small scale continental or world mapps" (ibid.), but that true or irreversible desertification—as distinct from reversible reduction in productivity—may be a function of settlement rather than (as is generally assumed) the over-intensive practice of particular food production technologies (cf. Spooner *et al.*, 1980).

Whether or not desertification is irreversible, it threatens reduction in productivity over vast areas which are already relatively low in productivity and poor in economy. However, because of their vast extent the total productivity of these areas is of great consequence for world totals—the more so because they are as yet relatively undeveloped and their potential for increased productivity may therefore be relatively high. As yet also their production is relatively unintegrated into larger economic systems. The actual figures, therefore, as marshalled for UNCOD, that one third of the land surface of the world containing (in 1976) 15% of world population, or 630 million, of whom already 8% or 50 million are threatened by desertification's productivity losses (see Hare *et al.*, 1977:339) may be a misleading understatement, because desertification is threatening the as yet unrealized and unknown potential of as much as one third of the world's land.

Desertification in its most widespread forms is generally represented as the immediate result of one or another of three practices: over-grazing, "over"-cultivation, or over-irrigation—in increasing order of investment required and of potential productivity per hectare, and decreasing order of area at risk.

Natural rangelands sustain pastoralism up to a certain threshold beyond which grazing and browsing are likely to eliminate not only individual plants but entire species and so cause modification in community composition as well as reduction in percentage cover throughout the range. Although less vulnerable species may take the place of those eliminated, they are always less palatable and so less useful to pastoralism (or any other form of food production) and invariably provide less protection against soil erosion. This threshold varies seasonally and annually with the variation in precipitation that is typical of arid and semi-arid climates. Population growth or loss of territory in many pastoral societies has led to increases in animal population per hectare to the point where that threshold is often and even continuously exceeded. Because of social and cultural factors pastoralists seldom adapt simply and directly to the ecological situation by reducing their herd size in tune with the reduced carrying capacity of the range (cf. Sandford, Chapter 3). Their technological adaptability to recurrent conditions of reduced carrying capacity has also in some cases been prejudiced or impaired by the effects of nationally and internationally sponsored projects to increase their productivity and integrate it into the larger economy, especially by making new watering points available which facilitate greater exploitation in good years and lead to herd growth (cf. Bernus, 1977). As ranges become more heavily overgrazed they first lose their more palatable species, then gradually overall cover is reduced leading to reduced absorption of precipitation, increased run-off and erosion. The conventional cure is reduction of the animal population to the point where the vegetation may recover, but this measure does not always produce a simple reversal of the process. Undesirable species may invade and take over the range before the earlier vegetation (which may have lost its seed base) can re-establish itself (cf. Conant, Chapter 6).

In the case of rainfed cultivation, the main hazard arises from the extension of the system on to marginal soils that will not support it. Opportunistic ploughing of such soils, which may produce a few good harvests in the short term, in the longer term leads to erosion. The wild vegetation produced by such soils often constitutes the better rangeland of traditional pastoralists. As a result of erosion the land is lost to both agriculture and pastoralism, and the pastoralists are thereby pushed back in denser numbers on to less productive rangeland which becomes further impoverished as a result. In addition to, or instead of, being extended, cultivation is sometimes intensified by the reduction of fallow periods—which similarly leads to the impoverishment and loss of soil and to long term reduction in productivity. These forms of "over"-cultivation have been encouraged not only by population growth but by increased economic demand and opportunity resulting from integration into a larger, cash economy, and also by the increasing availability of mechanical aids to labour such as tractors, which allow a much faster rate of growth in acreage than would be possible given simply the rate of population growth (cf. Schulz, Chapter 2).

A further important factor in over-cultivation is the reduction in the farmer's dependence on his land that often accompanies his incorporation into the cash economy. This shift and its implications in the historical study of the relationship between natural processes and human activities have been appropriately termed "the ecological transition" (Bennett, 1976). The economic opportunities of industry or urban life gradually provide viable social alternatives to rural life and individual farmers can afford to be less concerned about the possibility of decline in the productivity of their land. In this way the socio-economic reorganization that goes with development, may bring in its train, reduced ecological sensitivity in rural populations. It is worth noting, however, that the social security of these populations may at the same time be increased.

Over-irrigation, as a form of desertification, occurs in the land use system which is not only the most intensive and productive (though admittedly with the least—actual and potential—extent) but it is in a sense the most directly caused by industrialization. Unlike most other desertification processes over-irrigation is scarcely affected by local levels of population: it is a function of the large scale engineering of irrigation, in which the local operators do not control their own operation, did not develop it themselves and therefore have a less detailed "folk" understanding of the ecology of it. The investors, developers or organizers, whether in late prehistoric and early historical Mesopotamia or in the Punjab under the British in the nineteenth century, approached it as an economic enterprise, and now that the ecology is understood by the organizers it remains to work out ways to arrange the efficient operation of the system. In short, the limiting factor in agricultural production in many dry areas is water. Many dry areas can be transformed into agricultural miracles by the controlled application of water—which is possible, especially in large river basins, given large scale organization, engineering and above all industrial construction. But in order to be ecologically successful over a long period the flow of irrigation water must be engineered in such a way that it reaches the root zones in optimum quantities for the needs of the specific crops. Excess water that seeps out of the channels before reaching the crops, or sinks below the root zone of the crops, unless it can somehow be drained out of the basin, sooner or later causes trouble. However deep the original water table, excess irrigation water gradually builds it up until it approaches close enough to the surface to cause first salinization (since it invariably has a high mineral content) through capillary action, and finally waterlogging. The result is serious decline in yields and, in severe cases, total loss of productivity for indefinite periods.

Since this type of land use system is basically modern and industrial, generated by national or international investment, and integrated into the larger economy, loss of its production due to desertification is of more immediate economic significance than in the case of over-grazing or over-cultivation which are often unintegrated so that the potential value of their production in the larger economic context is difficult to determine. In the case of over-irrigation, therefore, figures are both more available and more meaningful.

This type of situation is dealt with in some detail in a case study from the Pakistani Punjab by Merrey in Chapter 5. Briefly here, before the development of the irrigation system, water table depths over most of the area now irrigated were about 24 to 28 metres. Of the 123 billion cubic metres of Indus water diverted annually into the irrigation system, only about 71·5 billion m³ are available at the heads of watercourses: the rest is lost through seepage. It has been estimated that from 5% to as much as 65% per mile is lost in the watercourses. Altogether, less than 30% of the water diverted from the rivers is stored in root zones for crop use. Historical data show that the water table has risen an average of 15 to 35 cm per year since modern irrigation was introduced. Further, in an area where the underground water has a salinity of 1000 parts per million, which is acceptable for virtually all crops, evaporation at a rate of 60 cm per year, which is a typical value where the water table is only a few feet deep, will raise the salt content of the top three feet of soil to about 1% in 20 years. This level is too high even for the hardiest crops. Not only, therefore, are environmental problems causing loss of cultivable land—the irrigation system is working at only 30% efficiency and this inefficiency is responsible for the loss. These processes are exacerbated by inefficiencies in actual irrigation and cultivation—that is, in application of water to crops. In the 1960s it was estimated that up to 40 000 additional hectares were being affected each year and, in the worst districts, 40–50% of the cultivated land was already severely damaged.[3]

A certain amount can be done to correct this situation by more engineering. Fields may be levelled, canals may be partially or wholly lined to reduce or prevent seepage; drains may be dug. Tube wells may be installed to pump the ground water back up to the surface for recycling where the mineral content is low enough. But these measures all increase both the capital cost and the running cost of the operation and reduce its profitability without increasing its efficiency. The basic aim—to provide the root zone homogeneously with just the right amount of water—remains elusive and can probably be solved only at the level of the individual operator or some social grouping of operators. Merrey deals with moves to approach the problem at this level. He shows that there has been little practical success so far, partly perhaps because so much rethinking of the relationship between government, official, engineer, and local operator or farmer is required.

Desertification, then, as conventionally defined, is a function of the complex interaction of a series of factors, which includes: topography, soil, vegetation, groundwater, wind, temperature, precipitation, and the nature and intensity of human activity. It is not new. It is not peculiar to any particular land use system, area of the world or culture. In fact it can accompany any land use system and has probably accompanied all land use systems to some extent at least in some periods. It is not qualitatively different from ecological change as a result of human activity in more humid areas. It has only recently been perceived as a serious problem— partly because the rate of global population growth has for the first time brought us to the point where it is not only conceivable that we shall run out of resources, but we have felt threatened by imminent failure to feed ourselves. It is important to note that the present consciousness of this threat derives at least partly from intellectual developments that have generated a new perception of the environment. But however infallible this perception may appear among planners and ecologists, it is not universal. Even where the problem is perceived to be immediate, the perceptions are not all uniform, and there are legitimate differences of opinion about the analysis of particular situations. It will be useful here to discuss the causes of desertification, bearing in mind that any causal analysis is likely to reflect a particular perception of the problem, and although one analysis may seem more persuasive than another, in the last resort what will determine the acceptance of one or another analysis in any campaign to combat desertification will be the political process of interaction between the various agencies, governments and social groups concerned. The most useful aim at the intellectual level of this volume is to seek to inform that political process by making available to some of the various parties to it the maximum amount of relevant information in its most digestible form.

What then are the causes of desertification? Arguments about ecological deterioration have been usefully grouped as:

"(a) *Structural* arguments. These lay the main blame on social and economic structures and relations (patterns of ownership, rights of use and control over resources).
(b) *Natural events* arguments. These see largely uncontrollable events, such as droughts or outbreaks of disease, as the prime causes of deterioration. We can also include in this category political developments whose origins lie outside the context of pastoralists and rangelands, and which are, therefore, similarly
(c) *Human fallibility* arguments. These lay the blame on the stupidity, ignorance, or short-sightedness and perversity of pastoralists, of governments, and of do-gooders.

(d) *Population* arguments. These see the main cause of the deterioration in the rapid growth of human and livestock populations" (Sandford, 1976:47).

In the terms of the conventional wisdom there has been a tendency always to assume implicitly that the immediate human cause of any particular symptom of desertification, such as over-grazing, was the significant cause, to look no further to pursue the reconstruction of a chain of causation, but to prescribe simple remedies in the form of management regimes and expect them to be easily and efficiently implemented. This tendency derives from the fact that the consciousness of desertification arose among specialists in the study of natural phenomena, who naturally sought to solve it in those terms. They were able to communicate their concern fairly readily to the community of the agricultural sciences, but they had less success with the communities of economics and the social sciences, whose first priorities were the efficient use of labour and social well-being respectively rather than the conservation of renewable natural resources. Although the dangers of desertification could have been presented in economic or social terms, the ecologists did not make a very convincing case in such terms. For this reason, and because of the underlying problem of the sectoral fragmentation of science there has been little interdisciplinary synthesis of desertification or transdisciplinary cooperation either to understand causes or to develop recipes for solutions. The communication or lack of it, between the disciplines or professions, with regard to desertification, has been exacerbated by the conviction of the ecologists, who discovered it, of the urgency of the situation—that something had to be done immediately in order to save a significant proportion of the world's resources not only for posterity but for the future well-being of the present generation, and especially for the good of the affected people themselves.

This element of panic is reminiscent of other similar campaigns also having to do with the global ratio of people to resources that have hit us since the mid-sixties. The first was population; then came food; more recently it has been energy. No time could be lost working out the ideal way of dealing with the situation. Early faint protestations from social scientists that simple imposition of management solutions based on experience in other areas which may well be ecologically comparable, but whose populations are socially and culturally non-comparable, were met with accusations of callous detachment.

Invariably, however, though they often seemed feasible at the time, we can now see that Western-devised management solutions for non-Western situations of desertification, or for that matter of "under-development", have not been successful, and we can now see that this lack of success should not surprise us. These management solutions are a form of technological change. It is now generally accepted that technological change does not occur in isolation from, and cannot be induced in isolation from, economic change and general social change. The development campaign generally has not fulfilled expectations because it has been technology-led, formulated in terms of technology and investment (and management) with the implicit assumption that the social relations that constitute the society in question will rearrange or re-form themselves in adaptation to the new exogenous techno-environmental conditions. It is now possible for the social scientist to join in the development effort—and therefore also the effort to combat desertification—because it has now become clear that in order to induce change successfully in any social system it is necessary first to investigate the dynamics of the existing system. It is unscientific to expect to change particular social practices without first ascertaining what generates the social formation underlying those practices.

The question then arises whether the plan to induce the change can be morally justified—especially when it is often based implicitly on assumptions about the interests of future populations. This is too large a question to be dealt with fully in this chapter, but its existence and importance must be acknowledged. For the time being we can assume that in the case of desertification many will argue that the attempt to induce change is morally justified by the prediction of ecological consequences based on the assessment of current ecological trends. However, there is still a moral problem in so far as the over-grazing or other practice that is the immediate cause may be a response not to local factors such as inefficient management of resources but to external factors such as market opportunities and the terms of trade for pastoral products. In this case do moral principles dictate a solution to the problem by imposition of a pastoral management regime (which is likely after all to be distasteful, if not a cause of serious hardship, to the local pastoral population)? or, for example, by administrative changes in tax structure that would alter the relationship between the pastoralists and the larger economy in which they are encapsulated, and possibly reverse the desertification trend by imposing some degree of hardship on other sectors of the population that use pastoral products and were benefiting from it?

In many cases of desertification the production system, for example pastoralism, may constitute a less significant pressure on the vegetation than the demand for fuel and for construction materials. Apart from forage the vegetation often has to satisfy the need for fuel for heating, cooking and in some cases also for processing milk into yoghurt and other products, and for roofing and the construction of animal pens. In an area of arid rangeland in northeastern Iran the average domestic consumption of firewood has been estimated at 5·3 metric tons per year per family plus as much as an additional 7–10 tons for milk processing (cf. Horne, Chapter 10). Historically the same area has also produced charcoal for urban markets. In other Middle Eastern rangelands, where no ligneous vegetation survives to satisfy even local fuel demands, animal dung is collected for fuel purposes. Perhaps the most surprising point is that wood still serves as the major fuel for heating for some cities, such as Kabul, which have cold winters. The distribution of fuel and construction materials is even more obviously a matter for central economic planning and political decision-making, and the "fault" therefore lies at the national level, not at the level of the user. It has to be generally accepted that desertification demands a "no-fault" approach, in which the interests of existing populations may not easily be set aside in favour of future generations, especially when the burden of cost could be more evenly spread among existing populations.

These simple examples enlarge the perspective on desertification by suggesting the moral, political, economic, as well as social and ecological, dimensions of any process that involves the interaction of human activities and natural processes in the modern world. If the problem of desertification is to be approached realistically, all these dimensions must be taken into account. Conceived in this way, the problem of desertification can be resolved only in the political process, but the political decisions necessary to resolve it are unlikely to be made for purely ecological reasons. We are likely, therefore, to have to live with increasing desertification for some time to come. Meanwhile, the goal of science should be to synthesize research on these various dimensions of the problem as conceived above, and seek to feed information from that synthesis into the political process.

In fact, however, there is little or no evidence either that current trends (where they are bad) are long term trends, or that direct intervention is likely to reverse them, and not enough notice is taken of contrary indications. For example, in at least two arid areas—northern Nigeria in the vicinity of Kano, and Rajasthan in northwestern India—population densities have exceeded 150 per square kilometre, and although there is definitely evidence of desertification, living standards are not deteriorating and disaster is not imminent (Hare *et al.*, 1977:340).

Some diagnoses of desertification are questionable. For example, erosion may not necessarily be disadvantageous. "In northern Libya many small dams were built specifically to trap the eroded sediment and to create deep, well-watered soils for agriculture. Roman agriculture could have expanded in a short period of favorable rainfall, much as agriculture expanded into the northern Sahel in the early 1960s. The Romans, like the Hausa, may have had to retreat in the face of a drought and what was, in effect, only a marginal increase in erosion" (Hare *et al.*, 1977:337, Vita-Finzi, 1969 who also cite other examples). The use of a natural factor, such as soil, as a yardstick of desertification, may disguise the assumption that a particular set of human social interests had priority—those that valued the present state of soil distribution over those that were happy to see it changed. The following example merits quotation in full:

"In the Valley of Nochixtlan in southern Mexico, many side slopes are ravaged by active gullies which remove the surface wholesale and leave the slopes bare of vegetation, fields or houses. Since the Spanish Conquest, an average depth of 5 m has been stripped from the entire surface area, producing one of the highest rates of erosion recorded in the world.* Set between the forested uplands and the agricultural valley floor, the area seems a wasteland which only drastic soil conservation measures could reverse.

Government experts share this view and have instituted conservation measures including the construction of low earth ridges to slow down soil movement. Few scientifically trained experts would disagree with their general perception of the gullying as a problem but the view from inside the valley is different. Gullies are seen not as a hazard but as a resource. By directing the flow of the eroded material, Mixtec farmers can annually feed their fields with fertile soil and can, with greater effort, extend their agricultural land by building new fields over a few years.

Over the past 1000 years, Mixtec cultivators have managed to use gully erosion to double the width of the main valley floors from about 1·5 km to 3 km; and to infill the narrow tributary valley floors with flights of terraces several kilometres long. Judicious use of gullying has enabled them to convert poor hill-top fields into rich alluvial farmland below, using the gullies to transport the soil. Thus before large-scale gullying began, the agricultural productivity of the valley area was less than it is today.

The difference between the 'outside expert' view and the inside Mixtec one rests on the farmers' greater experience and knowledge of the local situation. Their experience of the highly fertile and erodible local deposits, and their familiarity with the technical and social bases of controlling soil movement, are too particular to the Valley of Nochixtlan to be readily translated to other areas. Thus the concept 'gullies are good' is not part of the outside expert's portfolio. Nor could he be expected to know that intermarriage between the hill-top and valley bottom communities enables families to 'move with their soil' downvalley.

The Valley of Nochixtlan is an unusual case; usually different groups agree that soil erosion is a problem but disagree about how to solve it. This example is intended, however, to illustrate the importance of understanding local perceptions of the environment in the context of local resource use and social structure. But this is only the first, important step. In the example of Nochixtlan—as almost everywhere—both perceptions of the environment are valid, within their own contexts. For the farmers in Nochixtlan, gullies are an important agricultural resource. For the government authorities concerned with the area as a whole, gullies are also a problem—not for those farms who owners remain, but for the farms abandoned by their urban-migrating owners and no longer receiving replenishment and protection from the gullies. Thus, the national 'problem' is that of urban migration and rural depopulation, which is the higher-order one, and which is outside the scope of agricultural authorities and local communities" (Whyte 1977:11–13).

---

* The average erosion rate over the whole surface was in the order of > 10 mm per year over the last 500 years (M. Kirkby, 1972).

Analgous examples have been documented in the history of the Negev
(Evenari *et al.*, 1971) and of parts of Iran (Dennell, Chapter 9).

Any process of ecological change is likely to be perceived differently by
different social groups. Desertification studies so far focus on ecological
change resulting from human activity. If progress is to be made in the
campaign to combat desertification, their focus must be shifted to the
relationship between population and resources. But it will still be necessary to
develop a yardstick to measure desertification in those terms. "Flexibility"
may serve this purpose. A major factor in the social tragedy that followed the
Sahelian drought was the recent loss of flexibility in resource use. In arid and
semi-arid lands where precipitation is often more irregular and unreliable
than insufficient, flexibility is the most important aspect of any land use
system. Nomadism has survived for so many millennia in many such areas
because it is a strategy that gives first priority to flexibility. The value of this
strategy is still not fully appreciated (see Bernus, 1977). Flexibility is often lost
as a result of modernization and the introduction of new technologies.
Development increases investment and investment reduces flexibility. But
flexibility could be safeguarded as a matter of public policy—though such an
objective would require a major reorientation in development planning and
in government intervention in land tenure generally.

Over the last decade a few attempts have been made to develop research
programmes that would illustrate the interaction of social and natural factors
in desertification and provide better understanding of its dynamics in different
cultural contexts. The switch away from a focus on natural processes
irrespective of their social referents is slow, and data from these programmes
are still in the process of publication but examples may be found in Lamprey
(1978) and Novikoff *et al.* (1973–1976), as well as Part Two of this volume.

IV. THE PROSPECT

Desertification is real, but more complex and less well understood than even
UNCOD documentation suggests. However, for the most part it is probably
not irreversible and therefore not as critical as we thought. As a slogan, an
organizing concept, it is part of the international history of the late twentieth
century, taking its place alongside population, food and energy. But in the
final analysis it seems not to have the same force as the others. Some of the
figures on which its credibility depends have been shown to be exaggerated,
unrepresentative, misunderstood or mistaken (see Simon, 1980). The danger
now is that its credibility will decline further.

It is important that the campaign to combat desertification should not lose
credibility. But in order to maintain credibility, desertification must be put

over in such a way that the implications for counter-measures make sense in terms of the interests of the affected people as they see them. To achieve this end a great deal more rethinking is necessary. We need a yardstick and a set of indicators that will be applicable in different political and cultural as well as ecological contexts.

The first essential step in this direction is to discontinue the exclusive use of ecosystems as units of analysis. For example, if Rajasthan is providing meat and dairy products to the population of Delhi (Kates *et al.*, in UNCOD, 1977a:289–290) it is unrealistic both politically and ecologically to seek to combat desertification in Rajasthan simply by rearranging pastoral activities there. Central governments are responsible for the formulation and implementation of policy, and therefore also for the measures that influence the distribution of economic activity, developing some areas and marginalizing others. It is at this level of organization and planning that desertification can serve as an important concept to minimize the ecological degradation of dry lands.

Desertification needs to be rethought not only in terms of the articulation of ecosystems with human use systems but in terms of what we know about human adaptation. The concept of adaptation is commonly extended from evolutionary biology to explain the relationship between natural processes and human behaviour. But although this extension is often useful it does not rest on any sound theoretical basis and often leads to inappropriate expectations. While it is true that much human behaviour does make sense as adaptation to particular constraints or opportunities, the concept does not explain why people invariably choose to adapt to some conditions and not to others, and (despite elaborate recent attempts to explain the Jewish and Muslim pork taboo and Hindu cow protection) it by no means accounts for all behaviour.

The idea of adaptation is so ingrained in modern thinking that it is more practical to make it more explicit and seek to modify it, than to try to supplant and replace it. We may, therefore, continue to think of human behaviour as adaptive so long as we remember that it is adaptive to constraints and opportunities generally—natural, social and cultural—as they are selectively perceived by particular people in particular historical contexts. Since we do not know how to predict the selection in any particular case, the concept is reduced to practical insignificance in the present context.

Any attempt to put new life into the concept should focus on historical context: the integration of natural, social and cultural factors in the explanation of how people adapt selectively to different constraints and opportunities is best achieved through historical reconstruction. Ecologists have generally been ahistorical, but a growing awareness of historical context is an important part of the rethinking process in most disciplines. In ecology attention to historical context is necessary to show the essential difference between apparently similar ecosystems. For example, the history of interaction between human activities and natural processes in the arid and semi-arid rangelands of the United States, Africa and the Middle East explains why despite present superficial resemblances the imposition of one management regime on all of them leads to different results in each, because each is in fact the product of a historically particular process of co-adaptation between vegetation and herds. This type of co-adaptation and its implications are explored in some detail by Nyerges in Chapter 11.

An historical perspective in fact suggests that a certain degree of desertification (as conventionally defined) is the inevitable consequence of human activity. Desertification should probably, therefore be approached more positively as one aspect of an unavoidable compromise with population growth, bearing in mind that reduced natural productivity can be (and in some cases has definitely been) balanced by improved social productivity.

The reality of social factors cannot be overemphasized; they have their own dynamics and as yet we have little scientific understanding of their interaction with natural processes such as desertification. It is possible that systemic relationships generally cause social factors to intervene before desertification becomes severe—except in cases of critical external intervention. In any case ecological degradation can be accounted an evil only in relation to specific human interests—not in its own right (cf. Passmore, 1974: *passim*). But how can these human interests be evaluated to provide a scientific basis for the campaign to combat desertification? The answer underlying the theme of this book is that diagnoses and prescriptions of desertification processes should be measured carefully against sociocentric yardsticks—against the interests, that is, of the social groups concerned. Scientific ecologists are one such group. Local people may constitute several concerned groups and local and national government yet more. The assumed interests of future generations can be represented only by one or more of these groups. Final diagnosis and prescription will be a compromise between the interests of these groups. The compromise will be made, as always, in the political process. The fuller the socially relevant—sociocentric—information fed into the political process, the more satisfactorily it will work in the long term. The scientific objective should be to increase the flow of socially relevant information—not just from the sociocentric point of view of the scientist, but across the board.

To attack desertification is to select one symptom arbitrarily from a range of symptoms generated by a pervasive disease. For example, in the Sahel there is evidence that social disruption had set in before the drought (Bernus, 1977). The disease lies in the relationship between society and natural resources not only on a local but on a global scale. Desertification has in fact come into vogue only since that relationship has been perceived on a global scale.

Unfortunately, we are not prepared to deal with it on that scale because of the political and other divisions in society. We have a fair view of the natural resources on that scale and we must keep this ecological knowledge in mind as we search for social and political answers. In so doing we shall develop a new approach to development generally—a new approach that will be both scientifically more comprehensive and politically more acceptable.

The significance of desertification for the foreseeable future depends on the political process, and is a relatively insignificant ingredient in it. Despite a number of excellent scientific papers on the biological and physical processes involved (for example, Le Houérou, 1976 and 1977), there is no hard evidence that *overall* food production of environmental quality is decreasing as a result of it. But the overall political process could be helped and smoothed if the information deriving from the campaign against desertification so far, and from now on, is fed into it more efficiently. The campaign will have served its purpose if it has led to a situation wherein from now on ecological consultation will be built in to the development process. Though undeniable, desertification is not only insignificant relative to the overall socio-political problems of organizing production and distribution (while raising living standards and the general quality of life): it is unbeatable except within the framework of a programme to solve those problems. At the same time, it causes rearrangements of people and of rights to resources, which result in social, cultural and psychological costs that are not quantifiable, but change ideas and ideologies with unpredictable political implications.

To summarize, desertification may reduce the productivity of some natural resources seriously in the foreseeable future; but it is unlikely that a frontal attack on the problem, which attempts to rearrange human activities in direct management solutions, will have much impact. Desertification is intimately related to development and represents one aspect of the inadequacy of the development effort so far. The development effort is undergoing a process of re-thinking, in which the emphasis is shifting from purely economic criteria to criteria of social welfare, broader public participation and the local definition of priorities. The campaign to combat desertification can be effective only as a component of this general development effort.

Endnotes

[1] An earlier and shorter version of this chapter was published as The Significance of Desertification in Global Aspects of Food Production, edited by Drs M. S. Swaminathan and S. K. Sinha, London, Academic Press, 1982.

[2] These Transnational Projects, of which there were six—two ecological monitoring projects in Southwest Asia and South America, two greenbelt projects in the North African and Sahelian countries, a livestock stratification project in the Sahel and a ground water conservation project in Northeast Africa and the Arabian peninsula—

although they survived the Conference, began soon after to disintegrate. In an attempt to save some of them the Desertification Unit (which succeeded the UNCOD Secretariat, Nairobi) redefined them as sets of interrelated national projects. The twelve "Case Studies" were from Chile, India, Iraq, Niger, Pakistan, Tunisia (commissioned) and Australia, China, Iran, Israel, United States, and U.S.S.R. (contributed by governments). The four "Component Reviews" were prepared on desertification and climate, ecological change, society, and technology. There were also a number of country reports and other minor documents.

[3] These figures are based on Michel (1967), Spooner (1979) and Hasan (1976), which in turn are based on numerous other sources, including especially the reports of the Water Management Research Project, Colorado State University.

Part I
The Global Problem

The international campaign to combat desertification is concerned not with deserts, but with the development and intensification of desert conditions that impoverish human living conditions—in terms of the productivity of natural resources and of the quality of life—in areas of water deficit, whether arid, semi-arid or subhumid. In some forums it has been extended to include deforestation in areas that do not suffer from water deficit. Although desertification could be natural or anthropogenic, the campaign is for all intents and purposes concerned with correcting what are considered to be the avoidable results of human activity. Since the process is not confined to particular biomes and neither the related human activities nor the consequences for human well-being can be defined ecosystemically, it is a truly global problem: it potentially poses questions for many scientific disciplines about all major types of human activity. Work so far, however, has been concentrated in a relatively narrow part of this range. Part One of this book seeks to expand the sample by working out examples of under-represented approaches. It is divided into two sections: the first is disciplinary and deductive; the second empirical and inductive.   —Ed.

Section A
New Perspectives

The three chapters of this section develop new perspectives on desertification from the points of view of particular social sciences—political science, economics and archaeology—which have been under-represented in the general debate. Although individually they are not meant to be necessarily representative of the possible contribution of the authors' disciplines, jointly they demonstrate the importance of what would be gained from involving these disciplines, and others, in the research and planning process. Although they deal with particular problems in specific parts of the world—especially in the case of Chapters 2 and 3—their conclusions are immediately transferable to a more general level of application. —Ed.

2

Reorganizing Deserts: Mechanization and Marginal Lands in Southwest Asia

Ann Schulz

Clark University
Department of Government
Worcester, Massachusetts 01610, U.S.A.

Written from the point of view of political science, this chapter approaches the problem of explaining what has happened so far under the heading of desertification and the campaign to combat it by focusing on some of the relevant political interests. It shows that perceptions of ecological processes are to an important extent a function of those political interests. Since the most important political question in this context is food production, the ecological variables commonly discussed under the heading of desertification are re-evaluated as pawns in the politics of agricultural planning.   —Ed.

There is an irony in the desertification debate that demands careful attention. Precisely because desertification is a highly political issue involving the interests of powerful organizations and of state elites, politics rarely enters into how it is conceptualized. Instead, it is defined in ecological terms that obscure the influence of organized political interests over relevant public policy.

For many years, desertification was attributed primarily to geomorphological and climatological phenomena. It was studied almost exclusively by dryland ecologists. Their work was scientifically important. It also helped to preserve a limited definition of the problem. For this reason, in public policymaking, anti-desertification activities were routinely assigned to environmental agencies, rather than to planning bureaus or to ministries of agriculture, whose decisions directly affected land use. When social scientists began to play a role they were isolated within the bureaucracy and had little influence over agriculture. As a result, powerful agricultural interests have never been held accountable for the impact of their decisions on land quality.

International interest in desertification was provoked by the devastating Sahel famine of the early 1970s. The famine politicized food, elevating it to the status of a national security issue alongside arms and energy. In 1974 the World Food Conference publicized the need for a more reliable system of food distribution and for increased investment in agriculture throughout the developing world. Three years later, the United Nations Conference on Desertification (UNCOD) attempted to translate the Sahel experience into a problem of land degradation. Despite the evident relationship between the food production and distribution system and land fertility, the two conferences served to institutionalize desertification as an environmental issue quite separate from agricultural and land use policies at the national level.

The world's limited land resources raise the spectre of even more severe food scarcities in the future. The increase in global food production during the 1960s and 1970s resulted from bringing more land into production as well as from higher productivity per hectare (Brown, 1978:33). The demand for arable land will increase. By the end of the present century, the world's population is expected to exceed the level in 1975 by 58%. Most countries that have sizeable arid and semi-arid regions also have very high population growth rates (see Eckholm and Brown, 1977:20). Even now, in many Third World countries, lands that are presently used for food production are biologically less productive and have less regenerative capacity than those used to grow industrial crops. The high demand for food in the future will make it politically difficult to protect these fragile lands from overuse, and the prevention of desertification will become more critical.

The challenge of international policymaking is to initiate "concerted action" by governments and international bureaucracies to solve shared problems despite underlying political dissension (Rittenberger, 1978:24). UNCOD was able to gain approval for recommendations for land use in its final Plan for Action. But the recommendations will confront strong opposition from member states and from international development agencies before they can be implemented. Questions of agricultural land use generate far more political controversies than do climatological or local ecological phenomena. With future food scarcities, anti-desertification activities can be expected to provoke more intensified conflict.

One of the recommendations approved by the conference implied a choice between short-term production and long-term environmental protection. Directed toward land degradation in rain-fed farming areas, the recommendation called for "the establishment of legal limits to cultivation by tractor ploughing in marginal dry lands, which are ecologically better suited for grazing" (UNCOD, 1977b:25). The recommendation was based upon the assumption that rain-fed farming areas are particularly vulnerable to overuse, extensive clearing, and "excessive mechanical treatment".

Establishing limits on tractor ploughing directly challenges the interests of a complex agricultural establishment. Policy-makers in countries where land is scarce and population increase is high believe that mechanization increases productivity. For many years, international lending agencies also preferred to invest in mechanized farming projects. They were supported by machinery manufacturers who wanted to increase their markets and by farmers who hoped to avoid later problems by mechanizing. As a result of these over-lapping interests, mechanized farming spread dramatically during the 1970s.

The purpose of this chapter is to review the history of UNCOD, a milestone in articulating land use objectives, within the context of international agricultural policymaking. The apparently conflicting demands of short-term growth and long-term land conservation are particularly applicable to the countries of North Africa and Southwest Asia, where agriculture and desert exist side-by-side, where large areas are dryfarmed and where agricultural mechanization continues to proceed at a rapid rate. In four countries within this region—Pakistan, Sudan, Iran, and Turkey—the underlying principles of existing agricultural policy conflict with the Plan of Action's recommended limits on mechanized farming. But, unless progress toward a more equitable global distribution of resources reduces pressures on Third World policy-makers to use all available land to the fullest extent, conservation will continue to be seen as an unaffordable luxury. Furthermore, they are encouraged to mechanize by the multinational manufacturers of farm machinery with the backing of the governments of the exporting countries, and by many technical assistants and leaders who share a commitment to high technology agriculture.

I. POLITICS AND UNCOD

UNCOD had no impact upon the global distribution of resources, but rather was part of a trend toward the consolidation of North–South differentials. The declared goal of the Plan of Action approved at UNCOD in 1977 was to reduce the "diminution or destruction of the biological potential of the land". This definition of the desertification problem itself represented a significant shift from earlier formulations. During the 1950s and 1960s, several U.N. agencies initiated arid land studies, but it was the UNCOD secretariat and its consultants who identified social and economic causes of desertification. This broader orientation was confirmed by the inclusion of the United Nations Development Programme among the *ad hoc* group of agencies that planned the conference and by the selection of a director, Ralph Townley, who had earlier experience with social and economic issues as organizer of the World Population Conference.

Despite thorough planning, UNCOD confronted two overwhelming obstacles to producing strong, broadly-supported recommendations. One was the North–South confrontation which was becoming more polarized as UNCOD was being prepared. The other was the lack of involvement in the conference of international agencies, like the World Bank, whose resources, geographical reach, and involvement in agricultural development make their participation critical to any assessment of land use policies. Despite the significant progress toward identifying social causes of desertification, UNCOD was on the whole a scientific meeting about land degradation which skirted the issue of who is going to grow food where, how, and for whom.

The Sahel drought and the famine in Bangladesh in the early 1970s added a new sense of urgency to the new international economic order (NIEO) dialogue within the United Nations that began in the early 1960s, and focused more attention on the poorest of the poor countries—the so-called Fourth World. Upper Volta experienced the worst effects of the drought. Its delegation to the U.N. General Assembly introduced the resolution for the desertification conference on December 17, 1974. Poverty was a latent issue, if not a manifest one, in UNCOD from the beginning. During the African regional preparatory meetings, a proposal was approved to establish a special account to fund anti-desertification programmes. By 1977, "special accounts" had become a metaphor for the NIEO, in commodity agreements, in the International Fund for Agricultural Development and so forth. It represented the acknowledgement of all parties concerned that poverty underlay a variety of global ills.

In the final Plan of Action, the special account appeared as one of several sources of financing. The Plan also encouraged the diversion of funds from existing aid programmes to anti-desertification projects. A special committee, made up of public officials in the field of finance and development, met during the year following the conference and produced proposals for funding anti-desertification projects from sources such as the private capital market, equity funds mobilized by a public international corporation, and more radical possibilities for obtaining revenue from "the international commons", particularly the ocean and its minerals (United Nations Environment Programme, 1978:11).

The special account was formally approved by the conference and was accepted by the General Assembly on January 29, 1979, following a protracted debate. However, the OPEC countries, the United States and the British and Soviet governments all refused to contribute to it. It remained an account in name only.

Officials of the United Nations Environment Programme (UNEP) met with their consultative group in Nairobi in 1978, and in 1980, and attempted to elicit financial support for its proposed anti-desertification projects. The consultative group was as unproductive as was the move to establish a special account.[1] The failure of the Nairobi meeting resulted from inadequate planning coupled with a certain naiveté about the international funding process. The agencies that participated in the meeting as prospective donors (including the World Bank) were not committed to pledges ahead of time, as is the custom for such gatherings, and the projects that UNEP presented for funding were not tightly justified with the kind of cost-benefit analysis that the donors typically demand.

UNEP's plans also were constrained by the Western industrialized countries' reluctance to accept financial responsibility for environmental projects in the Third World. By the time that the Desertification Conference was convened in 1977, the North–South dialogue was at a particularly critical point. The U.S. delegation at the United Nations was overtly opposing Third World demands. Daniel Moynihan, ambassador to the U.N., articulated the U.S. position that Third World governments were too undisciplined and stifled growth incentives through their own heavy-handedness. It was they, according to Moynihan, not the industrialized West, that was responsible for the slow pace of developments.

The U.S. position on North–South issues carried over into the desertification debate. U.S. delegates to the conference emphasized the ultimate accountability of national governments for anti-desertification action. Recommendation 20, for example, reiterated the necessity of comprehensive national involvement in educational programmes (UNCOD, 1977). The position of the United States over the years immediately following the conference remained the same. The United States would take no major initiatives in multilateral international programmes so long as the governments of the affected countries failed to construct their own anti-desertification plans. The United States and the Mexican government organized joint anti-desertification consultations, but action on any further activities following the conference was left to a weak *ad hoc* inter-agency committee.

By 1980, the conference had produced neither a separate and distinct agency within the United Nations system, nor concrete projects within the UNEP desertification office. The industrialized states had been unwilling to fund any programmes. Third World representatives to the conference had been equally unwilling to create a new U.N. bureaucracy. After two decades of U.N. involvement in international development, many Third World leaders believe that the U.N. bureaucracy is an expensive luxury that does little to redress the fundamental grievances of the poor countries.

The secretary-general of the conference, Moustafa Tolba, was sensitive to the anti-bureaucratic position of the Third World leaders. When he met with the African delegates at a regional preparatory meeting, Tolba promised that there would be no new structures or programmes. Other members of the Conference Secretariat felt that Tolba conceded too readily to the conservative position of the member states. The World Food Conference produced two new agencies as a result of its determined political leadership.

Third World member states have been particularly critical of the oldest U.N. agencies that date from the colonial period and that have failed to identify their programmes with the movement for a new international economic order. The developing countries lobbied for UNCTAD, rather than FAO, sponsorship for the 1974 World Food Conference. In support of UNCTAD, Edmundo Flores, chairman of the Group of 77 delegation to the Food Conference, criticized the FAO as being "like the Catholic Church during the reformation. Its organization, its methods . . . no longer suit the changing times" (Weiss and Jordan, 1976:627). They failed to dislodge the FAO from the conference but succeeded in replacing it later with two new agencies responsible for international post-conference food programmes. The World Food Council was made up of political leaders from member-states rather than U.N. bureaucrats. The International Fund for Agricultural Development was able to obtain sizeable pledges from Saudi Arabia, Iran, and Kuwait.

A proposal was circulated after the Desertification Conference to give responsibility for implementing the Plan of Action to the United Nations Development Programme (UNDP), also relatively popular among Third World member states, and the U.N. Fund for Population Activities in conjunction with UNEP. In the end, UNEP received sole jurisdiction. Tolba, as the Executive Director of UNEP, hoped this would improve UNEP's weak position within the U.N. system. At the time, UNEP was threatened with the possibility of losing its independent pledging status, without which its budget would be funded solely through the division of pooled resources among several agencies. Financial independence is a critical part of bureaucratic influence in the United Nations as elsewhere. UNEP's success in acquiring the anti-desertification programme, however, probably damaged its viability. UNEP has no executing authority (it cannot implement projects), but can only recommend and fund them from voluntary contributions. UNEP has a small staff, in contrast to the nearly global presence of the United Nations Development Programme (UNDP) and the FAO.

The only desertification programmes that were kept outside the UNEP organization were those that applied to the fifteen-country Sudan and Sahelian region. A Sahelian Office was established shortly after the drought. It acquired a separate trust fund and was made jointly accountable to UNDP and UNEP. It is responsible for implementing the Plan of Action in the Sahelian region.

Policymaking is a continuing process and UNEP has responded to some of the shortcomings of UNCOD's approach to desertification. In March, 1980, a World Conservation Strategy was announced which was commissioned by UNEP and prepared by the International Union for Conservation of Nature and Natural Resources (IUCN). The strategy aims at broadening the concept of conservation from that of a "limited sector" to one that "cuts across and must be considered by all sectors" (Development Forum 1980:9). Among the objectives for national action spelled out in the document are the "advance assessment of the likely environmental effects of all major actions" and the "inclusion of non-monetary indicators of conservation performance in national accounting systems" (ibid.). This last recommendation is similar to the argument made by World Bank economists for including the costs of poverty directly in evaluations of investment project benefits. UNEP's persistence with this strategy could help to move desertification into the agricultural policymaking arena and to capture initiative in assessing land use policies before they are implemented. In order to do this, UNEP will need to move aggressively into the development field.

II. THE INTERNATIONAL POLITICS OF FARM MECHANIZATION

UNEP's lack of influence within the U.N. bureaucracy is compounded by the range of international public and private organizations whose actions affect land use and land quality. Promoters of the World Conservation Strategy envision gaining access to international development agencies to encourage conservationist views, but that will not be easily accomplished. Early U.N. programmes for arid lands have been dominated by scientists and have generally not succeeded in involving economic planners or political decision-makers. The problem is equally an ideological one. Even were political leaders to be involved, the view that environmental protection is only tangential to growth makes it unlikely that leaders would adopt conservationist policies (Enloe, 1977).

The assumption of unlimited growth potential is common, even within the United Nations. In 1977, for example, FAO Director-General Tsouma announced, in Moshi, Tanzania, that around him he could see abundant resources, awaiting exploitation—resources that could make the region self-sufficient in food. The distribution of land and its influence on land degradation are secondary in the FAO's more technological approach to agricultural growth. Such a position is politically attractive because the assumption of abundant land to be developed in the Third World avoids North–South conflict over the international distribution of food and land resources.

Emphasizing the land's capability to grow more crops skirts the issue of inequality and limited land resources at the risk of promoting the overuse of fragile lands. Desertification is a politically marginal issue relative to high-technology agriculture in the international bureaucratic hierarchy. The anti-desertification programme was placed within UNEP, a weak agency without executing powers, which has had difficulty keeping the issue alive. Agricultural modernization, by contrast, has attracted public international development agencies and private multinational enterprises. Where the two issues clash, the modernizers greatly out-number, out-organize, and out-spend the environmentalists.

The effects of mechanized farming upon land that already is of poor quality were becoming clearer as, by the end of the 1970s, mechanized farms covered more of the semi-arid lands in the non-industrialized world. A good indication of the recent growth of mechanized farming is the increase in numbers of tractors in use. In Sudan, for example, tractor numbers nearly *tripled* between 1967 and 1973 (Lees and Brooks, 1977:44). Tractor-driven ploughs can quickly remove topsoils from fragile lands. Also, farmers who have machinery can actively farm larger areas, contributing to land degradation by (1) extending their operations into fragile or semi-arid land areas and (2) by pushing herdsmen from stronger grazing areas into more fragile ones.

Most Third World countries import the agricultural machinery that they use. The export promotion policies of the industrialized countries and of development funding agencies encourage mechanized farming. The World Bank is one of the development organizations that typically has supported agricultural projects with high mechanization requirements. Through the Turkish Topraksu agricultural development agency, the World Bank and the European Investment Bank funded two state irrigation projects (accounting for 80% of all Topraksu irrigation expenditures) that demanded increasingly large investments in machinery and equipment. That portion of the Topraksu budget devoted to machinery and equipment increased 800% between 1965 and 1971 (Mann, 1972:6). The outside funding agencies' projects dominated the state budget. It was assumed that Topraksu funds would be made available on a priority basis for these two individual projects. Other smaller-scale activities indicated that the same objectives could have been ac-complished without the burden of heavy machinery imports.

Joint projects between state planners and international funding agencies often select the most expensive alternatives in terms of import costs and domestic subsidies. Tractor imports into Pakistan from the United States were promoted by Massey-Ferguson, AID, the World Bank and the Harvard Institute for International Development (HIID) which found both radical and conservative regimes in Islamabad receptive to mechanized farming. The fruits of the advisor's influence were a doubling of Pakistan's tractor imports between 1968 and 1972. In fact, Pakistan's government subsidized U.S. tractor sales so heavily that they were far cheaper to buy there then they were in the United States (Herring and Kennedy, 1977:28).

Advisors from industrialized countries foster farm mechanization in Third World countries in two ways: by setting the agenda for agricultural policy (the HIID economists in Pakistan functioned as a "shadow" planning commission) and by providing financial support for investments in farm machinery. Machinery producers promote sales directly. Their international operations are extensive and their experience with mechanized agriculture tends to be highly valued (sometimes overvalued) by agricultural planners in the importing countries (Bhatt, 1978:30, 45).

The preference of international development agencies for investments in mechanized farms has been clearly defined. According to an economist with the International Finance Corporation (IFC), agribusiness provides potential investors, of which the IFC is one, with a variety of investment alternatives. Most favoured by the IFC are investments in the "well-defined, commercially-oriented part of (an) overall project" (Lowe, 1977:26–28). Lowe feels that the IFC is in a unique position to provide the loan and equity finance needed by high-technology agriculture. Because traditional sources of credit are limited to smaller and shorter-term loans, project appraisals at the IFC reportedly do consider the higher infrastructure and overhead costs of high-technology projects but at the same time most project sponsors are agriculture equipment manufacturers.

By contrast, investments in improving land are less attractive. Land enrichment projects hold the least interest for the IFC because such "infrastructure . . . is already available in developing countries", while "provision of this infrastructure in a developing country can be very costly" (ibid.:28). It remains to be seen whether a February 1980 agreement by major international development banks to consider environmental impact will come to include enrichment projects. The IFC statement acknowledges reluctance.

As in the case of Pakistan, the organizational network that fosters high-technology agricultural production within developing countries characteristically includes private and public agencies. The Sudanese government has been actively trying to attract foreign enterprise, and the promotion of Sudan as a "breadbasket" for the Middle East accompanied the initiation of several large agribusiness projects there. In 1974, the government Mechanized Farming Corporation (MFC) approved a livestock and cattle project proposed by the Arab-funded Triad Investment Company. Technical assistance for the project was provided by the U.S.-based Arizona–Colorado Land and Cattle Company (Lees and Brooks, 1977:156). Another project, the

Simsim Mechanized Farming Project, was jointly funded by the IBRD and the Sudanese government, and implemented by the MFC.

The cost of the Simsim project was divided among land development (clearing and roads), machinery and equipment, and advisory and administrative costs. Two-thirds of the costs were for foreign exchange, the IBRD's share, making it an expensive project from the standpoint of the Sudanese investment alternatives.

The Simsim project promoted private, mechanized farming on lands where soil fertility had been a perennial problem. Sudan is part of the drought-affected region of northern Africa, and its delegates took an active role in UNCOD, but Sudanese environmental officials do not participate regularly in planning agricultural projects. Environmental input would almost certainly lead to less enthusiasm about high technology farming projects. When the environmental effects of the Simsim project finally were evaluated under the auspices of the United Nations University (UNU), a consultant concluded that the "rate of return to the economy seems to be roughly equal to the private rate, as long as long-term environmental consequences are excluded" (Thimm, 1978:18). According to the UNU study, if environmental consequences were taken into account, the IBRD appraisal of a 17% rate of return would have to be revised downward to 5%!

The political significance of Sudan's strategic position on the volatile Horn of Africa further explains the upsurge of interest in its agricultural development, and of U.S. and Saudi advisers in gaining access to the Sudanese bureaucracy. For the United States and the conservative states of the Arabian Peninsula, Sudan had become a valuable buffer against the pro-Soviet regime in Ethiopia, the radical Libyan regime, and the liberation movements active on the Arabian Peninsula. In this connection, it is interesting to note that farm labourers for the Simsim project were drawn from Eritrea and from the Tigre province in Ethiopia, both of which were in rebellion against the Ethiopian government. The immediacy of international strategic interests created an environment that was favourable to outside involvement in the Sudanese economy and Sudanese officials seized the opportunity to initiate visible projects. According to one interview, even a U.S. weapons manufacturer, Raytheon, was studying the possibility of investing in Sudanese agribusiness.

Mechanized farming is well established in all these countries with the support of influential government agencies and of private and public international organizations. Land quality is also a serious problem in all these countries. However, the high priority of food production and the capital and organizational backing of mechanized farming make it unlikely that environmental costs will be seriously considered in selecting agricultural investment projects.

III. STATE AGRICULTURE AND LAND DEGRADATION IN IRAN

States elites and development planners also reap political benefits from mechanized agriculture. In Iran, before the revolution, the bureaucracy supported high-technology farming over the protection of fragile lands. Although only one tenth of Iran's land is arable, the threat of further land degradation did not significantly affect agricultural policy.

In fact, public policy contributed to the process of land degradation in several provinces. In 1968, Iranian agriculture was at a critical transition stage in the process of mechanization. According to one view, it was vitally important that Iran not respond to the incentive to mechanize in the same expansive manner as had Turkey (Bowen-Jones, 1968:585). The ecological result of using tractors on fragile lands could be "catastrophic" (ibid.).

Nonetheless, the number of tractors imported into Iran increased ten-fold between 1962 and 1972 (Aresvik, 1976:164). Population pressure on Iran's limited range lands raised concern about overgrazing. The government responded to the perceived land degradation by placing legal limits on the numbers of grazing animals per hectare. The incentives to overgraze that led to the initial problem were not dealt with and tension between state regulations and economic survival remained.

Soil salinity, water shortages, and other arid land problems placed physical limits on agricultural productivity in Iran. The government's agricultural and good policies reinforced the effects of geography and climate. The Shah was committed to industrializing Iran and, although he initiated land redistribution, his regime allocated little capital to the agricultural sectors. With the massive oil revenues that were available, it was possible for the government to meet rising demand for food with imports from the United States, Australia, New Zealand, and elsewhere. The Iranian food market became increasingly important for U.S. farmers and for the balance of payments. In 1973, Iran was the seventh highest recipient of credits from the Commodity Credit Corporation (see Schultz, 1979:184). By 1977, Iran's bill for imported food was $2 billion.

Public investment in agriculture generally favoured large-scale mechanized agribusiness enterprise. One foreign agribusiness investor pointed to the benefits that he enjoyed: "They (the Iranian government) develop the water first and we come in and farm. It's an attractive arrangement" (George, 1977:150). Legislation governing farm land irrigated from dams was passed by the parliament in 1959. It gave the government the authority to lease all below-dam agricultural land to large-scale commercial agricultural enterprises. The legislation specifically was designed to permit agribusiness to move into southwestern Khuzistan (a semi-arid region). The project was fostered by the head of the U.S.-based Development Resources Corporation, David Lilienthal, who enjoyed the personal backing of the Shah. Small peasant farms below the dam were bought out with the assistance of World Bank and IFC capital (Anon., 1976). Forty-thousand acre farms were set up by (Iranian–U.S.) joint stock companies to produce cash crops. The farms were large and unmanageable and the government finally bought out most of the U.S. investors.

Iran's contribution to UNCOD, a multidisciplinary study of desertification in Turan in northeastern Iran (see Part Two B) showed among other things that mechanization increased the economic dependency of those who farmed marginal lands and, in some areas, led to intensified use simultaneously with less investment in the quality of the land (Iran, 1977:12). Farmers spent scarce capital on machinery rather than on restoring the land in order to maximize short-run productivity. By contrast, the vast agricultural project in the southwest of the country was undertaken without any preparatory impact studies despite the magnitude of the project and the consequent potential for land degradation of equal proportions.

The Turan investigations indicated that land degradation resulted from processes more complex than nomadic overgrazing, a conventional explanation for desertification. Adequate protection for fragile lands would necessarily involve planning for "desert *and* related non-desert areas" and would "apply the same ecological, as well as economic, standards to both" (ibid.:8).

Implementing such a recommendation would require radical changes in the political structure of agricultural policymaking and in the attitudes of decisionmakers toward investments in land. Farm mechanization in the agricultural region north of Turan took place without regard to its effect on the quality of the land. During the 1930s, Reza Shah, founder of the short-lived Pahlavi dynasty, took over most of Gorgan as crown land that he could personally control and develop. He initiated mechanized, cash-crop farming—cotton and wheat—under the assumption that mechanization would increase agricultural productivity, while simultaneously increasing state control over an area where nomadic and settled populations had been in conflict for many years. Assisted by public funds, tractors were introduced through the local agricultural office, while profits from the land reverted to the Pahlavi family (Okazaki, 1968).

Reza Shah's son, the late Mohammad Reza Pahlavi, allowed the crown lands to revert to private ownership, giving local bureaucrats and military officers options to buy the land. The new owners enjoyed several incentives to continue with mechanization. The central government's Plan and Budget Organization gave them loans to buy tractors. They also had access to cheap labour from Baluchistan, where a drought was pushing people off the land. Individual farm profits were high because the state was underwriting the costs of infrastructure and machinery. The transformation of Gorgan's agricultural economy from animal husbandry and small farms to large mechanized farms also helped to undermine the Turkmen tribal economy and society and centralize control of the region under Persian rulers in Tehran. Few of the landowners would hire Turkmen workers, who were left with reduced grazing lands and without alternative income. Resentment over the usurpation of Turkmen grazing land still was sufficiently alive in 1979 to foment rebellion against the new Islamic regime in Tehran, with the support of leftist Fedayeen.

According to Okazaki the long migration of farm workers from southeast Iran to Gorgan's cotton fields illustrates the economic interdependence of Iran's provinces, which are more isolated in appearance than in fact. This is also true of Gorgan's farms and the isolated agricultural communities on the edge of the central deserts. The extension of cash-crop farming in Gorgan meant the expansion of cotton and wheat production, both relatively drought-resistant crops, compatible with mechanized farming techniques. Although Iran can produce cotton neither efficiently nor cheaply, by the 1960s cotton processing came to dominate the regional economy. Half of the cotton is processed in local factories. The remaining half is exported as lint with very low returns. Cotton mills in Semnan and Gonbad-e Kavus provided employment for villagers unable to find land to farm. Even Turan was eventually affected by the expanding cotton economy. The cumulative effect of the original mechanization of farming through direct state intervention was to increase pressure on marginal lands. In Turan, this was aggravated by government licenses which were issued to commercial, non-resident herders to use local grazing lands (Spooner, 1979), and now 80% of the animals grazed in Turan belong to non-residents.

State intervention in agriculture in Iran has been consistent in its support for mechanized farming. Where local organizations were able to resist state intervention, as in the shrine lands around Mashhad, small-scale food crop farming typically was preserved (Flower, 1968). When Islamic leaders took over the state apparatus in 1979, they announced plans to emphasize small and medium-sized farming over agribusiness (largely in the hands of foreign corporations). But they simultaneously promised to increase mechanization. Mechanized farming still appeared to offer the best means of increasing food production, reasserting state control over the countryside, and getting food to the politically volatile cities.

The stability of the new Islamic regime was threatened by the food shortages that it inherited from the monarchy and that were worsened by the revolution itself. Throughout post-revolutionary Iran, control over land is the central issue in determining who would dominate the countryside. During the interregnum many landless labourers took over vacant lands under the authority of the Islamic precept that land should not lie vacant and unproductive. Farmers in Turan began to ignore the state's restrictions on the protected lands and extended their cropping boundaries (Spooner, personal communication). These political realities make it unlikely that anti-desertification will be a high priority in Iran in the near future.

IV. CONCLUSION

The scarcity of farmland in Iran and the rush of international investors to Sudan are the reality confronting United Nations policymakers. UNEP and other anti-desertification programmes operate alongside, and often in competition with, international agricultural interests and powerful state bureaucrats. The political objective of short-term food security unites the interest of state elites, investment bankers, the producers of agricultural machinery and other inputs, and the providers of technical assistance. Agricultural investment decisions in Turkey, Sudan, Pakistan, and Iran all conform to a pattern in which the appeal of modern farming technology and the availability of foreign capital encourages political leaders to implement mechanization policies. In the process, supportive bureaucracies are created which have ties to international organizations and to the mechanized farms at home—a powerful mechanization lobby. Mechanization gains further momentum from the opportunities that it offers for state elites to control the agricultural sector.

In one very fundamental respect, desertification is an international North–South issue. The need to expand food production makes it more difficult for political leaders to consider the environmental impact of their agricultural policies. Industrial crops, providing foreign exchange, also compete for scarce land. In light of the extremely unequal global distribution of food, a strong argument could be made that any short-term decreases in food production imposed by land protection measures should be shared by the wealthier countries. Much of what is produced on these lands is consumed in the developed world. In as much as a global agricultural system clearly exists, we also have a global environmental system.

UNCOD was unable to gain support for a desertification fund. The organization of the conference and the environmental focus of the background work did not contribute as much as it might have to making the link between the global agricultural and environmental systems that would have justified the fund. It should be possible to develop guidelines for a "land use tax" from cost-benefit studies like those undertaken in Sudan and Turkey which would provide "shadow prices" for agricultural products in which the costs of revitalizing land would be included.

The country examples in this chapter also illustrate the need to integrate anti-desertification planning into national agricultural development pro-grammes. They indicate how strong is the appeal of mechanization and how closely agricultural policy is identified with state security. UNEP's new World Conservation Strategy indicates a sensitivity to the need to link land protection to agricultural development plans. A useful next step would be to review UNCOD's recommendations on crop rotation, irrigation, and other agricultural practices within their public policy contexts, as we have looked at mechanization, to determine appropriate targets for anti-desertification programmes.

Desertification is an international problem, a national one, and a local one for the most marginal farmer and herdsmen. UNCOD and its aftermath does not prove the futility of international action in a world of nation-states. An international agricultural establishment very effectively influences decisions at the nation–state level. UNCOD's lesson is the need to use international organizations to articulate alternative land use policies clearly with their attendant costs and benefits. Mechanization is just one process that can contribute to land degradation and that nonetheless has continued unchecked because investors have been allowed to discount its costs. This is where international environmental agencies can make a contribution, confronting directly the tension between the environment and powerful international organizations and state bureaucracies.

Endnote
[1] Most of this recent history of UNEP's desertification activities comes from interviews with former UNEP officials and from conference participants.

3

Archaeology and the Study of Desertification

R. W. Dennell

University of Sheffield
Department of Prehistory and Archaeology
Sheffield S10 2TN, U.K.

Archaeology is perhaps of all the academic disciplines that may be included under the heading "social science", generally thought to be the furthest removed from practical application. However, a notable inadequacy in most of the work on desertification so far is lack of sufficient historical perspective to allow realistic assessment of present trends in relation to the past. Archaeology can make good this lack. This chapter reviews the competence of archaeology in relation to the problem and holds out the promise of setting desertification in a more realistic perspective.   —Ed.

Investigations into the causes, processes and even diagnosis of desertification are frequently hampered by the lack of a long-term time-scale. The reasons for this are not hard to find. In many dry regions, it is only in the last few decades that the attempt has been made to compile the type of environmental and demographic data that were being routinely collated in parts of Europe as early as the 18th century. Studies of long-term processes—whether of climate, vegetation, geomorphology or human settlement—were undertaken in temperate regions long before they were attempted in dry lands, partly perhaps because deserts posed few research problems that could not be answered in greater comfort and more easily elsewhere. The methodologies used in these studies often proved inappropriate when transferred to dry

43

lands, and this problem further impeded the understanding of similar processes in deserts. As a result of all these factors, the data for placing present-day dryland conditions in a long-term perspective are either unavailable, or woefully inadequate.

The lack of a long-term outlook on dry lands has two serious consequences for our understanding of desertification. The first is that it is often difficult to distinguish short-term from long-term trends. Climatic fluctuations are a case in point. Before weather forecasting in dry lands can be reliable on even a ten- or twenty-year basis, we need to place the trends recorded by meteorological observations over the last fifty years or so against those of the last thousand years or more. Only then will we learn whether or not the frequency and severity of droughts (and floods) in recent times have been "normal" for dry lands. If, as has been suggested, they are atypical, and if they occur on a cyclical basis, we need to know how often and how regularly they happened in the past, and for how long they persisted. Recent events in the Sahel confirm this point. Had it been realized earlier that conditions of extreme aridity were episodic, the dangers of increasing stocking levels during moist decades earlier this century might also have been realised long before rainfall began to decrease (Winstanley, 1973). Given the vulnerability of dry lands to the effects of minor climatic oscillations, historical climatology clearly becomes far more than a mere academic exercise with no relevance to the contemporary world. The same point will be made in this paper with respect to archaeology: the more that is known about the way man has used deserts in the past, the better we might be able to evaluate the causes of desertification in the present.

The second adverse result of having only a short time-perspective on dry lands is that it is often difficult to identify the causes of environmental change, particularly as each component of the environment may change over a different time-span. Increased rates of soil erosion, for example, can result from several factors, such as changes in the intensity and amount of rainfall, in stocking rates and cropping practices, or in the type of material being eroded. Each factor, along with its effects on other components of the environment, has to be assessed before appropriate remedial action can be taken. Often, the data on record extend back too short a time for all these to be properly evaluated. Epirus in western Greece provides a convenient example of this point. Like other Mediterranean lands, it has suffered extensive soil erosion, and this has been widely attributed to the consequences of goat-grazing over the past six millennia or so. However, geomorphological investigations indicated that the area had experienced soil erosion long before the inception of goat grazing, and indeed, over most of the last fifty thousand years. This led the agriculturalist, Hutchinson (1969), to reverse his previous judgement about the extent of ecological damage caused by the goat, and to conclude that in this instance goats could not be regarded as the sole causal agency of soil erosion.

At first sight, it might seem that prehistoric archaeology has little to contribute to either problem, or to any others about desertification. After all, for the general public, much of the fascination of the stone tools, pot sherds and ancient metalwork that fill museum cases lies in their remoteness from present-day concerns, and the material unearthed in excavations would strike few as particularly relevant to contemporary problems. However, whilst there is some substance to popular notions of what archaeologists do, it would be erroneous to claim that the only value of their activities is to enhance our appreciation of past societies and to provide opportunities for occasional distraction from the present. Prehistorians have in fact a great deal to contribute to our understanding of the problems currently affecting dry lands and many of their inhabitants.

The main value of prehistoric studies in this field lies in the fact that prehistorians are the only people who investigate human behaviour on time-scales that can encompass the phenomena studied by both the social and natural sciences. Social scientists such as social anthropologists and economists can only rarely obtain information on the ecology of dry lands that extends back more than a few decades. Although historians are more fortunate in that they can study processes over centuries, and even over a few millennia in rare cases such as Egypt or Mesopotamia, much of their information is piecemeal, second-hand and difficult to verify. Archaeological data provide the only evidence of human activity from periods before reliable written or oral evidence.

Natural scientists are more fortunate than their colleagues in the social sciences in that they can obtain direct evidence for processes that may operate over millennia or even millions of years. However, whilst they can place present-day conditions within a long-term perspective, they often find it difficult to evaluate the effect that man may have had upon the processes that they study. The way that human activities are likely to have modified the development of plant and animal communities, or the geomorphic processes of erosion and deposition are some obvious examples.

In the context of desertification, archaeologists can provide a useful long-term perspective on human activities in dry lands that can supplement the short-term studies conducted by historians, anthropologists and economists, and can enhance our understanding of the long-term processes that concern botanists, zoologists and geomorphologists. This chapter reviews what they can contribute.

I. LONG-TERM STUDIES (10^4–10^7 Yr BP)
OF THE PLEISTOCENE

Archaeological problems on this time-scale are chiefly concerned with human evolution. The earliest part of this immensely long period is marked by the separation of hominids from the ancestors of modern apes in the Miocene, and their earliest tool-making activities in the late Pliocene and early Pleistocene. Major areas of research in later periods are the evolution of *Homo sapiens*, his adaptations to the environmental conditions of the last glaciation, and the colonization of the Americas, Australia and peri-glacial regions. The Pleistocene is a period relevant to many disciplines—notably climatology, geomorphology, botany, zoology and ecology—and archaeological studies of early man impinge on several types of studies in the natural sciences.

In what ways can archaeological studies of desert regions during the Pleistocene be useful to our understanding of the contemporary process of desertification? The first is that Pleistocene studies depict a simpler situation than the present. Since man can for the most part be excluded as a major component of Pleistocene ecosystems (but see below), archaeological studies on this time-scale can show us how desert regions functioned before the additional complication of man-induced environmental changes. It might be argued that this type of study need not be confined to the Pleistocene. Although this point is to some extent a valid one, studies of this period have several advantages over those of earlier periods. One is that both the quality and quantity of information on the Pleistocene is far superior to that of other geological periods; others are that the plant and animal communities are much the same as today's, and that complications such as continental drift and phyletic replacement can be excluded.

If man was an unimportant part of Pleistocene environments, it might be thought that investigations of deserts on this time-scale are best left to natural scientists. However, archaeologists have an extremely useful role to play. Pleistocene archaeological sites provide many disciplines with their main sources of dating material, and much of their palaeo-environmental evidence in the form of pollen, faunal remains, soils and sediments. Often, the results of such investigations are of mutual benefit to archaeologists and other specialists. A good example can be found in the recent multi-disciplinary studies of the Egyptian Sahara (Wendorf *et al.*, 1976), where archaeological sites provided the main evidence for faunal and vegetational changes associated with datable deposits. In return, palaeo-ecological analyses ensured that the record for prehistoric occupation could be placed within a detailed climatic and environmental context. Recent studies of the Thar Desert (Allchin *et al.*, 1978) and the Mungo area of Australia (see Jones, 1979) are equally noteworthy examples of how the Pleistocene history of dry lands can be documented by a judicious combination of archaeological and palaeo-environmental expertise. A further example is provided by recent taphonomic investigations concerned with showing how living animal populations became fossil assemblages, and have proved immensely useful to palaeo-ecologists wishing to model the composition and habitats of extinct animal communities, as well as to ecologists studying mortality rates of modern animal populations. Much of the stimulus to this type of study has come from archaeological investigations into the environmental contexts in which early hominids lived and died, and into the food and tool debris they left behind (see, for example, Behrensmeyer and Hill, 1979).

The second way that archaeological studies of the Pleistocene can be relevant to our understanding of desertification lies in assessing the extent to which human groups at low levels of population and technological efficiency can modify their surroundings. In more specific terms, this problem can be defined as one of evaluating the role of late Pleistocene human populations in causing vegetational and faunal changes. One of the most interesting suggestions raised by this type of enquiry is that the concept of a "natural" desert landscape, as one wholly unaffected by human activities, may be illusory for much of the period during which modern man has been living. This viewpoint obviously impinges uncomfortably on those who envisage hunter-gatherers as living in a "natural" world, in contrast to pastoralists and farmers who were supposedly the first to modify their environments.

This type of investigation is clearly more sophisticated than ones which simply view archaeological sites as convenient sources of data for various specialists. So far, the effects of late Pleistocene populations on their environments have been studied in two ways. The first has been to consider the causes of vegetational changes. Perhaps the most interesting speculations on this matter have come from Australia, where it has been suggested that the aboriginal populations of Tasmania and much of the mainland may have altered the vegetation over the last 20 000 years or more through the deliberate burning of scrub in order to promote the growth of young shoots and to facilitate the capture of game. This "fire-stick farming" (Gould, 1971) reduces the probability that the Australian vegetation was "natural" before the arrival of Europeans with their sheep and cereals. Indeed, it is even possible that the Australian vegetation was being artificially modified long before the generally accepted emergence of modern *H. sapiens* some 35 000 years ago. At Lake George near Canberra, a pollen profile has been obtained which apparently covers the last interglacial (probably *c.* 100 000 and possibly 150 000 BP) and which, it is claimed, shows evidence of anthropogenic burning (Singh, in press). Although this claim has yet to be confirmed by independent data, it becomes more reasonable when set against other recent suggestions (Sagan *et al.*, 1979) that man may have been altering the earth's albedo by burning vegetation over the last half million years.

In the Old World, a few studies have indicated that late Pleistocene

populations may well have been modifying their surroundings. In addition to
the well-known evidence for early Holocene burning of woodlands in
northern Europe (see e.g. Mellars, 1975), we should note the evidence for
scrub burning from the Upper Nile some 14 000 years ago. South of Aswan, a
layer of charcoal could be traced in sections along the Nile for about 200 km;
this was thought to be too extensive for a natural fire (Clark, 1971). When
considered alongside other archaeological evidence that the late Pleistocene
populations of the area were intensifying their quest for food by using grain
resources, small game and fish, the suggestion seems more reasonable. We can
also note the point made by Wendorf *et al.* (1979) that human groups may
have been altering the vegetation of the Egyptian Sahara as long ago as
18 000 bc to encourage the growth of cereals.

A less direct way of investigating the effects of palaeolithic societies on their
environments has been to examine the causes of Pleistocene animal
extinctions. The main arid area where this problem has so far been
investigated is again Australia, where the "megafauna", comprising several
types of large marsupials, died out after the arrival of man. As in other areas
where large-scale faunal extinctions occurred at the end of the last glaciation,
the debate has been intense as to whether man, climatic change, or both, were
responsible. A decade or so ago, opinion tended to attribute these extinctions
to human agencies (e.g. Mosimann and Martin, 1975), but now seems to have
swung back in favour of climatic causes (e.g. Gillespie *et al.*, 1978; Grayson,
1977; Jones, 1979), with man perhaps delivering the *coup de grâce* to many of
the larger mammals.

The question of whether or not late Pleistocene populations modified the
vegetation—and, as a result, the fauna—of arid lands is more than an
academic quibble, and has some serious implications for many conservation
programmes. If man modified dryland environments both before as well as
after the introduction of domesticated plants and animals, it would seem
naive to hope that dry lands could be restored to their "natural" state by
simply trying to make them revert to their pre-agricultural conditions. In
order to re-create the "natural" state in which they existed before being
altered by man, it would probably be necessary to change them back to their
condition of some 20–30 000 years ago. This goal is clearly impossible, since
the climate then, at the height of a glaciation, was completely different from
today's; many species have also become extinct. From an archaeological
viewpoint, a "natural" environment under present-day or comparable
climatic conditions is unlikely ever to have existed in areas inhabited by man
since the late Pleistocene. The concept of a "natural" environment in fact
embodies something quite different, but rarely recognized as such: it is simply
an environment that contemporary attitudes perceive as sufficiently different
from the present as to *seem* uninfluenced by man. In other words, it is a set of

subjective attitudes that determines a "natural" environment, and not a group of objectively defined environmental parameters.

The point is illustrated by an example from temperate regions. In highland Britain, there are large expanses of moorland that are generally regarded as "natural", and these are protected both by legislation and public concern. Yet these moorlands are probably no more "natural" than the fields and hedgerows around them. As pollen analysis and archaeological investigations have shown (see e.g. Moore, 1975), moorlands were usually created by prehistoric over-grazing and forest clearance which together removed the tree cover, and encouraged the formation of peat and its associated grass and sedge communities. To restore these areas to their "natural" state would now be impossible because the soils have changed irreversibly; it would be even less practical (or even desirable) to restore the fauna to its pristine composition by re-introducing the bear and wolf.

What is at issue here is that the concept of a "natural" arid landscape under modern, post-Pleistocene climatic conditions, is not just simplistic but misleading. It would seem more appropriate to think in terms of a "desired" environment, whether or not it existed in the past, than one that is spuriously "natural". A consensus definition of a "desired" environment would probably embody several attributes: i.e. that it contains a high diversity of species; is self-regulating and thus requires little supervision; is a suitable refuge and breeding area for rare and endangered species; and is also aesthetically pleasing. The archaeologists' contribution is to help show how this type of "desired" environment differs from previous environmental conditions over the last ten thousand years.

II. MEDIUM-TERM STUDIES (10^3–10^4 Yr BP) OF THE HOLOCENE

Studies of this time range cover the origins and dispersal of agriculture in both the Old and New World; the spread of pastoral economies across the Near East, the Sahara and eventually into southern Africa; and also the rise of literate urban communities in the major river valleys of the Nile, Tigris–Euphrates and the Indus. For the moment, we can leave aside the problems connected with studies of early civilizations, and instead concentrate upon those associated with early agricultural use of dry lands.

There are two ways in which archaeological studies on this time-scale can contribute to discussions on desertification. The first is to help improve the reliability of the evidence for climatic change in dry lands over the past few millennia by distinguishing environmental changes that were induced by man from those brought about by a change in climate. In practical terms, this contribution is useful as it will ultimately help to improve our knowledge of how and why dryland environments have changed over the last few millennia, as well as the reliability of climatic forecasting in these regions.

A. Holocene Environmental Change in Dry Lands: Man-Induced or Climatic?

Although the changes in the earth's climate since the end of the last ice age some 10–12 000 years ago have received a prodigious amount of scientific attention, they have proved extremely difficult to elucidate. Three reasons for these difficulties are worth considering before we turn to the problems of unravelling the climatic record of dry lands.

In the first place, the climatic changes within the present interglacial have been very subtle compared with the profound contrasts between glacial and interglacial conditions that characterize the study of the Pleistocene climate over the last million years. Holocene climatic changes have been short-term, minor and regional, and their detection requires a record of climatic change that is sufficiently sensitive to monitor shifts of perhaps only 1–2°C in mean annual temperatures over a few centuries. Secondly, our "signal" of climatic change over the last few thousand years is usually a compound one that is difficult to break down into its components. A common example is where an environmental change could have arisen from an alteration in temperature, precipitation, or a combination of both. In only a few instances can one of these variables be safely eliminated. For example, changes in the width of tree-rings in dry lands is usually determined by variations in precipitation; while the growth of glaciers in areas such as New Zealand can be attributed to fluctuations in temperature, as precipitation is not a limiting factor.

The third reason why Holocene climatic change is so difficult to record is that almost all the evidence is indirect, being derived from a record of environmental change. Unfortunately, climate is only one of many variables that can alter the environment. The link between a climatic change and an environmental response is often a tenuous one. For example, the expansion of forest over much of northern Europe at the end of the last glaciation may have been caused by a change in climate, but its composition and the rate of its expansion were determined by the speed with which each tree species could colonize new areas. Similarly, the disappearance of reindeer from western Europe at the end of the last ice age is popularly attributed to its intolerance of warm climates but probably owed more to the replacement of its staple lichen by grasses that were in turn more successfully utilized by red deer.

In other cases, environmental change could have occurred entirely independently of the climate. Perhaps the classic example of this state of affairs is provided by vegetational succession, whereby the vegetation of an area can pass through several stages before reaching its final "climax" community that is thereafter stable. A second example that is of far greater importance in the context of this paper concerns the environmental consequences of human activities. Once man had developed crop cultivation, pastoralism and, in later periods, technologies such as metallurgy that required large-scale usage of fuel, he was able—either deliberately or inadvertently—to alter his environment. (As noted already, the environmental effects of pre- and non-agricultural populations might not have been negligible.) How long ago man's agricultural activities had begun to leave a permanent mark on the environment is still unclear. The steppe vegetation of Syria and South West Iran may have been altered by sheep and goat grazing by 6000 BC; by 4000 BC, the expansion of cereal cultivation and stock keeping over much of temperate Europe had already affected many areas of woodlands. The developments of smelting techniques in the following millennium throughout much of Europe and the Near East is also likely to have had widespread repercussions on the vegetation and rates of soil erosion. Although the effects of these activities are still little known, it nevertheless seems reasonably clear that the signal of climatic change embodied in the environmental record is increasingly swamped by the background noise of human activities for much of the last 5000 years over a vast territory from the Indus to the Atlantic, and from Scandinavia to the Sahel.

These problems have considerably impeded our understanding of how and why different environments have changed since the beginning of the Holocene. This is particularly true of most dry lands since the quality and quantity of data are generally insufficient to show what parts of the environment have changed, and still less why. These uncertainties have given rise to two major lacunae in our understanding of desert regions. The first is a general problem of whether we are dealing with processes of desertization, or desertification. Present opinion is far from unanimous. Some years ago, Raikes (1967) performed a useful service by attacking the prevalent customs of ascribing Holocene environmental changes to climatic factors at the expense of human agencies. Since then, it has become apparent that both are important: Holocene climate has not remained constant, and the environmental consequences of human activities have also fluctuated considerably. Doubts as to which process is the more important permeate many discussions of dry lands.

Studies of the Sahara and Thar Deserts are appropriate examples of these doubts. Some (e.g. Lhote, 1973; Raikes, 1978) have claimed that the Sahara's aridity can largely be attributed to human activities. Adherents of this view can cite the depictions in rock art of the Early Holocene of animals such as hippopotamus that are now either extinct or very restricted, and later representations of pastoral communities with large numbers of cattle. It is also possible in arguments of this kind to cite recent models of biological feedback mechanisms in desert environments, which show how over-grazing can increase the regional albedo, rates of soil erosion and surface run-off, and depress the reliability and amount of rainfall (e.g. Charney et al., 1977; Watkins et al., 1974). On the other hand, opponents of this viewpoint have argued that man adopted agriculture in the Sahara against a background of increasing aridity, and thus responded to a declining supply of game and wild plants by using different resources that could be more closely husbanded (Shaw, 1976). In the case of the Thar Desert, Bryson and Barreis (1967) suggested that its present aridity is very much a human artefact. Overgrazing in the third millennium BC—during the time of the Indus civilization—led to soil erosion and the release of large quantities of dust into the atmosphere; this in turn reduced atmospheric subsidence and restricted the influx of monsoon rains. Unfortunately, archaeological evidence from the area is inadequate to confirm or rebut this model. In addition, Harshvardhan and Cess (1978) examined some of the mathematical features of this model and found them unsatisfactory.

Debates of this kind highlight the inadequacies of much of our data on the environmental changes that have occurred in dry lands since the end of the last ice age. In many areas, the causes of environmental change over the last 5000 years cannot be confidently attributed to climatic, human or other factors. An unfortunate consequence of these uncertainties is that our understanding of the long-term causes of desertification are weakened if we do not know the extent to which present desert environments are a climatic or human product. Until this issue is resolved, it remains unclear whether the past human record in these regions has been one of marked adaptability in the face of climatic change, or one of increasingly irreversible environmental damage.

Although the present state of research is far from satisfactory in many desert regions, the next few years should see some marked improvements in our knowledge of climatic and environmental change over the last ten millennia. One of the most promising ways of documenting climatic change would seem to lie in the analysis of lake sediments, (for example Maley, 1977; McClure, 1976), since lake volume and chemistry are largely dependent upon temperature and precipitation. The value of lake studies may also be considerably enhanced if the techniques of isotopic analysis that have been so successfully developed for studying Pleistocene climatic change from ocean-floor sediments (see for example Imbrie and Imbrie, 1979; Shackelton and Opdyke, 1976), can be transferred to inland bodies of water (Shackelton, 1980). Once the main climatic variables of temperature and precipitation are established, environmental changes affecting lakes and their environs can be monitored by noting changes in sediment, fauna and pollen. This allows discrepancies between the climatic and environmental records to be assessed in terms of human activities such as pastoralism, crop cultivation and smelting; clearly, at this stage of investigation, archaeological participation becomes essential.

B. Desertification and Human Activities

The last-mentioned point leads us to the second problem raised by our poor understanding of the past history of desert regions, namely that we cannot yet reliably distinguish between the long-term effects of different kinds of human activities. This is important as the past history of dry lands has often been cited in political arguments about the causes and remedies of desertification. Arguments of this nature usually take one or two forms. The first is to use historical evidence in nationalistic arguments that assert the superiority of one ethnic group (usually the one in power) over others (usually neighbouring or previous power groups). Recent changes in the way history is viewed in parts of North Africa provide one example out of many that could be chosen. When Libya became an Italian colony, the Romans were often extolled as an ideal example of how desert regions could be made fertile by a judicious combination of investment and wise government. A generation later, we find that the Romans are now being cited as a prime example of how colonial exploitation can all but destroy the agricultural potential of their overseas possessions. The sad truth in this instance—and in most of the similar arguments echoed throughout many dryland areas—is that the evidence of land use over the last two millennia is so poorly documented that it can be used to support any claim.

The second type of political argument that invokes historical data can be termed ecological imperialism. Here, the preferences and prejudices of groups living outside dry lands towards particular resources or technologies can be superimposed on dry land development programmes. The goat is a good example in this context. West Europeans and North Americans tend to hold this animal in very low esteem on the grounds that it wreaks ecological havoc. When this outlook is combined with historical generalizations to the effect that the goat has destroyed much of the original agricultural potential of the Near East and North Africa, it is easy to see how its elimination is often regarded as one of the main remedies for desertification. Yet the evidence that the goat has browsed its way over the last ten millennia through a lush and fertile Eden is far from impressive, and is certainly open to other interpretations. After all, this animal has been an integral part of subsistence economies in the Near East, the Mediterranean and North Africa for some eight millennia, and it would seem dubious that a resource would persist for so long if it were as much of an ecological vandal as is frequently claimed (cf. Nyerges, Chapter 11). What needs to be considered carefully are the long-term effects of other human activities on scrub vegetation—notably the collection of brushwood for domestic fuel and industrial charcoal—as well as the long-term ecological consequences of other animals (for example, sheep and cattle) at different rates of stocking, and different types of crop cultivation. Whilst the goat has undoubtedly played a part in the process of desertification, it would be unreasonable to single it out as the primary cause without a detailed knowledge of the ecological consequences of other agencies.

The long-term impact of intensive agricultural systems on dry lands is another topic that is relevant to both desertification and archaeology, and which is at present poorly understood. As usual in archaeological debate, we can easily find views on the merits and drawbacks of ancient irrigation systems that are diametrically opposed. The older and possibly more prevalent view sees the ancient civilizations of Egypt, Mesopotamia and the Indus as examples of how dry lands can be developed, given a clear-sighted and stable government, and the appropriate technology. In the case of Iraq, the collapse of the Abbasid dynasty under the impact of the Mongols, and the subsequent abandonment of the irrigation networks, can easily be invoked as an instance of how exogenous social factors can destroy an otherwise stable agricultural system. The political implications of this type of interpretation are fairly obvious: given stable government and sufficient injections of capital, similar systems can be made to work today. A clear example is demonstrated by Adams' (1962) work in Khuzistan on Sassanian irrigation systems, and the recent development plans for the area.

The alternative view draws attention to the ecological consequences of large-scale agricultural systems in antiquity, and their fragility. This point was made forcefully many years ago by Jacobsen and Adams (1958) for ancient Mesopotamia, and by Raikes and Dyson (1961) for the Indus civilization, and more recently by Gibson (1974), again for Mesopotamia. All these studies stress the short-term benefits of large-scale, centrally-controlled irrigation systems, the long-term adverse effects on soil fertility, and the increasingly difficult administrative problems of maintaining, repairing and regulating such systems. This argument also has its political connotations, not the least being a salutary reminder against the construction of large-scale irrigation projects simply because they are large-scale, and not because they are necessarily the most appropriate.

III. SHORT-TERM STUDIES (10^2–10^3 Yr BP) OF THE RECENT HISTORY OF DRY LANDS

Archaeological studies on this time scale can make several useful contributions to our understanding of desertification. Some are the same as those mentioned in the previous section. Much, for example, remains to be learnt about the climate over the last millennium, and archaeological data can be as useful in helping to elucidate this topic in dry lands as they have in much better documented areas such as northern Europe. Another contribution that archaeologists can make with short-term studies is by documenting the types and environmental consequences of activities such as charcoal-burning, smelting, pastoralism and crop cultivation.

Whilst these contributions are of value, they are not the main ones that archaeologists can make to our attitudes about present-day desertification. Archaeological studies of the last millennium could be especially informative in two main ways. The first is in showing the background to present-day and recent desert societies; and secondly, in helping to explore the role of exogenous political and economic factors in causing desertification.

A. The Historical Background of Contemporary Dryland Societies

As is well known, the history of many dryland communities before this century is very poorly documented, and much of what passes for knowledge is often little more than hearsay. This is as true for those regions with a long tradition of literacy—such as the Near East and much of North Africa and India—as for those like the Australian interior that were prehistoric until the last century. Because the documentary sources are often so meagre and imperfect, archaeological evidence is essential if we are to learn about the inhabitants of dry lands in pre-colonial or pre-contact times. Knowledge of their recent history is of value to those directly concerned with desertification today in two ways.

In the first place, the absence of evidence for change amongst dryland populations has often encouraged the view that no change has in fact occurred. Popular views of the Australian aborigines, the Kalahari Bushmen or the Bedouin nomads as the last survivors of an immensely ancient way of life are perhaps some of the most blatant examples of this outlook. It also manifests itself in the contrast often drawn between modern conditions and an earlier "traditional" way of life. The former is frequently seen as an embodiment of change, and even "progress"; the latter as static, conservative and not infrequently as "backward" and "primitive". In point of fact, many desert communities have changed profoundly in the last thousand years or so, and their history is no less dynamic than that of many societies living beyond the desert margins. Although the "traditional" way of life recorded by many ethnographers can depict a way of life that had changed little for centuries, it can also represent merely a short transitional period of change that was occurring at the time of European contact, and which was mistaken for the final phase of a long period of stasis. An obvious consequence of this point is that knowledge of the recent history of dryland populations can do much to place their present social and economic behaviour in perspective.

One recent study which is highly relevant in this context is by Schrire (1980) on the San Bushmen of the Kalahari, who have often been cited as one of the few extant examples of the hunting–gathering way of life that characterized Pleistocene economies. As Schrire points out, the San have lived in close proximity to herders for several centuries. Her analysis of historical records suggests that these people have frequently switched from hunting–gathering to herding—either for clients or by obtaining their own stock—as and when opportunities permitted. In much the same way, the Australian aborigines no longer seem the epitome of an unchanging people in an unchanging landscape, as they were presented only a few decades ago. On the contrary, recent archaeological research indicates that aboriginal society changed markedly over the 40 000 years before European arrival, and was changing at the time of contact (see, for example, Mulvaney, 1975; Jones, 1977).

Agricultural societies of both the Old and New World desert regions are as likely to have changed at least as much as those of hunter–gatherers over the past millennium. In many instances, evidence for these changes and their environmental consequences will depend upon archaeological evidence. In the Old World, we can note the spread of camel husbandry for herding and transport over much of the Near East and North Africa, as well as *qanāt*[1] irrigation across North Africa and into Spain. The origins and dispersal of both are still shrouded in obscurity: the dromedary was probably first domesticated in the Arabian peninsula in the first millennium BC, and the camel in Central Asia around the same time (Afshar, 1978). The spread of *qanāt* irrigation from Iran probably resulted from the eastward expansion of the Arabs after the seventh century AD. There is also much still to be learned about the introduction, diffusion and nutritional consequences of New World domesticates—notably potato, tomato, maize, peppers and sunflowers—across Europe, North Africa and the Near East. Nor should one overlook the consequences of inadvertent introductions such as prickly pear (*Opuntia ficus-carica*) which is now used as a fodder plant and for its somewhat sickly fruit. Although the adoption of American cultigens by the Old World was arguably the most important agricultural development over the 5000 years before the advent of mechanization, their rate of diffusion and effects on indigenous cropping systems have been sadly neglected. Indeed, the date of their introduction is still so poorly established that the case for pre-Columbian maize in Africa is difficult to confirm or refute (see, for example, Jeffreys, 1975).

In the New World, we can see the mirror image of the processes which so profoundly affected Old World agricultural societies over the last few centuries. The introduction of sheep, pig, horse, wheat and barley into the Americas following the Spanish conquest of Mexico and Peru had enormous repercussions on indigenous cropping systems. The same can be said of Australia, where the introduction of the rabbit was one of the least useful results of European contact. As in the Old World, much has still to be learned about the diffusion of these resources, and their impact on the landscape and peoples of the arid regions of the Americas.

The second way in which an historical perspective upon dryland populations is valuable is that it enables us to assess the extent to which they are now adapting to the consequences of their own past as well as to present-day conditions. The influence of the past upon the present is not always readily apparent to the observer, but it should not be ignored. An ethnographic example is provided by Marx's (1977) analysis of how the different subsistence strategies of two Bedouin groups living in the same environment are explicable by reference to their different political experiences over the last century and a half. Archaeological instances can also be cited. As argued elsewhere in this volume (Dennell, Chapter 9), the present-day irrigation farming systems in Tauran (north east Iran) occur within and are derived from an earlier system of land use that probably dates from the 10th century AD. Consequently, the decisions by modern farmers about what and where to irrigate are influenced by what had happened several centuries earlier, as well as by prevailing conditions of weather, market prices and local demand. A similar situation exists in parts of coastal Peru, where many of the irrigation systems date from pre-colonial times (Farrington and Park, 1978). As in Iran, decisions about water-management are consciously made in the light of contemporary conditions, but also unknowingly influenced by previous centuries of land use.

B. Desertification and Exogenous Factors

Perhaps the most challenging aspect of desertification that archaeologists could help investigate by studying historically documented periods is the extent to which it has been caused by exogenous political and economic factors. The circumstances in which the activities of those inhabiting deserts were influenced by the demands and decisions of those living elsewhere are not of course confined to the last millennium. The earliest instances probably arose some five millennia ago when centralized urban policies along the major river systems of Mesopotamia, Egypt and later the Indus extended their influence and sometimes control to those communities living beyond the limits of irrigation farming. The same theme was later repeated on a much larger scale when the Romans annexed North Africa and Syria, and again with the expansion of the Arabs after the 7th century AD.

Instances of desert regions being drawn into cultural and political systems extending far beyond their margins become both more numerous and better documented within the last millennium or so. From the last 1000–1500 years we can note the brief flowering of the Saharan Empires which controlled vast areas from the Sahel to the Sudan; the complex and fragile web of towns and roads that spanned the vast terrain from Cairo to Kabul; the growth of Mecca as the pilgrimage centre of the Islamic World; the emergence of ports around the Arabian peninsula and Persian Gulf that linked others along the East African coast and India; and the growth and gradual disintegration of the Ottoman Empire, with its vast fringe of desert territories extending from Mesopotamia and Arabia to Morocco. Similar examples further removed from the Mediterranean and Near East can also be cited: the vast areas of desert which lay within Mogul India, or within the Aztec, Inca and later Spanish Empires in Peru; and even within the Russian Empire after its eastward expansion into Central Asia after the 16th century.

These examples provide a fascinating range of circumstances under which dryland populations have been drawn into wider political and economic systems in recent times. As might be expected, the relationships between dryland and other populations have also varied enormously. Urban and rural populations beyond desert margins generated a variety of needs that could be supplied from dry lands—for example, meat, milk products, wool, minerals, ivory and even slaves. In some cases, they also generated capital that could be invested in dry lands in the form of irrigation systems, for example, and in addition depended upon communication networks that often traversed desert areas. In return, the more populous and fertile areas on the margins of deserts provided opportunities for wealth, power and prestige that the inhabitants of deserts could obtain through marriage, commerce, treaty or, on occasion, raiding. The political devices that were used to regulate relationships between those in deserts and beyond also took on a variety of forms. Military methods ranged from direct annexation to the maintenance of garrisons along the desert margin, to occasional punitive raids and shows of strength, or to arming client buffer states or other political groups. Intermarriage, bribery and intrigue were other devices that were often used, with varying degrees of success. Finally, the desert peoples with whom outside populations interacted also varied enormously—from the hunter–gatherers of the Kalahari, Australian interior and American South-West, to the loose confederacies of nomadic pastoralists of Arabia in the final years of the Ottoman Empire, to the highly organized but short-lived kingdoms of the Zulu and mediaeval Sahara.

What we have in the historical record is a vast amount of information about the extent to which desertification was caused by external pressures on desert environments. The record of the last thousand years or so should enable us to determine whether the process of desertification became prevalent only after desert peoples were incorporated—either directly or indirectly—into economic and cultural systems which linked them to outside areas. If the environmental consequence of these large-scale systems on dry lands was desertification, it would also be useful to know under what political and economic conditions it was most pronounced; what type of dry lands was the most vulnerable; and what part climatic fluctuations played in exacerbating or moderating the effects that these extraneous pressures had on the environment.

Not surprisingly, given the complexity of the subject as well as the amount and variety of data, such enquiries present a challenge for future research rather than a topic of present discussion. Whether or not it proves fruitful will depend upon two conditions being met. One is that it is multidisciplinary, involving the collaboration of historians and natural scientists. Secondly, it should utilize a model which can study both desert environments within an ecological framework, and their human populations within a cultural framework which itself integrates them with the wider political and economic systems to which they belonged.

Archaeologists have an important role to play in such investigations. In the first place, they have an advantage over historians in their ability to study long-term processes, and can thus place the record of the use of dry lands over the last millennium within a wider perspective; secondly, they are familiar with the use of environmental data and are thus able to see human societies as an integral part of the landscape. Finally, much of the evidence needed for this type of study on trade, agriculture, mining and settlement patterns is archaeological, because of the paucity of detailed documentary sources for many dry regions.

IV. CONCLUSION

This chapter has outlined some ways in which archaeologists could add a useful dimension to the study of desertification. Throughout, I have deliberately tried to build as much as possible upon the types of problems that they already study, and the techniques that are currently employed. What is proposed is largely a re-organization and re-alignment of existing research strategies: that archaeologists should widen their horizons to consider present day conditions; and that those concerned with present conditions in dry lands should take greater account of the past.

The point can be made another way. Archaeologists have long drawn upon the natural sciences to place ancient societies within an environmental context; and as a routine field procedure, they attempt to obtain as much information as possible about the present-day climate, environment, subsistence and settlement of the area in which they work. It is, after all, only in this manner that they can create a base line against which to compare the behaviour of people in the past. However, the flow of this information has often been one way, and in the direction of the archaeologist. What is suggested here is that their perspective on the past use of dry lands has much to offer our understanding of present-day conditions, and is too valuable an asset to be casually dismissed. Perhaps the most obvious instance where the proposed type of "applied archaeology" has produced exciting results was the pioneering work some years ago by Evanari and his team (Evanari et al., 1973) on the Iron Age run-off farming systems of the Negev desert. Here, initial curiosity about these led to a formal analysis of the mechanics of run-off farming, which in turn prompted ideas on how knowledge of how the area was utilized in the past could be used to the benefit of the present. This research is not the only one than can be cited, but serves as one example of how the study of the past and present can reinforce each other.

Given that interdisciplinary voyages are rarely contemplated without some prospect of scholarly enrichment, it would be reasonable for archaeologists to ask how they could profit by studies of contemporary desertification. Potentially, present-day and recent trends in dry lands have much to offer our understanding of the past. For example, the city and territorial states of the Ancient World often incorporated or interacted with desert areas, and thus their destinies were intertwined. Many of the instabilities, weaknesses and concerns of these states may be illuminated by a greater understanding of what was affecting adjoining desert regions. Hall (1976), for example, has suggested that the expansion of the Zulu kingdom in the late 18th century, with all its dramatic consequences upon the people of southern Africa, may have been initiated by a period of drought which reduced their pasturage and encouraged an expansionist policy. If—as has been suggested in the previous section—desertification has been a persistent feature throughout at least the last millennium, it would seem worthwhile for classical and historical archaeologists to consider how events in "peripheral" desert regions affected both those within them and the populations on their margins. Similarly, an appreciation of the vulnerability and fragility of many desert environments to human and climatic pressure might well illuminate many aspects of the history of dry lands in the remoter periods of prehistory.

Endnote
[1] Qanāt technology is explained on p. 147.

4

Pastoral Strategies and Desertification: Opportunism and Conservatism in Dry Lands

Stephen Sandford

Overseas Development Institute
10–11 Percy Street, London W1P 0JB, U.K.

There have been serious attempts to involve economists in the desertification debate. But the UNCOD Secretariat was mainly concerned with cost-benefit analyses that would back up the Plan of Action and justify the associated Transnational Projects. Little thought has been given to the potential role of economists in clarifying the human dimension of the process. This chapter is essentially economic in approach, and in treating one of the major causes of desertification—overgrazing by traditional pastoralists—shows how environmental damage is just one of several variables that the pastoralist justifiably bears in mind in making management decisions. It provides a framework for weighing the costs of overgrazing against those of avoiding it, with important implications for the campaign and for pastoral development generally. —Ed.

In the dry areas of the world not only does rainfall fluctuate, in a fairly regular fashion, from season to season within the year, but there are also wide and unpredictable differences in rainfall between one year and another. These interannual differences in rainfall cause corresponding fluctuations in the amount of livestock forage which grows naturally on grazing lands. One choice of strategy which pastoralists in dry areas have to make is how to cope with these interannual fluctuations in the availability of forage. This chapter contrasts two pastoral strategies, opportunism and conservatism. Initially, and for simplicity of exposition, the discussion proceeds as though the choice lies only between pure opportunism and pure conservatism. This is a caricature since, as will be seen, neither of the extremes is, in practice, attainable. Later in the chapter various factors are considered which will

61

influence pastoralists to adopt strategies that lie on a range between the two extremes.[1]

An *opportunistic* pastoral strategy is defined as one which varies the number of livestock in accordance with the current availability of forage. Such a strategy enables the extra forage available in good years to be converted directly into economic output (milk, meat) or into productive capital in the form of a bigger breeding herd. The economic output may be immediately consumed or it may be exchanged for easily storable wealth, such as money or jewellery, that can be re-exchanged for consumables when needed. In bad years livestock numbers are reduced. In most cases where an opportunistic strategy is actually attempted livestock numbers in bad times are reduced too little and too late and as a consequence ecological degradation may occur. One can, therefore, distinguish an *efficient* opportunistic strategy as one where livestock numbers are varied at the *appropriate* time. If one thinks in terms of a "proper use factor", (a range management term for the proportion of forage produced which can safely be consumed by livestock; see Stoddart *et al.*, 1975:205), then an efficient opportunistic strategy is one which ensures that the proper use factor is adhered to.

A conservative strategy is defined as one which maintains a population of grazing animals at a relatively constant level, without overgrazing, through good and bad years alike. A conservative strategy implies that during good years livestock numbers are not allowed to increase to utilize all the additional forage available.

Differences in strategy are better revealed by behaviour than by statements of intent. Few pastoralists, other than commercial ranchers with their own land, intend to reduce the size of their herds in bad times. Such reduction is usually forced on them. The differences in behaviour that reflect conservative and opportunistic strategies are differences in breeding practices and in slaughtering, selling and purchasing livestock, and they may affect the species, age and sex composition of the herd, as well as the absolute numbers of livestock and the quantities of produce that are derived from them. It would also be possible to define opportunism and conservatism in terms of stability of *output* or of *income* derived from livestock. In this chapter the emphasis is on stability in the *number* of livestock because of the relation between this and desertification. There is some connection between stability in livestock numbers and in income but changes in livestock prices between good and bad times may prevent the connection being very close.

As will be shown later, some pastoralists pursue a conservative and some an opportunistic strategy. Among range scientists there seems to be little disagreement with the view that a conservative strategy presents less risk of degradation to the environment and should therefore be adopted. This view finds expression in the concept of a single estimate of an area's "carrying capacity", which for the rest of this paper is defined as the maximum number of livestock that an area will support for a complete year without causing deterioration in range productivity (that is, the proper use factor for each type of vegetation is adhered to), and in the recommendation that the number of livestock should not exceed this single-valued carrying capacity. Two such scientists have written recently:

> "Carrying capacity is limited by [the] harshest period during the climatic cycle. For instance the carrying capacity of the Sahelian desert areas would be limited to the number of animals able to maintain themselves during the driest year of the drought" (Box and Peterson, 1978:37); and "all development schemes must be examined against historical or simulated time periods long enough to reflect a valid statistical sample of climatic variance to insure that carrying capacity at any site will not be exceeded" (ibid.:43).

In spite of what they sometimes say, range scientists who advocate conservatism do realise that whatever level of carrying capacity they select, there will always be some years which are so dry that almost any figure they choose will turn out to be too high. They tend, however, to pay little attention to calculating risks. For example, Stoddart *et al.* (1975:335) have written "usually 65–80% of average forage production is a safe base for calculating grazing capacity" but no argument to support this estimate is advanced in terms of either variability of rainfall or of effects on production or range degradation. Instead of selecting a single level of carrying capacity (x animals) which will never be excessive, one needs to estimate the degree of risk that any particular level selected will prove too high. If we select a figure, say x_1 (e.g. 200 animals) as our estimate of the carrying capacity, there will be a particular degree of risk or probability (P), say P_1 (e.g. 0·10—or 1 in 10 years[2]) that there will not be enough forage in a given year to feed as many as x_1 animals. If we select a smaller value of x, say x_2 (e.g. 150 animals) there will be a lower degree of risk, say P_2 (e.g. 0 ·05 or 1 in 20 years) that this lower level of carrying capacity will be excessive in a given year. But for every degree of risk of the level selected, x, being too high there will be a corresponding risk, of the order of 1 minus P, that x will be too low and that there will be more forage than there are animals to eat it. If the level selected is too high, there will be losses from *over*stocking in the form of forced slaughter, sales at knockdown prices, deaths from starvation and environmental damage. If the level selected is too low then there will be costs of *under*stocking in that the available forage will not be fully utilized and the benefits, in terms of meat, milk and wool, will not be reaped which could have been if the area had been stocked with more animals.

In deciding which strategy to follow, and, if a conservative choice is made, what level of carrying capacity to select, one of the considerations to be taken into account is the long term average balance between the losses from overstocking and the costs incurred by understocking. The next section of this paper will look at one side of this balance, at the costs incurred by understocking, and will try to indicate a general method for estimating how these costs can be calculated. Concentration on only this side of the balance can be justified on the grounds that it has in general been neglected in the past and that the costs of understocking can be extremely high. Subsequent sections of the paper will discuss in more general terms the factors which will tend to make pastoralists veer towards conservatism or opportunism and the implications for government development programmes.

I. THE COSTS OF UNDERSTOCKING

In order to calculate the potential costs|of understocking an area, one needs to be able to forecast natural forage production in each of the years ahead and the quantity of livestock products into which this forage will be converted at different levels of stocking. Firm predictions are not available, so instead one must rely on the assumption that forage-yields in the future will follow roughly the same pattern as in the past so that one can project the future from past experience. The assumption here is that future yields are part of the same "population" as past yields. Those who believe that desertification is taking place will deny this. The justification for the assumption at this stage is that conservatism is being contrasted with an efficient opportunistic policy. The definition of an efficient opportunistic policy excludes desertification. Later in the chapter the possibility of desertification is readmitted. Only in a tiny handful of places in the world have annual natural forage yields actually been measured over a number of years, and in any case not enough is known about the statistical distribution of forage yields for one to calculate how big a sample of past yields would be adequate to estimate the parameters of a "population" of yields. However, it is known that in drier parts of the world the annual yield of natural forage is very substantially determined by the amount of annual rainfall. In many areas a fair amount is known about annual rainfall so that an estimate or reasonable guess can be made about its average value, variance and statistical distribution. By putting these pieces of information together, it is possible to generate information about forage yields.

Figure 1 shows an example of the close relationship over a number of years between forage yields and annual rainfall. The illustration is drawn from the USA but it is also true of other dry parts of the world. In general the relationship between annual rainfall and annual forage yield is approximately linear (see, for example, Le Houérou and Hoste, 1977) at any rate in drier areas (say less than 800 mm of average annual rainfall in tropical Africa)

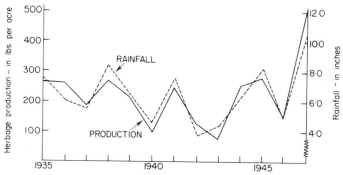

Fig. 1. Herbage production (measured in October) in Utah and rainfall in previous twelve months. (From Hutchings and Stewart, 1953.)

where moisture availability is always likely to be a constraint on forage production. Figure 2 illustrates the linearity of the relationship with data from Mali. This close and linear relationship between rainfall and forage provides a device for generating representative but synthetic data for natural forage yields from our knowledge of an area's rainfall parameters.[3] A brief diversion into elementary statistics in a number of steps is necessary to demonstrate this.

For many dry areas of the world data about average (arithmetic mean) annual rainfall are now available from recording stations; elsewhere average

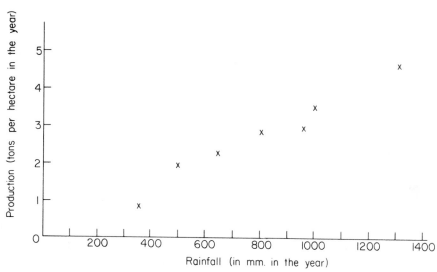

Fig. 2. Herbaceous production and rainfall at seven sites in Mali in 1974. (From Diarra and Breman, 1975.)

rainfall can be roughly estimated from the appearance of the vegetation. For the present purpose an estimate of the coefficient of variation[4] of the annual rainfall of an area is more important than an estimate of its average value. In general the lower the average annual rainfall the higher the coefficient of variation (for example, see Breman, 1975:251) although there is no universal formula relating one to the other. Nevertheless within one region of a country it will often be possible to guess the coefficient of variation fairly accurately from a knowledge of average annual rainfall and in some countries the coefficient of rainfall variation has been mapped. Even in quite dry regions it may range from as low as 10%, or even less, to as high as 50%. One can also roughly predict the general shape of the statistical distribution of the annual rainfall of a particular place in individual years around its long term average. The shape of this distribution is often roughly "normal" (in a statistical sense,[5] Hastings, 1965) although in very low rainfall areas its shape is closer to normal when the logarithms of the true (untransformed) values are mapped rather than the transformed values themselves (Griffiths, 1958:4). (Such a statistical distribution is known as "log-normal".)

One can now proceed to the next step. With knowledge of the coefficient of variation of annual rainfall for a given place, of the statistical distribution of annual rainfall, and of the properties of a "normal" statistical distribution, it is possible to predict in what proportion of years the annual rainfall total will fall within particular ranges of values expressed as percentages of the area's long term average annual rainfall.[6] For example where the statistical distribution is normal and the coefficient of variation is 30%, then in 5% of years the annual rainfall will be less than (approximately) 50% of the long term average, in 45% of years it will be between 50% and 100% of the average, and in 50% of years it will be greater than the average figure. If the coefficient of variation is 40%, then annual rainfall will be less than 50% of average in about 11% of the years. If the coefficient of variation is as high as 50% (still with a normal statistical distribution) then annual rainfall will be less than 50% of average in about 16% of cases. Suppose, however, that the statistical distribution of annual rainfall is not "normal" but "log normal"; then where the coefficient of variation (of the real, i.e. untransformed values) is 50%, only in about 12% of years will the (untransformed) value of annual rainfall be less than 50% of average.[7]

The general hypothesis here is that in dry areas the annual yield of natural forage is linearly related to annual rainfall and that the graphed relationship passes through the origin or intersection of the x and y axes. If this is correct, it is possible to translate statements about the proportion of annual rainfall values that fall into certain ranges of values in relation to long-term average annual rainfall into identical statements about forage yields. If the coefficient of variation of annual rainfall is 30% and the statistical distribution thereof is normal so also will be the coefficient of variation and distribution of annual forage yields, and thus the proportion of annual values falling into certain ranges in relation to the long term average annual yield will also be the same.

One more step needs to be taken before it will be possible to provide a general model for calculating the costs of understocking. This step is to relate the annual yield of natural forage to the number of livestock that can feed thereon and to the annual value of livestock products that they produce. This step is more difficult than the previous one since it lacks a basis of empirical evidence. At this point a number of simplifying assumptions will be made; some of these are a good deal less realistic than others and the distortions they impose on the general model will be discussed later. The assumptions are:

(a) Under both conservative and opportunistic strategies livestock are fed only on natural forage and eat as much of this as they want (i.e. "*ad lib*" feeding).
(b) The quantity of livestock products produced in an area varies in direct proportion to the number of livestock kept therein; i.e. a 10% increase in the animals kept leads to a 10% increase in all products. Coupled with assumption (a) (*ad lib* feeding) this implies that the age and sex composition of the herd, and the calving and mortality rates, remain fairly constant through time even when the absolute number of animals varies.
(c) Livestock numbers can be varied at will and immediately upon a decision to do so. Adjustment to numbers comes by natural increase, by buying from outside the area, by slaughter and by selling off; in short an efficient opportunistic policy is possible.
(d) Grazing cannot be "carried over" from year to year, i.e. it cannot be stored as hay nor will overgrazing or undergrazing in one year affect, favourably or adversely, productivity (through changes in composition, cover, vigour) in subsequent years.
(e) The price of livestock and livestock products remains constant through time.

The assumptions made in the last paragraph make it possible to describe the value of output from a pastoral area as being linearly dependent on the number of livestock kept[8] and, if an efficient opportunistic strategy is followed, also on the annual yields of forage and on annual rainfall. If a conservative strategy is followed we can roughly calculate the value of output from the number of livestock; provided that the particular conservative strategy involved is defined in terms of the risk of the selected carrying capacity proving too high.

Figure 3 illustrates this point. The horizontal axis represents time, a "typical" sequence of twenty years. The left hand vertical axis represents annual rainfall and so also annual forage production. If all this forage is fed to livestock this vertical plane also, by the arguments of the previous paragraph,

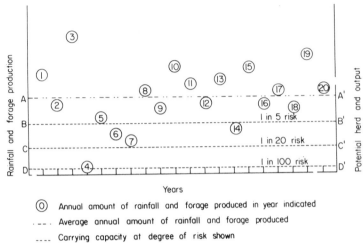

Fig. 3. Carrying capacity, risk and variability in rainfall and forage production.

represents the number of livestock that can be kept and the value of livestock output. The right hand vertical axis is, therefore, labelled "potential herd and output". The points, indicated by numbered circles, represent rainfall, forage and potential output in each of the years 1–20, the number is the circle indicating the year. The distribution of the points approximates to a "normal" distribution. The average values of the annual figures is indicated by the level of the line A – A'. The other lines, B – B', C – C', D – D', represent the values below which there is a 1 in 5, 1 in 20, and 1 in 100 risk respectively of the annual values falling. A pastoralist—the term hereinafter includes a pastoral society which, either through the communal ownership of livestock or by regulation of individual holdings, pursues a collective policy— pursuing an efficient opportunistic strategy will adjust each year the number of his livestock to the amount of forage available as indicated by the figures for years 1, 2, . . . 20. The herd size will, therefore, show very large fluctuations between successive years; the actual output will always be equal to potential and the *average* value of actual output will be at the level indicated by A – A'.

A pastoralist pursuing a conservative strategy will decide on the level of risk that he is prepared to accept of his selected carrying capacity proving too high. For example he may decide on a risk of 0·2 (1 in 5 years). This means that he will not permit his herd to be larger than can be supported by the amount of forage which grows in 80% of years, the level indicated by line B – B' in Figure 3. In years 1, 2 and 3 his herd size will be at B – B', and forage available in those years in excess of that level will be wasted; in year 4 he will have to reduce his herd substantially, in year 5 it can be restored to $B - B'$ again, but in years 6 and 7 it will have to be reduced again; and so on. The value of his annual output can never exceed $B - B'$ and the average value of his output will be slightly below the level indicated by $B - B'$ because of the downward adjustments he has had to make in years 4, 6, 7 and 14. Similar processes of reasoning apply to pastoralists who select other risk levels, for example 1 in 20 or 1 in 100.

The potential costs of conservatism can now be demonstrated with some quantitive illustrations. In Table I six variants of a conservative strategy are compared in terms of the average value of output that will arise from each of them in relation to the average value of output that would arise from pursuing an efficient opportunistic strategy. The results are expressed in terms of the "cost of understocking"; for example if the average value of output that would arise from pursuing an efficient opportunistic strategy is 100 and the average value that arises from a particular variant of a conservative strategy is 70, then the cost of understocking of that variant is 30 $(100 - 70)$. The six variants of a conservative strategy are defined in terms of the levels of risk of the carrying capacity selected proving too high in any particular year. The levels, in order from most to least risky, are 1 in 5 $(P = 0·2)$, 1 in 10 $(0·1)$, 1 in 20 $(0·05)$, 1 in 50 $(0·02)$, 1 in 100 $(0·01)$ and 1 in 200 $(0·005)$ respectively. In order to illustrate the effects of different degrees of reliability of rainfall, comparisons are made for three areas with coefficients of variation of rainfall ranging from 20–50%. Such coefficients would be typical for dry regions (Barry and Chorley, 1968, p. 357), as evidence from Mali (Breman, 1975, p. 252) and Botswana (Sandford, 1977a, p. B6) in particular demonstrate. The figure, for example, in the top line of the centre column of Table I indicates

Table I. The Costs of Conservatism and Understocking

(Costs are expressed in terms of output foregone because of understocking, relative (%) to output attainable (100%) under an efficient opportunistic strategy)

Variant of conservative strategy	Cost of variant where coefficient of variation of annual rainfall is:		
(expressed in terms of the risk—1 in n years—of the level of carrying capacity selected proving too high)	20%	35%	50%
1 in 5 years	17	29	42
1 in 10 years	26	45	65
1 in 20 years	33	58	82
1 in 50 years	41	72	100
1 in 100 years	47	82	100
1 in 200 years	52	90	100

that, in an area where the coefficient of variation of annual rainfall is 35%, a conservative strategy which selects a carrying capacity which will prove to be too high, on average, in one in every five years, will achieve a level of output of only 71% of what would be achieved under an efficient opportunistic strategy, i.e. the cost of this variant is 29% $(100 - 71)$.

The figures in these columns of Table I slightly underestimate the costs of conservatism, in that they are based on the calculation that the average value of output of each variant is equivalent to what would be produced if the pastoralist was actually able to keep a constant number of animals at the level of carrying capacity selected. In practice, as discussed in relation to Figure 3, even with very conservative variants a pastoralist will have to destock in some very bad years, and in consequence the actual number of animals kept will, on average, be less than that indicated by the selected level of carrying capacity; and the actual average level of output achieved will, in consequence, also be less. The underestimation of costs is greater for the variants of higher degrees of risk.

The figures in Table I indicate that very conservative (i.e. low risk) strategies have extremely high costs in all areas in terms of potential output foregone, and that in areas with very unreliable rainfall (high coefficients of variation) even moderately conservative strategies are very costly. The figures in the table are not, of course, derived from measurements of actual events. They are simply the mathematical consequence of combining the various hypotheses and assumptions with a table of the normal distribution function. If, instead of a "normal" statistical distribution of annual rainfall values, a "log-normal" distribution had been assumed—as would have been more realistic for areas with high coefficients of rainfall variation—the results would not have been much different for the more risky variants of a conservative strategy, but would have exhibited lower levels of cost for the less risky variants. For example in the right hand column of Table I, corresponding to the place with a coefficient of variation of 50%, the cost of the six variants, with a log-normal distribution would have been (the figures in brackets are those for a normal distribution as shown in Table I): 44 (42), 56 (65), 64 (82), 71 (100), 75 (100) and 78 (100)% respectively.[9] Although there are substantial differences between the costs arrived at by using these two different assumptions about the statistical distribution of annual rainfall nevertheless in both cases the costs are very large.

Three substantial objections can be raised against the procedures and arguments set out so far, especially to the assumptions (a) to (d) of a few paragraphs earlier. The first of these arguments is that not only do pastoralists fail to practise opportunism efficiently by destocking too late after environmental damage has been done, but efficient opportunism is, in fact, impossible (see assumption c) because it is not possible to restock fast enough after a drought. The question of destocking will be discussed later. Restocking after drought can be through purchase or by the natural reproduction of the herd. Purchase may not always be possible in remote areas away from markets or where a whole region is hit by drought and then tries to restock throughout in one season. There simply may not be enough suitable animals available for purchase. Upward adjustments of stock numbers by natural reproduction is dependent on animal biology. Under traditional pastoral systems of Asia and Africa the long term maximum natural growth rate of a camel herd is between 1% and 7%, for cattle about 11% and for goats as high as 45% per year (Dahl and Hjort, 1976). In the short term slightly higher growth rates may be possible; I have used these long term rates in conjunction with fifty years of actual annual rainfall data for Ghanzi in Botswana where annual rainfall averages 420 mm and the coefficient of variation is 40%. I calculated that a semi-efficient opportunistic cattle-keeping pastoralist who, whenever necessary, adjusted his herd size downwards immediately, but who could only build up his herd again through its own natural reproduction and not by purchase, would, on average over the last 50 years, have been able to keep only 62% of the number of animals that would have been kept by a perfectly efficient opportunist who had been able to purchase stock. This is roughly the same average level of performance as a conservative with a risk selection of 1 in 6 years. If, however, this semi-efficient pastoralist had kept goats rather than cattle he would, because of their faster rate of reproduction, have been able to achieve an average stocking level of 85% of that of the perfectly efficient opportunist.

The second substantial objection relates to the assumptions (a) and (b) p. 67, that animals under both conservative and opportunistic strategies eat "*ad lib*" and that the quantity of livestock products varies in direct proportion to the number of animals. The second of these assumptions largely follows from the first, for if animals are fed *ad lib* in bad years the only way one can exploit the increased forage available in good years is by increasing the number of animals.[10] If, however, animals do not get as much to eat as they want in bad years then in good years they will increase their intake of forage (per head) and also their yield of products. In this way the quantity of products from livestock will vary more than proportionately to their numbers. In typical conditions in dry areas where animals are underfed, only about 10% of total gross energy intake of livestock feed is available for producing useful products other than dung (Sandford, 1978:9). With better feeding this proportion, and in consequence yields per animal, can probably be doubled. There is some force in this objection insofar as it asserts that the model proposed in this chapter will be a poor predictor of actual performance under present conditions. It is not, however, a valid objection to the procedures proposed insofar as they relate to comparisons of potential output, and in particular the objection cannot be used to give general support in favour of a conservative strategy. In dry regions under conditions of extensive grazing half-starved animals will always eat the forage which is available to them. One cannot, therefore, argue both that under a conservative strategy the livestock will not overgraze the vegetation in excess of "its proper use factor" in a bad year and also that the same livestock will be able, in a subsequent good year, to expand their individual appetites (and their yields) to take advantage of the more abundant forage then available to them.

The third substantial objection relates to the assumption (d) on p. 67, that over- or under-utilization of the forage in one year will not be "carried forward" either in the form of hay or as a factor affecting potential productivity in subsequent years. In practice, with the exception of certain areas in India, hay is not carried forward as a reserve from year to year for both technical and economic reasons. The issue of overgrazing, and consequent decline in productivity, or of deliberate undergrazing as a way of restoring productivity, has been evaded so far in this paper by explicit concentration on the costs of *under*stocking and through the device of defining efficient opportunism in terms of adhering to a proper use factor; this device simply defines away the possibility of desertification or of improving productivity by deliberate understocking. Of course in the real world loss of productivity through overgrazing does occur. This loss is the main plank in the conservatives' platform. However, the heavy costs of conservatism indicated in Table I suggest that pastoralists' frequent reluctance to embrace a conservative strategy may be better founded than non-pastoralists in the past have realized. The balance between the losses arising from overgrazing and the costs of understocking is not universally such as to suggest that conservatism should always be the rational choice. More attention needs to be paid to finding some alternative way, such as efficient opportunism, to avoid the harm done by overgrazing.

These three objections do not to any significant extent invalidate the arguments and procedures set out so far in this paper. To the extent that the first two objections do have some force, they indicate that the procedures are likely to be more useful as an analytic than as a predictive tool. Too much stress should not be put on the precise figures of Table I; obviously modification of some of the assumptions will alter the figures somewhat, and they should be looked on as orders of magnitude rather than as precise estimates. The arguments of the paper, and the substantial costs from under-stocking, indicated in Table I, that arise from pursuing a conservative strategy, bring out three points very strongly. The first is that advocating a conservative strategy based on carrying capacity is meaningless unless it is attached to some estimate of the risk of the selected carrying capacity proving too high or low. The second point is that the precise degree of risk selected has very large consequences in terms of the average value of output which pastoralists will command—and hence on their standards of living. The third point is that there are advantages in opportunism which mean that it should at least be considered by planners and scientists as a serious alternative to conservatism—which is seldom done.

II. THE FACTORS WHICH INFLUENCE CHOICE OF STRATEGY

The discussion so far has been in terms of a formalized model of behaviour. What strategy do pastoralists in fact follow? Douglas Johnson (1979:31) claims that all persistent cultures in dryland ecosystems follow an opportunistic one. Heathcote (1969:320), talking of capitalist pastoral enterprises in the U.S.A. and Australia, says "Pastoralists on both sides of the Pacific very soon claimed that they had to take what they could while they could if they were to survive on the range. Opportune use of the fleeting resource seemed to them the only solution". In contrast Almagur (1978:54), for example, has written of an East African people: "The Dassanech must, therefore, keep their animal population stable, and do this by regular slaughtering"; and there are many references in the literature (e.g. Monod, 1975:118) to the delicate balance between pastoral man and his environment, which, it is said, has only recently been disrupted by advances in veterinary medicine and by outside pressures. No attempt will be made to list here the pastoral groups following either conservative or opportunistic strategies. Rather it is assumed that there *is* a diversity of behaviour along a range between more opportunism and more conservatism and an attempt will be made to identify the factors which are likely to influence behaviour towards one or the other.

Three sets of factors may influence people in their choice of strategy, causing them to veer towards conservatism or opportunism. The first set is the physical characteristics of their environment. A relative high degree of variability in rainfall over time, as we have seen, increases the gross cost of conservatism and will influence people towards opportunism. It is, perhaps significant that maps which indicate the relative reliability of rainfall on a world scale, (e.g. in Barry and Chorley, 1968) appear to show that central Asia and Africa, whose pastoralists seem most reluctant to embrace conservatism, have less reliable rainfall than other dry regions in the world.[11] However, if within the general region used by a particular individual or group of pastoralists different places tend not to suffer the same variations in weather at the same time, then the effects of variability in weather can be mitigated by the migration of livestock and the costs of conservatism reduced by this means.

The nature of the soils, topography and vegetation will also determine the extent to which over- or under-grazing in one period affects the supply of forage in subsequent periods, i.e. leads to the sort of "carry over" effects discussed previously. Erodible soils on steep slopes will demonstrate such effects more than stable soils on flat plains. Some plants are better able to withstand heavy grazing than others. Moreover, although the general response of forage yields to rainfall, as we have seen, is linear and immediate, there are some differences; annual grasses will show wider interannual fluctuations and more immediate responses than perennial grasses or deeper-rooted browse (Cook and Sims, 1975).[12] The greater the carry over effects, negative or positive, the less will be the benefits and the greater the costs of opportunism; in contrast conservatism will be more attractive.

The second set of factors influencing choice of strategy are social, political and economic ones. If, for example, pastoralists are mainly keeping camels, which have a very low potential rate of natural population growth, then, where restocking after drought through purchase of animals is impossible, pastoralists are likely to opt for conservatism since opportunism will be impractical. If the main species of livestock kept are sheep or goats, with high potential rates of natural population growth, opportunism will seem preferable.[13] Where exchange of one species of livestock for another is possible, the loss or forced sale of most of a camel herd during a drought can be compensated by exchanging the remaining camels for small stock which, when their numbers have grown rapidly after drought, can be re-exchanged for camels (Dahl and Hjort, 1979:28). The existence, as a consequence of economic, political or social factors, of a reliable mechanism for exchange will favour opportunism. This is also the case where, in the event of a disaster to his livestock enterprise, a pastoralist can rely on relief from within his own community or from government, or on paid employment in some non-pastoral enterprise, to meet his subsistence needs and to restart his pastoral enterprise after the drought is ended. Droughts are not the only crises to affect pastoralists. Wherever political conditions are unsettled and stock theft occurs on a large scale, it is likely to be difficult to convince pastoralists of the merits of conservatism. The same holds true where government veterinary services are unreliable and heavy losses from disease may also occur.

A third set of factors influencing choice of strategy will be people's attitudes—their aims and values. If pastoralists' main concern is the assurance of their survival, either as individuals or as a coherent social unit, or if government or business is concerned with the need to secure reliable supplies of meat from dry areas, then the choice of strategy will veer towards conservatism. If, however, an ideology of performance and competition in respect of income, wealth or status predominates, then opportunism is likely to seem preferable; for conservatism not only reduces overall long term average income[14] it also reduces the opportunities for some individuals to advance relative to others.

A society may be concerned to minimize differences in the wealth and status of individuals within it; and the fluctuations in total livestock numbers that are a consequence of an opportunistic strategy are often identified as the occasions on which the rich get relatively richer and the poor poorer. In confirmation of this view a recent survey among Maasai pastoralists in Kenya (see Table II and Campbell, 1978) found that during the drought of the first half of the 1970s among cattle herds of different sizes the correlation between

Table II. Losses during drought as a % of pre-drought herd numbers

	% Died	% Sold	% Decline
Smallest 25% of herds	38	15	52
Next 25–50% smallest	33	17	50
Largest 10% of herds	27	15	42

high loss and small herds only occurred where, even before the drought, the herd size was below the family's subsistence needs. Once above this minimum the proportional losses during drought were not markedly different between herds of different sizes and the overall distribution of cattle wealth was much the same before and after drought.

The processes by which fluctuations in herd numbers may cause inequality need clarification. The poor man's herd is likely to suffer more in bad times when it is too small overall to feed him and his family. He will therefore have to pursue some other occupation which means that he cannot take livestock to where pastures are best. The poor man's herd is likely also to suffer worse where there are benefits in management to be obtained by running separate herds of different classes of stock, e.g. milch cows, dry cows, immatures, and where it is too small to justify the amount of herding labour required for such splitting of the herd. On the other hand the poor man's herd may do as well as, or better than, the rich man's where, in a drought, much labour is required for drawing water or cutting browse, and where the amount of labour required is directly proportional to the number of stock owned. However although the fluctuations inherent in pursuing an opportunistic policy may provide many of the occasions for social differentiation, it may also happen that the initial adoption of a conservative strategy by a pastoral community may involve the allocation of stock quotas to individuals. If the size of these quotas is proportional to the current livestock holdings of the individuals, then the allocation will make permanent any existing patterns of inequality and will provide a further important occasion for social differentiation.

The three sets of factors, physical, socio-economic and attitudinal, are obviously related. The physical characteristics of the environment, for example, as well as the availability of reliable markets and relative prices, will determine the species of livestock kept; and pastoralists' relative concern for security or pre-eminence will be affected by their current physical and economic environment as well as by their past history. But reducing one factor or set of factors to another provokes such controversy that it seems preferable to treat them separately.

III. THE IMPLICATIONS FOR GOVERNMENT DEVELOPMENT PROGRAMMES

If an opportunistic strategy is to be efficiently conducted in a way that takes advantage of the high production of forage in periods of above-average rainfall without causing damage to the environment in bad years or without the system being in constant danger of collapse, it requires devices that permit the timely adjustment of stocking rates, so that, as bad years succeed good, livestock are removed from pastures before they cause damage and are replaced later as fast, but only as fast, as the recovery of the vegetation should permit. The remainder of this chapter will look at some possible implications for government-initiated development programmes if efficient opportunism is to be possible. This does not imply that efficient opportunism is always a more desirable policy to pursue than conservatism, but rather that too little attention has been paid hitherto to making opportunism work better, and this imbalance needs to be redressed.

If pastoralists are to destock at the right time they must be both able and willing to do so. For them to be able to do so almost certainly requires that they sell their extra animals, since not even in the case of rapidly reproducing goats is mere adjustment to recruitment by lowering fertility likely to produce a quick enough response. The bottlenecks which will restrict rapid marketing are likely to be the physical facilities for transporting and processing animals, the availability of increased working capital to finance the holding of animals between first sale and final consumption, veterinary regulations which inhibit the free flow of animals out of pastoral areas, and the size of the final market. With the possible exception of working capital these factors are likely to militate against systems in which a high value product is being produced and marketed. High value products, whether they are breeding or fattening stock, are liable to require not only stringent veterinary regulation, but also the transport of animals by vehicle rather than on the hoof, processing facilities, such as feed lots, slaughterhouses, cold-storage and refrigerated transport with high fixed capital costs in relation to output, all of which make the provision of spare capacity to be used only occasionally very expensive. High value final products, especially fresh and chilled meat, in contrast for example to canned meat, also tend to require very rapid turnover due to physical deterioration, and so require a bigger instant expansion in the size of the final market. The case of working capital is more ambiguous. Lower quality animals may take longer moving to a processing point on foot, and have a higher working capital requirement in this respect; however this may be counterbalanced by the greater speed with which they can be despatched, e.g. without a fattening period. But the lower rate at which canned and frozen meat can be released on to the market in turn implies a higher working capital requirement.

When the final market for animals culled rapidly under an opportunistic strategy, is a small domestic market, it will be very difficult to expand the demand of this market sufficiently to accept the necessary 3–4-fold increase in flow[15] without a collapse in prices. This is particularly the case because other parts of the agricultural sector which normally purchase pastoral off-take will simultaneously, because of the drought, be short of cash and livestock feed. Only where there is access to a large urban or export market, where the supplies from the pastoral area concerned are small in relation to total supplies, is a very rapid expansion of sales likely to be feasible. A temporarily very high off-take rate also implies that animals of a sex and age not normally marketed will be sold; again this implies a market mainly interested in low quality and low value supplies.

A number of studies, mainly of pastoralism in Australia and the U.S.A., have indicated that where pastoral enterprises have a high level of fixed costs (especially interest on debt) which do not vary with the number of livestock, there is a reluctance to reduce stock numbers at the appropriate time. Such reduction would probably lead, whatever happens to the condition of the vegetation, to bankruptcy so that the enterprise could only profit by taking the risk of overstocking and hoping for a change in the weather. The implication is that, if an opportunistic strategy is being pursued, fixed costs should be kept low and financed by equity or grant rather than by fixed-interest loan capital.

To make pastoralists willing to destock in good time requires a shift in the balance of incentives between destocking and retention. Not only do those who try to destock at the onset of drought normally face collapsing prices at the time they sell, but also they know that if they do not retain sufficient animals to provide them with basic maintenance they will face rising prices for other forms of food as they become more dependent on these. They also know that those pastoralists who do come through a drought with a number of animals in excess of their own subsistence requirements will be able to obtain very large rewards, in terms of power and wealth, in return for their subsequent distribution of these animals to those in need of them. Proposals for a government-run or -guaranteed livestock bank of animals (Goldschmidt, 1975) with which to restock after drought run into the difficulty that if the spare feed existed wherewith to keep a "bank" of adequate size alive, then the bank itself would not be needed. There are several possible interventions that a government can make to tilt the balance of incentive in favour of timely destocking. It can, at the onset of drought, support livestock prices at sales for a limited period whose terminal date is announced well in advance. It can, for example, publicly guarantee a reasonable price for all classes of stock for the next three months; with the implication that prices will collapse thereafter. Of course there are difficulties in judging whether a period of sub-normal weather is going to continue and turn into a serious drought. But if government miscalculates and good rain comes it will probably not be called on to fulfill its guarantee in respect of many animals.

Government also needs to ensure that a reliable supply of grain, for human consumption, is made available in pastoral areas at moderate and stable prices. It can facilitate the movement of animals, after a drought, from stock surplus to stock deficit pastoral regions to enable the restocking of the latter when the vegetation has recovered to the point where such restocking is desirable. Unfortunately the resilience of the vegetation is often under-estimated and it seems likely that governments will probably restrict the flow of livestock into deficit areas for longer than is appropriate under an opportunistic strategy. Finally government can initiate, or support where they already exist, measures for the compulsory and unrewarded transfer of livestock within a pastoral society from individuals who are relatively well-off in respect of livestock to those who are relatively poor. Some pastoral societies have traditional institutions for such redistribution which governments are destroying without realising that they ever existed. The conflict between the interest of the individual and of society at large can be resolved by such measures which communalize the use of livestock instead of by the ones, usually recommended, which individualize the use of land, but which the climatic conditions render unsuitable in dry regions.

A government can also attempt to influence the species and genetic make-up of a livestock herd by its policy in respect of licensing or encouraging the introduction of new breeds. In the past governments have often discriminated against goats not as part of a conscious choice of a conservative strategy but as a result of probably misguided notions of goats' destructive effects (see Nyerges, Chapter 11). But within a species also, a government can choose to disseminate strains with relatively high or low fertility. Again they already often do this, although the choice of strategy often seems to follow from the choice of breed rather than vice versa.

IV. CONCLUSION

This chapter has classified pastoral strategies on the basis of response to variations in rainfall and consequent forage yields and has sought to show that although neither of the two extremes—pure conservatism and pure opportunism—is wholly attainable in practice, the distinction between them is analytically useful, and that the choice of where, in the range between the two unattainable extremes, a pastoralist should take up his own position is an important issue which requires explicit consideration. It has also shown that the choice of an opportunistic strategy for a pastoral region or society may require, if the strategy is to be efficiently implemented, that a government take measures which are quite different from those normally carried out in development programmes. Considerable emphasis has been placed on the importance of identifying the factors which will tend to make either a relatively conservative or a relatively opportunistic strategy more appropriate in particular circumstances. This emphasis rules out the adoption of any universal prescription which might tend to favour either one or the other.

Endnotes

[1] It is always difficult to be certain where one's ideas come from. In writing this paper the author benefited from three other papers, at that time unpublished: (i) Dahl and Hjort, 1979; (ii) Johnson, 1978; (iii) Fukuda, 1976. He is particularly indebted to Gudrun Dahl for several ideas jointly developed during an assignment as consultants, and to his colleagues at the Overseas Development Institute for comments on earlier drafts. Remaining errors of fact and logic are the author's.

[2] Jackson (1970) has pointed out that one should not unthinkingly proceed from statements about long term probabilities, e.g. 0·1 meaning 1 in 10 years *on average* to thinking that such long term probabilities apply to particular short periods of years, e.g. to the *next* ten years.

[3] In anticipation of the following three paragraphs one can note that a random variable which is a linear function of a normally distributed random variable is itself normally distributed (Chao, 1974:185). Moreover, one does not need to know exactly how many kilogrammes of forage are produced by a millimetre of rainfall, provided that the graphed relationship between the two passes approximately through the origin (i.e. some rainfall, however little, produces some forage and no forage grows without rainfall). For in that case where rainfall in any year deviates (plus or minus) by x% from average annual rainfall it will be associated with the same x% deviation of forage yield in that year from average annual forage yield. One should note that some of the linear functions reported relate average annual forage production to average annual rainfall and others relate production in a particular year to rainfall in a particular year. Figure 1 gives data of rainfall and forage in a particular year for different places.

[4] The coefficient of variation of a variable is defined as $\dfrac{\text{stand deviation}}{\text{arithmetic mean}} \times 100.$

[5] For a description of the "normal distribution" and its characteristics and properties see any elementary statistical textbook, e.g. Croxton and Cowden, 1955, Ch. 23.

[6] It is assumed that annual rainfall in any one year is independent (in a statistical sense) of rainfall in other years. This is an assumption still under debate by climatologists. The assumption is necessary for the formal validity of the arguments of this paper although a moderate degree of "non-independence" would not greatly affect the practical implications of the arguments.

[7] This is not a statement based on *a priori* considerations about the relationship between untransformed values and their normally distributed logarithms. It is derived empirically from an analysis of the actual data for one weather station in Botswana.

[8] In this paper "value of output" means *gross* (value of) output before deducting fixed and variable costs. *Net* output (= gross output minus costs), which can be loosely described as "profit", tends to vary more than proportionally to livestock numbers, forage supply and rainfall. Under reasonably realistic additional assumptions the general model of this paper can be applied to the calculation of net output but with a greater degree of complexity than is appropriate for present purposes. See also note [14].

[9] The apparent tendency for the value of the difference between the results derived from the two assumptions to reach a maximum at the "1 in 50 years" and then to decline again at yet more conservative variants is a freak arising from the impossibility of "negative" rainfall. The major problem with the assumption of normally distributed annual rainfall is that, when coupled with estimates of high coefficients of variation it leads to assigning positive probability to obtaining negative (i.e. less than zero) rainfall. This is the main reason for preferring the assumption of log normal distribution.

[10] One cannot force-feed range animals in the manner of geese for *"paté de foie gras"*, nor is it possible to restrain range animals from eating what is available.

[11] The maps are on too cramped a scale to allow more than the formation of a tentative impression.

[12] Noy-Meir (1975) discusses the plant characteristics which make some grazing systems inherently unstable even where exogenous shocks such as poor rainfall do not disturb them. Such systems may actually benefit from fluctuating grazing pressure.

[13] It may be more possible to pursue an opportunistic strategy using sheep and goats; but, as Noy-Meir (1975:470) points out they may, through their greater efficiency as grazers, create the very instability which requires opportunism to cope.

[14] Whitson (1975) used a programming model to provide a hypothetical example of the trade-off between security and profit. In round terms, he found that, under his chosen assumptions, to reduce the standard deviation (as a measure of instability) of profit by $x\%$ one has to reduce long term average profit by $2x\%$.

[15] If normal marketed off-take is 15% of the total biomass of livestock kept on the ranges while this biomass is stable in size, then to reduce grazing pressure by 60% over 2 years will require a roughly threefold increase in the normal market flow. Typical average marketed off-take rates in pastoral Africa are 5–10% for cattle and 20–30% for sheep and goats.

B. *Case Studies*

The following three chapters are case studies illustrating problems from three of the types of human activity that are most obviously related to desertification (whether as cause or effect)—canal irrigation, traditional pastoralism, and resettlement. They illustrate in markedly different ways the importance of understanding the dynamics of people's thought and behaviour before designing any intervention in the name of development.   —Ed.

5

Reorganizing Irrigation: Local Level Management in the Punjab (Pakistan)

Douglas J. Merrey

University of Pennsylvania
Department of Anthropology
Philadelphia, PA 19104, U.S.A.

From the sociological point of view desertification, and environmental degradation generally, is—like underdevelopment—a problem of organization or management. The first of these three case studies focuses specifically on the management problem as it has been recognized in a particularly serious example of one type of desertification. Besides its value in demonstrating the importance of the relationship between the ′structure of social relations and the organizational requirements of a technology, it shows how misleading the concept of adaptation can be in dealing with development problems.　　— Ed.

Discussions of the problems facing irrigated agriculture in Pakistan usually begin with an irony: despite the favourable climate, fertile land, hard working farmers, and possession of the largest integrated irrigation system in the world, agricultural production is very low. It is low by any standard— in relation to similar situations in other countries, to the potential demonstrated year after year on demonstration plots and—even more so—to the needs of Pakistan's rapidly growing population.

Before World War II the area which is now Pakistani Punjab was a major exporter of wheat. By the time of Independence in 1947 exports had ended.

Output *per capita* continued to decline until the mid 1960s. By the early 1970s, with the adoption of high yielding varieties of wheat, *per capita* productivity had returned to the levels of the late 1940s, but during the 1970s it stagnated. In most years up to 1978 Pakistan had to import about one quarter of its wheat requirements. Although the figures have improved significantly since 1978, mainly as the result of the introduction of improved rust resistant varieties, yield per hectare remains low.

Although the irrigation system had originally been built in order to transform drylands into highly productive farmland, the productivity of large areas has been either destroyed or significantly reduced by salinity and a rising water table. A considerable proportion of the population now lives under the threat of this type of desertification. Some surveys show that more than half the approximately thirteen and a half million hectares of irrigated land in the Indus Basin are affected by varying degrees of waterlogging and salinity (Clyma *et al.*, 1975a; Malik, 1978). The situation is regarded so seriously that newspapers carry learned and often passionate articles by experts and frequent editorials urging more action.

There is considerable controversy concerning the seriousness of waterlogging and salinity in Pakistan, and whether the situation is deteriorating or improving. Two factors are responsible for the controversy: first the relevant information is inadequate, inconsistent, and subject to different interpretations; secondly, since such large funds are involved in the various programmes for controlling the problems, the whole question has become a political issue.

The process of deterioration is generally blamed on the "twin menace" of salinity and waterlogging. In order to reverse it several programmes have been launched to reclaim waterlogged and saline land by means of drains, or high capacity tubewells to lower the water table (referred to as the Salinity Control and Reclamation Programme, or SCARP), and to flush salts out of the soil. Until recently, all the irrigation projects in Pakistan emphasized large-scale capital-intensive construction: SCARP tubewells, link canals, and dams—culminating in the Tarbela Dam Project on the Upper Indus River. This dam is billed as the largest earth-filled dam in the world. It is financed by huge foreign loans and is designed for both irrigation water control and the generation of electricity. It is important to note that these projects were based on research that was in turn based on certain assumptions about local level water management. The loss of water in watercourses before it reached the crops was assumed to be minimal, and due mostly to evapotranspiration. It was not thought to be adding to the water table. In particular, it was assumed that most of the water delivered to tertiary irrigation ditches (called watercourses) reaches the root zones of the crops.

In the early 1970s research teams from Colorado State University, in cooperation with Pakistani organizations, especially the Mona Reclamation Experimental Research Project (supported by funding from the United States Agency for International Development), began to explore the possibility that a major cause of the twin menace should be sought at the farm level.

They started by measuring watercourse and field application efficiencies. These studies demonstrated a wide variety in the delivery efficiencies of watercourses, but on the whole they were shown to be substantially lower than had previously been assumed. Overall, delivery efficiencies of watercourses seem to be less than 60%—often substantially less; that is, 60% or less of the water entering the watercourses reaches the fields. Unlevelled land, fragmented plots, and lack of knowledge about plant–soil–water relations reduce the efficiency of water use even further. Early studies in a SCARP area demonstrated that wastage of water in these areas is especially high. SCARP tubewells pump the ground water into watercourses where it mixes with canal water, increasing the available irrigation water, while lowering the water table. However, because of poor watercourse construction and maintenance, most of the water returns to the groundwater, minimizing the effectiveness of the SCARP programme in reducing waterlogging. Overall, Pakistan's irrigation system is estimated to be operating at less than 30% efficiency; that is, less than 30% of the water diverted from the rivers is stored in root zones for crop use.[1]

These findings led to the development of pilot projects designed to improve the efficiency of water delivery and usage in order both to increase agricultural productivity and to reduce waterlogging and salinity. A key component of these projects was an attempt to induce local water users to co-operate in reconstructing and maintaining their joint watercourses. This component has also proved to be a major obstacle: it is quite difficult, though not impossible, to organize farmers to co-operate on a short term improvement project, but it is even more difficult to induce them to continue co-operating for maintenance and management of the watercourse on a longer term basis. This chapter presents a case study of a reconstruction project on one watercourse and identifies the impediments preventing its successful completion. It also summarizes the results of a larger survey of organizational problems on improved watercourses. The basic argument is that a major source of the severe technical problems of Pakistan's irrigation system is ineffective organization of management especially at the local level; and attempts to improve the system so far have been hindered by the failure to recognize this social dimension of the problem.

I. THE "INDUS FOOD MACHINE":
HISTORY AND DEVELOPMENT PLANS

The history of the development of Pakistan's irrigation system is not long, but it is complex. The best recent account is that of Aloys M. Michel (1967), which draws on older histories of irrigation in Punjab plus British and Pakistani administrative and technical documents, and post-Independence research. Only a brief summary is possible here.

A. Environment

The Indus plain is a vast piedmont alluvial plain; the Indus River system has a mean annual flow of 175,156 million cubic metres in Pakistan, twice the mean annual water production of the Nile River (Johnson *et al.*, 1977). The climate is predominantly arid and semi-arid: in the northern regions annual evaporation averages about 152 cm while in the south it is about 190 cm; annual precipitation ranges from 50 cm in the north to 7·5 cm in the south. The combination of low and unpredictable rainfall and sub-tropical arid to semi-arid climate makes irrigation a necessity for successful agriculture. Since the major portion of the modern system, as well as the research reported here, are centred on what is now the Province of Punjab in Pakistan, the remainder of this paper focuses on this area.

The Punjab plains are crossed by six rivers, the Indus, Jhelum, Chenab, Ravi, Sutlej, and Beas. Since the implementation of the Indus Waters Treaty of 1960 between India and Pakistan, Pakistani Punjab has been utilizing the waters of the western rivers—the Indus, Jhelum, and Chenab; the bulk of the other rivers' water is utilized by India. The rivers are silt-laden, and because of the deposition of heavy silt particles, the beds are usually higher than the flood plains. The flood plains were (and still are) subject to flooding during the summer monsoon. The land between the rivers includes areas in the central portions that are above the flood plains. These high areas are called *bār*. Before the modern irrigation system these bār were covered by grassy and woody vegetation. They were exploited by semi-nomadic people with large herds of camels, sheep, goats, and cattle. These people also engaged in some rainfall agriculture, and cultivated small parcels of land irrigated by wells with Persian wheel, an endless chain of pots worked by a gear and shaft mechanism and powered by yoked animals. These wells were often 15–30 m deep.

B. Development of the System

The British began planning canal projects even before the annexation of Punjab in 1849. The first modern canal in Punjab, the Upper Bari Doab

Canal, was opened in 1859 and irrigation commenced in 1861. Thereafter the British continued building increasingly sophisticated and large-scale canals, with stock-taking interludes between them, until the end of their rule. Since 1947 Pakistan, with the aid of international donors, has completely remodeled and expanded the system: aside from several new canal projects, two huge dams (Mangala and Tarbela) have been completed, and link canals constructed to carry water from the western to the eastern rivers and canals to replace water retained by India. According to official figures, there are some 62 790 km of canals in Pakistan; and there are 88 000 watercourses, irrigation ditches that carry the water to the farmers' own lateral ditches and to their fields. These average over three kilometres in length. Of 14 million hectares of cultivable land with access to river water for irrigation, about 10 million hectares are irrigated and cultivated every year. It is no surprise that the system has been referred to as the "Indus Food Machine" (Johnson *et al.*, 1977; Planning Commission, 1978:3).

The British engineers who built the system had no previous experience in building irrigation works; and when they began, they had little theoretical knowledge of hydraulics, and knew little about ground water hydrology and the like (Michel, 1967:50–51). Furthermore, modern construction technology—machinery and building materials—was not available at first. By trial and error and experiments, they developed many of the basic formulas and techniques now used throughout the world.

The system uses barrages to divert water from the rivers into the canals. The canals are designed for continuous operation at or near full capacity, some year-round (perennial canals), some for only the summer season (non-perennial). They are designed to maintain a "stable regime"—that is, silting and scouring ultimately balance in the main canals (Michel, 1967:61). But the amount of flow cannot be regulated on demand and it may sometimes be interrupted, during floods or repair work. Water flows continuously from canals into distributaries, through concrete modular outlets (*mogha*) into watercourses (which often have several branches), and finally into farmers' laterals and onto the fields (see Fig. 1). The mogha is designed to deliver a fixed quantity of water when the canal is flowing at full capacity. Each watercourse commands from 60 to 250 hectares of land, generally cultivated by from ten to over 150 farmers. The Irrigation Department is directly responsible for the operation and maintenance of the barrages, canals and distributaries, to the mogha. It also lays out the route and commanded area of the watercourse, but its operation and maintenance are the joint responsibility of the farmers cultivating land in its commanded area. Individual farmers (or small groups) build and maintain small ditches (laterals) to carry water from the watercourse to their own fields.

Each farmer has a right to water proportional to the size of his land holding.

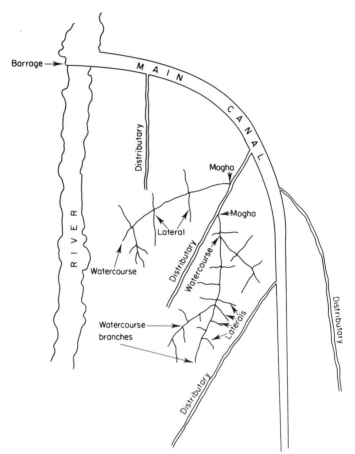

Fig. 1. Schematic diagram of canal system.

Water is distributed on a time basis: each farmer is allotted a period of time to take water, usually on a weekly rotation basis. Originally the watercourse rotations (*wārābandī*) were devised by mutual agreement among the shareholders. Shareholders who cannot agree on a rotation system may apply to the Irrigation Department to establish a fixed legal one; most watercourses have now been converted to this "*pakkā wārābandī*". Since it is a continuous flow system farmers receive the same share every week regardless of needs—which leads to periodic over- and under-irrigation. Trading of water among farmers is illegal since it causes waste, but it is commonly practised.

The system was originally designed to operate with a minimum of human regulation or interference. For example, water is regulated at the head of the canal, but it is not possible to vary the flow into watercourses. Aside from engineering considerations, undoubtedly the British knew that recruitment of competent and responsible people would be difficult and a flexible system of water distribution would lead to uncontrollable abuses. They were also concerned to keep operational costs at a minimum since they were interested in recovering their investment quickly. These considerations also underlay the policy of minimal local intervention: farmers were expected to build and maintain their watercourses and settle disputes among themselves. The Irrigation Department retains considerable residual power, set out in the "Canal and Drainage Act of 1873" (Jahania, 1973). This power is used only when the shareholders appeal to the Irrigation Department. Similarly, the extension of irrigation included no instructions to the farmers on irrigation techniques. Farmers were left to their own devices (Johnson *et al.*, 1977:1257). The major method continues to consist of the flooding of small basins. There is no adequate means of communicating information from the users to the higher level managers, or even from the top down. Finally, no efforts have been made to organize farmers locally on either a formal or informal basis to manage the watercourse. The watercourse is a "collective" or "public good" which benefits all farmers using it, but there is no mechanism to insure that each contributes his share to its maintenance (Olson, 1965; Lowdermilk *et al.*, 1978 (II):119–20).

The British had several motives for building the canal system. In the beginning there was an idealistic and enthusiastic desire to extend irrigation to demonstrate the benefits of European science. A decisive motive for the first canal was to give employment to potentially disruptive Sikh army veterans. Another more important and lasting motive was to improve the agricultural value and thus the revenue-producing capacity of the newly annexed lands. Yet another motive was fear of famine (Michel, 1967:65–66).

The earlier canals were mostly designed to improve agriculture in already settled areas. Later projects emphasized settling new "waste" lands; they involved not only canal building, but laying out of new villages, cities, roads, railroads, etc., and distribution of land to settlers. The British hoped to reduce famine in India by making Punjab the "granary of India" and to relieve overpopulation in the eastern districts of Punjab by settling farmers from these areas on new lands.

C. Characteristics of Modern Indus Basin Development Plans

Michel (ibid.) provides a detailed account of all the programmes that have been proposed to solve the problems of the Indus Basin irrigation system, and the results of their implementation, up to the mid-1960s. In this section I do not discuss these programmes *per se* but focus on certain common characteristics of the proposals and policies. Since the Independence of Pakistan and especially since the signing of the Indus Waters Treaty in 1960 with India, a series of distinguished panels and study groups have produced long and detailed reports on the problems of the Indus Basin and their solutions. Perhaps the most influential reports have been the so-called "Revelle Report" (White House, 1964) and the "World Bank" report (Lieftinck *et al.*, 1969). A more recent panel, of which Dr Revelle was a member, has reviewed the Revelle Report's recommendations, and progress (or lack thereof) so far, and offered a series of research guidelines and topics (Planning Commission, 1978).

While these panels have suggested a wide variety of solutions to the problems of Pakistan's irrigation system, all of their recommendations share several basic characteristics. The first is that all of them have emphasized engineering and technological programmes: large-scale tubewell projects, dams, and link canals and the like. The White House (1964) report does also recommend a massive integrated extension programme to get farmers to adopt modern technology, but this programme never materialized. The second characteristic is that all of the advice offered has required massive capital outlays, mostly of foreign origin. Michel discusses Pakistani criticism of the SCARP tubewell programmes made on this basis in the early 1960s— criticism that was not heeded and that Michel himself dismisses (1967:470–72). The third characteristic is that the implementation of the proposed projects requires continuous input from foreign "experts". That is, foreign consultants, financed by major donors to Pakistan, have advised Pakistan to adopt highly capital- and foreign-expert-intensive solutions, many of which have been and are being implemented. Michel discusses this point and defends the use of foreign consultants, but his arguments are not entirely convincing (ibid.:357–64).

The final major characteristic of the advice given in the Revelle and World Bank reports concerns the administration and organization of the system. Neither panel included any kind of social scientist other than economists. No research data on organization were available—but there is no indication in the reports that this was perceived as a handicap. Both reports were written after the Water and Power Development Authority (WAPDA) had been set up to execute the large scale water, power and reclamation projects envisioned by Pakistan's planners; in fact Michel suggests that one motivation for establishing WAPDA may have been to attract foreign aid (ibid.:350). The few pages devoted to advice on organizational matters (White House, 1964:179–84; Lieftinck *et al.*, 1969 (2):186–91) suggest further centralization of planning and management at the highest levels, and better co-ordination of the various organizations involved. No serious consideration seems to have been given, either by the foreign advisors or by the Pakistani planners and administrators, to the problem of the relationship between the users of the irrigation water and the managers of the system; all assumed that the problems and their solutions were ones the planners understood best, and could impose from above. One short paragraph in the Revelle Report entitled "Long Range Goals", does suggest that "a central hope for the future should be the gradual emergence of associations for farmers . . ." (White House, 1964:184), but this obviously was for the future, *after* the system had "developed"; it was not recommended as a strategy for development.

More recently, for the first time in Pakistan's history, research has begun to focus on local level water management. This research has identified local level waste and mismanagement of irrigation water as a key constraint on improving agricultural production as well as a cause of waterlogging and salinity. Research on a similar irrigation system in India suggests that certain characteristics of the administration of the system itself result in farmers' uncertainty about their water supply, and this is a major constraint on productivity (Reidinger, 1974; Gustafson and Reidinger, 1971). All of this research has resulted in ·a recognition of the importance of local organizational factors, and recommendations for forming local farmer organizations (Water Management Research Project, 1976; Planning Commission, 1978).

Many of these recommendations, however, retain the "engineering mentality" of earlier advisors; that is, farmer participation and organization are viewed as being of the same order as technological problems: the function of research is to discover the appropriate organization of design—"the solution"—and introduce it. Just as the problems of rising water-tables, salinization, and inadequate supply of irrigation water are "solved" by installing a network of tubewells, so, it seems to be assumed, inadequate organization can be solved by "installing" a new farmer organization. Furthermore, only the formation of local organizations with vaguely defined but limited responsibilities have been suggested; there has been no consideration of the dynamics, the adequacy or the consequences of the present organization of the irrigation system.

Most of the high-powered recommendations,. and the policies pursued to date, then, share the same characteristics: an orientation toward purely technical solutions, designed and implemented from the top down, with the financial and advisory aid of foreign organizations; and an assumption that the "experts" know best what the problems are and how to solve them. Although the major reports and recommendations are thick and "comprehensive", none have seriously addressed the most fundamental problem of all for the future of the Indus irrigation system: how should it be organized? What should be the role of its users in the management of it? What have been the consequences of the present organizational structure? Policy based on faulty assumptions about the goals, values, ability to co-operate, and behaviour of local users is bound to fail. For example, if local irrigation associations were established in Pakistan, and the legal framework for these organizations were to specify "Western" rules and procedures such as decision-making by majority vote and some version of Robert's Rules of procedure, these organizations would probably not work in the way envisioned by the planners because such procedures are inconsistent with local users' decision-making patterns as well as with the prevailing stratified socio-economic structure of rural society. The remainder of this chapter illustrates the potential of an approach from social science towards the problems of designing and evaluating programmes to involve the farmers in local irrigation improvement projects.

II. WATERCOURSE RECONSTRUCTION: A CASE STUDY

Gondalpur[2] is a village in central Punjab, on the Chaj Doab, the area between the Jhelum and Chenab Rivers. This area has traditionally been called *Gondalbār*, because historically the Gondal "tribe" dominated the area. In the flood plains among the rivers, intensive agriculture based on flooding by the river, inundation canals, and wells, has been practised for centuries. Being located above the flood plain, Gondalpur had no canal or inundation irrigation before 1904. The vegetation of this semi-arid area consisted of various small and deep-rooted trees, which provided fuel, fruit, and fodder, and a variety of grasses, on which the ancestors of the present inhabitants raised large herds of cattle. At the time of the first British survey in 1857, there was one Persian wheel well irrigating 7 ha; irrigation was also practised during the monsoon by catching runoff in a low place, and planting primarily millets. By the 1880s, thirty years after British rule was established in the area, there were three Persian wheel wells. The wells tapped a water table 15 to 25 m deep, so that even with good oxen or buffaloes, only about one fifth of a hectare could be irrigated in a 12 h turn.

Informants claim that animals, not land, were wealth: a man's standing in his community depended mainly on the size and quality of his herd. Agriculture was meant to supplement a diet which was based on dairy products and meat. The British land records show that the short-lived previous regime of Ranjit Singh had imposed a head tax on animals, and no tax on land, in this village. The British discontinued this policy and imposed a moderate (in their eyes) land revenue. This meant that land had to be registered in individuals' names—an innovation. Informants say that their ancestors regarded this as an unfair burden and some sold or gave their land rights to others for almost nothing. Stories are told of how people in nearby villages punished their enemies and servants by having land registered in their names.

During the 49 years between the first British land settlement and the arrival of the canal water in Gondalpur, however, there was a substantial rise in population and a gradual extension and intensification of agriculture (see Table I). Population grew far more rapidly than did the extent of cultivable land, mostly as a result of immigration. There was also a fairly large-scale transfer of control over land to outsiders—and a concomitant increase in tenancy. The Lower Jhelum Canal was officially opened in 1901, but its water did not reach Gondalpur until the 1904–1905 winter (*rabī*) growing season. Its impact was immediate: scores of hectares of land came under cultivation during both the summer (*kharīf*) and winter growing seasons. Former cattle keepers and part-time farmers became full-time farmers, either on their own land or as tenants on others' land. The area available for grazing animals declined, while the number of animals increased, so that even a few years before the canal was introduced most farmers had begun devoting a substantial percentage of their land to growing fodder for their animals. Other changes since the introduction of canal irrigation include: a further rise in population; increasing fragmentation of land holdings; major changes in diet; increasingly intensive agriculture; and a rise in the water table of 12 to 20 m,

Table I. Changes in population and cultivated area in Gondalpur since 1857

Year	Population	Areas of crops harvested in hectares			
		Rainfed	Well irrig.	Canal irrig.	Total
1857	67	18·8	7·3	0	26·1
1890–91	310	87·1	17·2	0	104·3
1901–02	568	69·3	18·2	0	87·5
1905–06	NA	4·5	0	194·4	198·8
1910–11	565	0	0	233·7	233·7
1921	767	2·0	0	315·1	317·1
1931	758	133·4	0	139·7	273·4
1951	914	40·5	0	242·6	283·1
1961	1117	25·9	0	313·9	339·8
1972[a]	1246	36·5	0	384·3	420·8
1977[b]	1450	21·4	5·2	356·2	382·8

Sources: All data are from unpublished village records except the 1961, 1972, and 1977 population figures. The 1961 and 1972 population figures are from the District Census Handbooks for those years; the 1977 population is based on a census carried out by the author and his wife in February–March 1977. NA means "not available".

[a] The figures for area harvested are 1968–69 figures, the closest ones available to the 1972 population figure.

[b] The figures for area harvested are for 1975–76.

so that today nearly everywhere it is less than six metres below the surface, and in some it is less than a metre and a half. A large low-lying tract in Gondalpur has become waterlogged and an adjacent previously productive area is now saline and unproductive. The major crops today are wheat and fodder crops in the winter, and rice, sugar cane, fodder crops, and some melons and cotton in the summer. Most of the land is double-cropped every year.[3]

A. Watercourse Social Organization[4]

The dominant landowners in Gondalpur, the Gondals, are divided into four named *birādarī* "brotherhoods" which are local co-resident groups based on a

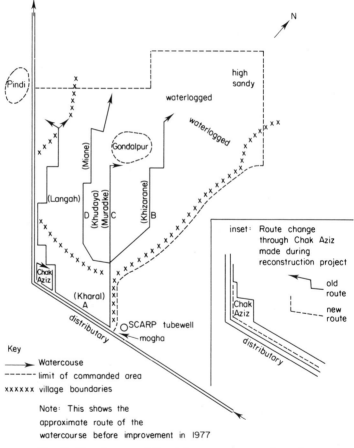

Fig. 2. Sketch map of Gondalpur watercourse branches.

combination of patrilineal descent and marriage (see Alavi, 1972). The biradari are concentrated on different watercourse branches (see Fig. 2). Table II shows the relationship between biradari and watercourse branch. The Gondal biradaris are Khizarane (branch B), Muradke (branch C), the Khudaya and the Miane (branch D). The first three named so dominate particular branches that these are known by their names. Besides the Gondals, families belonging to other groups, or $zāt$[5], also have land on various branches. A few Awan have very small holdings on branches B and C; two Bhattis have some land on branch B, as do three Sayid families; and some Muradke and Khudaya have land on branch B, though most of this land is either waterlogged or saline, or too high for irrigating. On branch D, aside from the Khudaya, a few Pindi farmers also have land, as do the religious leaders of Gondalpur, the Miane. The *numbardār*[6] and his family, who have

Table II. Biradaris involved in watercourse reconstruction[a]

| Biradari | Number of | | Watercourse[b] | Position on |
	Households	People	branch	branch
Gondal-Khudaya	11	70	mainly on D; a little on B.	Head; Middle; Tail
Gondal-Khizarane	21	105	B	Head; Middle; Tail
Gondal-Muradke	7	43	C; a little on B.	Head; Middle; Tail
Gondal-Miane	5	36	D; a little on A	Middle; Tail
Langah	5	36	A	Middle
Awan	11	47	B and C—very small holdings	Middle
Bhatti-Rajeane[c]	18	78	B (2 households)	Middle
Sayid	3	25	B	Head; Middle
non-Gondalpur biradaris:				
Kharal (Chak Aziz)	3	?	A (a little on B)	Head
3 Pindi biradaris[d]	under 10	?	A & D	Tail on both

[a] This is not a complete list of all biradaris in Gondalpur; only those having land irrigated by the watercourse reconstructed are listed. Figures are based on 1977 complete household census.

[b] See Fig. 2.

[c] Only two households of this biradari have land on this watercourse. There are 7 Bhatti biradaris in Gondalpur with a total of 90 households and 416 people as of 1977.

[d] These biradaris did not play an important role in the improvement project—their major holdings are on other watercourses; they generally acted together on this project.

relatively large holdings (20 to 30 ha) are Khudaya; the Miane holdings are also relatively large (about 10 ha for each of three households) while the other two Gondal biradaris are mostly small farmers (2 to 8 ha). Bhattis outnumber Gondals in the village as a whole as do the Massali labourers, but the former are mostly tenants and the latter landless and poor.

At the head of branch A, where it passes through Chak Aziz, are four related households of Kharal zat. One is a very large landowner (about 122 ha), having bought much land in a nearby village; his brother has about 20 ha on branch A; and their half-brother's two sons also have about 20 ha between them. Though these two brothers often quarrel with each other, they did not during the watercourse project. I shall refer to them collectively as "the step nephews". Following the Kharal, on Gondalpur land, branch A irrigates the land of several very small farmers (0·5 to 9 ha) of Langah zat. The members of this zat, though poor, have marriage relations with the Khudaya, Muradke, Kharal, and a large Pindi landlord. Some of the land belonging to the Miane is irrigated after the Langahs by branch A, and finally, at the tail, branch A irrigates small portions of the relatively large holdings of several Pindi families.

B. Watercourse Conditions Before Reconstruction

At the time of the study (1976–1977) the level of maintenance of all the branches on the watercourse was extremely poor. A SCARP tubewell had been installed at the head of the watercourse in the mid-1960s, doubling the amount of water flowing through the watercourse. As is generally the case in the SCARP areas, the intensity of cultivation increased substantially as a result of the increased water supply. Most land is now double-cropped each year on this watercourse. However, the capacity of the watercourse was not increased; further, for some years after the installation of the tubewell, there was no perceived shortage of water. According to informants this led to a decrease in maintenance efforts, atrophying the already weak sanctions enforcing participation in watercourse cleaning. Further, fragmentation of plots has led to increased numbers of "illegal" (that is, not sanctioned by the Irrigation Department) cuts in the main water channels. The watercourse, on all branches, was choked with grass, bushes and trees; leaked through rat holes, thin banks, and at junctions; and water remained standing in many low sections after irrigation. On branch A, since the Chak Aziz land is relatively high, the Kharal owners actively sabotaged efforts to clean the head of the watercourse. Silting raised the water level, and thus their ability to irrigate their high land; but it blocked a large percentage of the water from reaching the middle and tail farmers.

This lack of maintenance, combined with increasing pressure to raise production (in part limited by the water supply) had created considerable dissatisfaction with the condition of the watercourse by 1976.

C. The Reconstruction Process

In response to this dissatisfaction, I was instrumental in arranging for the Mona Reclamation Experimental Project to choose this watercourse for an experimental improvement programme.[7] In this programme the Government supplies technical advice and supervision, and materials such as concrete outlets (*nakkā*); the farmers are responsible for supplying all labour for the earthen improvements, masons for installing outlets, concrete sections, and culverts, and for subsequent maintenance. Some Gondalpur farmers had heard about the success of the improvement programme in other villages; according to a survey they were aware that the losses from their watercourse were high and they were eager to improve it.

The improvement programme on this watercourse undoubtedly faced more problems than is usual on a single watercourse; but this makes it an important case to study as all of the problems encountered characterize other watercourse reconstruction efforts to various degrees. A description of all that happened during the six months of active improvement work would constitute a book in itself; a brief summary will show the kinds of problems faced by the project. At a farmer meeting in June 1977 two committees were set up; one, for branch A, included a Kharal representative from Chak Aziz (the youngest of the two step nephews), a Gondalpur Langah, and the Pindi numbardar. For the main branch and branches B, C, and D, one Khudaya, one Khizarane, and an Awan were chosen. The branch C Muradke refused to take part in the improvement programme on their branch and therefore had no committee member. There were several reasons for their refusal: they did not perceive much of a water shortage; they preferred to continue cutting their watercourse freely; and since they were angry at their Awan relatives over unrelated issues they opposed any programme the Awan supported.

Work began on branch A—but on the same day as an announcement of land allotments under the land consolidation programme in Gondalpur; therefore, only Chak Aziz shareholders were present at the work site and they successfully pressured the government engineer to start work on a new route for the watercourse, parallel to the distributary around their village (see insert on Fig. 2). This route had been discussed previously and opposed by the middle shareholders, but now it became a *fait accompli* and they could not oppose it. Since the old route had passed through the step-nephews' land and another Kharal's courtyard, while the new one is on government land and higher than the old one, the Kharal benefited substantially from this change.

Over the next few months, work continued fitfully, on branch A, and the engineer had branch D and B work begun even though he had not yet done a survey to indicate the route, width, and depth. The farmers on B and D noticed that their water supply was *reduced* as a result, leading to considerable tension between them and the engineer. At a meeting with the farmers the engineer accused the farmers of not co-operating with him and gave them an ultimatum—to follow his instructions without argument or he would abandon the project. The farmers were angry but agreed to his demands. These branches were then surveyed and the work redone.[8]

A number of disputes broke out among the farmers (aside from a series of continuing disputes between the farmers and government officials):

1. On branch D, two Khudaya, the numbardar (supported by the Miane), whose lands were at the head and middle, and his paternal cousin, a watercourse committee member most of whose land is at the tail, disputed over how far towards the tail the improvement work should go. The numbardar and Miane wanted the work to stop about 300 m short of the cousin's land, so that no improvement work would be done on the section through their land. When the tail cousin refused to co-operate unless his demands were met the numbardar agreed, though the Miane continued to protest and refused to co-operate on the work.
2. The Miane, near the middle and tail of D, continued to dispute with Khudaya over how far the improvement should go, and over the route of the watercourse. The engineer, based on his survey, wished to straighten it. Since it skirted the edge of the Mianes' land and over the years had shifted, increasing their land, moving it would have reduced their land slightly. It was straightened, finally, but over their continuing protest.
3. On branch B, the Khizarane leader frequently argued with Muradke, Khudaya, and Sayid shareholders over the division of the work.
4. On branch A the Pindi shareholders and Miane were lax about doing their share of the work, leading to conflict with the others and long delays in completing the section.
5. The Langah committee member and the Kharal member disputed over route changes in branch A demanded by the Kharal, and division of work shares; because of his weak position, the Langah pursued these issues more with the engineer than the Kharal directly. In every case, the Kharal won, because both the government officials and other farmers feared the consequences of the Kharal not co-operating, given their strategic position on the watercourse.
6. The Kharal step nephews, who had traditionally taken "unauthorized" water from the main branch, successfully sabotaged the work on that branch, including preventing the removal of trees and straightening the route. There

seemed to be three reasons for their obstructionism: they realized taking illegal water from the main branch would be more difficult; they would lose a little of the land they had occupied if the watercourse were straightened; and they were jealous. They opposed any programme that would benefit the weaker Gondalpur people, perhaps fearing they would become independent.

7. The Kharal demanded, and by threatening to sabotage the project obtained, extra nakkas and double-sized culverts for their land; but even after getting them the two step nephews in particular continued to sabotage the work.

A project that was expected to be completed in less than two months was not finished in December 1977, the sixth month, when I left; and when I returned in May 1978 I discovered that some sections still have not been reconstructed, especially in the middle and tail sections of A and B; some of the sanctioned nakkas had not been installed, and several of the installed ones had been damaged; and no cleaning or maintenance had been done. All the branches were choked with weeds and silt and leaked from new unauthorized cuts in rebuilt banks. Even in October 1978, the normal watercourse cleaning in preparation for the winter season had only been haphazardly done.

The sections completed up to December 1977, immediately after reconstruction, did not leak, and farmers enthusiastically reported up to five times as much water reaching their fields as before. However, by October 1978 the water delivery had drastically declined to only slightly above pre-improvement rates. The watercourse sides, because of both poor construction and very poor maintenance, had deteriorated considerably and were leaking badly; much water remained standing in the ditch after irrigation; and many farmers felt discouraged about the prospects of maintaining even the present level of efficiency.

III. PUNJABI CULTURE: THE GAME OF IZZAT

There is no doubt that one source of the problems faced by this project is the relationship that developed between the farmers and the government officials supervising the programme. Although some of these engineers and extension workers have rural backgrounds, their education has seemingly made them unfit for rural work. Possessing a formal degree, and a respectable position in the government bureaucracy, they are "officers". They create barriers between themselves and their clients by wearing Western clothes, speaking an urban dialect, and doing all they can to create the impression they possess a superior knowledge and position which ought to be respected. When the clients assert themselves, and refuse the officer the respect (read obeisance) he claims, conflict arises, and the officer's low opinion of his clients is confirmed in his mind. This kind of relationship between government officials and farmers is not confined to Pakistan.[9]

Another factor was that the potential benefits of the programme were not perceived as equally distributed (see Doherty and Jodha, 1977). Indeed equal distribution of benefits in a watercourse reconstruction programme is impossible to achieve because of differences in size of landholdings, differences between owners and tenants, and most crucial, the relatively greater benefits accruing to farmers with land at the tail than to those with land at the head of the watercourse. Even if the benefits of reconstruction were distributed equally, one could argue that any rational individual will minimize his contribution toward such a collective good because he cannot be denied its benefits even if he does not invest in the project (Olson, 1965). However, the active attempts by the step-nephews to sabotage the programme, even on other branches in order to prevent others from benefiting, and the disputes that developed among persons whose benefits were about equal, suggests these factors are insufficient as explanations of the problems encountered.

A major source is to be sought within the social organization and culture of rural society. Punjabi rural society is characterized by a set of values and structural mechanisms which—in relation to their irrigation system—*encourage* conflict, make it endemic and unavoidable, and thus tend to discourage co-operation on a long term basis. These values may have been adaptive before the irrigation system but have continued to operate even though they appear maladaptive under present conditions.

The ancestors of the Gondalpur farmers, who were cattle herders and part-time farmers, were probably not permanent residents of Gondalpur before the British settlement. This settlement awarded permanent rights that had not existed before. One characteristic of pre-British Gondal society was relative mobility of individuals and families: larger local groups were unstable, as people were free to move and often did move with their animals. The type of situation now known as the "tragedy of the commons" (after Hardin, 1968), in which individual herdsmen continue to increase the size of their individual herds even after the carrying capacity of commonly owned grazing lands had been exceeded, did not arise because people were able to leave for less crowded, if not greener, pastures. It seems likely that under these conditions it was not necessarily recognition of overgrazing *per se* that triggered dispersion, but rather a high incidence of social conflict maintained dispersion.

The most fundamental concept, or theme, in rural Punjabi culture, in terms of which much of Punjabi behaviour can be understood is *izzat*.[10] Izzat may be glossed as "honour", "esteem", "reputation", "status", or "face". It is a "limited good" (Foster, 1965): one acquires it only at others' expense. As in a zero-sum game, the success of one person is a threat to all the other players, a characteristic that generates competition and jealousy. For example, when government officials agreed to a very reasonable request for a double-width culvert for truck access to one of the Kharal's brick kiln, his step-nephew demanded a double-width culvert for himself. Informants said his izzat was at stake: if he got less than his step-uncle he would lose izzat. Government personnel, not accepting the rules of the local izzat game, rejected his demand, which led to further problems with the man.

All men wish to avoid losing izzat, but many men also attempt to increase their own izzat or reduce others'. One acquires and increases one's izzat by several different strategies. First, one must have the ability and, more importantly, the willingness to use force. There is a famous Punjabi saying, "Whoever holds the stick owns the buffalo". This does not necessarily mean force is resorted to frequently; it is enough to create the impression that one is willing and able to use force, and in times of tension, much calculation and speculation revolves around this issue. The Kharal step-nephews were feared because they had demonstrated their willingness to use force in previous fights. The Bhattis of Gondalpur, mostly tenants and poor, in the past also had a high izzat for the same reason. On the other hand, the Khudaya numbardar, despite land holdings, his official position, and several adult brothers, had less izzat than he might have had because it was known he feared violence. This was not an unreasonable fear since his father had been murdered in 1962.

A second means of acquiring izzat is possession of influence with government officials, and willingness to use it for one's supporters and against one's enemies; the Kharal step-nephews, some Pindi landlords, and a recently deceased poor and landless Bhatti leader before his death, all had a substantial amount of izzat from this source (as did the author). A third source is willingness to entertain guests lavishly, whether they are government officials or relatives at a wedding—even to the point of bankruptcy. The deceased Bhatti leader mentioned above kept himself bankrupt but high in izzat by this strategy.

Success in competition, whether in organized games such as *Kabadī* or in a stick fight, is another source of izzat. *Winning*, not a valiant loss, is the key. Another source is generosity, not to the general public, but towards individuals (who are obliged then to render support).[11] Finally, successful one-upmanship, including revenge for a previous defeat or insult is important. For example, disputes are often taken to the police; and the person or group that can avoid jail or being beaten by the police, while getting the opponent punished, and spend the least money, "wins". Such cases often become very long, involved, and expensive; but they continue even when people are aware that after so much trouble and expense they will have nothing tangible to show.

In order to improve izzat, *tāqat* (strength, power), is needed, but tāqat alone is insufficient; it is also necessary to use this power to help clients or defeat enemies. The richest of the Kharals has less izzat than one would predict from his wealth and government contacts because he is unwilling to use his position in this way. A person whose tāqat and izzat are increasing attracts followers and allies who hope to benefit; but he also attracts the jealousy and fear of others who are likely to band together behind the scenes to plot strategies to limit or reduce him. If a group (such as a biradari) or several brothers become too powerful, efforts are made to sow dissension and thus weaken their unity; because individuals' primary loyalties are to themselves and each one assumes this to be true of others, efforts to divide groups, or even two brothers, often succeed.

People informally recognized as leaders are supposed to work for the benefit of their followers as a group; but more often than not, such persons keep their own interests in mind first and attract clients by aiding *individuals* (against their enemies or with the police for example) who are then obliged to them. Only infrequently do leaders work for the benefit of a group or community as a whole—and even when they do, others may accuse them of seeking only their own benefit.

Opposition is often expressed verbally in terms of issues, but in fact the issue is nearly always a pretext: men oppose or support decisions and programmes based on their perceptions of their competitors' position. For example, even though all farmers were suffering the exactions of a corrupt tubewell operator, they did nothing because, informants explained, if one man or group proposes petitioning for his removal, others will oppose, not out of love for the tubewell operator, but to prevent the others from utilizing the issue to gain some advantage, or to pursue some long-standing grudge. This can be carried further: the non-cooperative behaviour of the Kharal on branch A during the watercourse reconstruction was interpreted by informants as based on a desire to prevent others from benefiting—even if it means foregoing their own potential benefits. There is a Punjabi saying, "If my neighbour's wall falls it is good—even if it falls on me". Opposition is never legitimate in the western parliamentary sense: it is always personal (or interpreted as personal) and aimed at weakening others or strengthening one's own position.

There is a strong ethic of loyalty to one's kinsmen; one ought to be prepared to make sacrifices for their benefit. Marriage within the biradari—siblings and cousins exchange children—is intended to cement their affections and relationships. Divisions within the community, in Gondalpur and other villages, are usually between biradaris; this was the case for most disputes over the watercourse reconstruction programme. There is a feeling of a biradari's izzat, which must be protected from others' attacks; and if a man's izzat suffers at the hands of a member of a different biradari, all of his close kinsmen may unite in opposition to the "enemy".

Nevertheless, despite the emphasis on loyalty to one's kinsmen, tensions among biradari members are always present; patrilateral cousins and brothers often have tense and competitive relationships and do not completely trust each other. One's brother's or cousin's personal izzat is not necessarily one's own; hence a man in apt to be jealous of and feel threatened by a brother's success. Tension is also generated among biradari members by joint potential rights in land. One of the worst cases of conflict in Gondalpur history, resulting in two murders and three executions, occurred within the Khudaya biradari over land; one branch attempted to deprive another branch of rights to some land. Tensions built up and the latter finally took action, by murdering the numbardar and his brother. The amount of land involved was in fact not great; the real issue was izzat. If the second group had allowed itself to be deprived of the land, its members' izzat would have been severely damaged.[12]

During the improvement process there was much petty conflict among biradari members over work shares and the like; the Kharal are seriously divided, and the Khudaya only slightly less so. The Awan and Muradke, though separate biradaris in one sense, are closely intermarried, yet at the time of this project they were involved in conflict over several issues which prevented them from co-operating on the project.

The sense of community within the village is real, but also intertwined with izzat. In opposition to outsiders villagers will act together in a stick fight or a competitive game, to preserve the izzat of the village. However, co-operation within a community to achieve a mutually beneficial goal is very difficult as people fear others may benefit more than they, or the leaders will gain undue influence. In some villages there are leaders who are sufficiently trusted (or feared) to insure that farmers co-operate to maintain their watercourse, but this is not true of most communities, and is not a permanent characteristic of any community.[13]

IV. SOCIAL ORGANIZATION ON OTHER WATERCOURSES[14]

During 1978 I collaborated in a study of the social organization of ten reconstructed watercourses in Punjab. We deliberately chose our sample so as to include several "problem" watercourses and several "model" watercourses. We also chose watercourses for which a maximum period of time had elapsed since improvement (the range was four months to two years), and which represented several different agronomic areas of the province. The purpose of the study was to identify those sociological characteristics of rural society that both promote and inhibit effective co-operation on watercourse rehabilitation and maintenance. The results complement the intensive research reported above. Mirza and Merrey (1979) provide a detailed discussion of the methods and results of the research; here only a brief summary of the conclusion is possible.

We discovered that both the ease and completeness of the actual reconstruction process, and the quality of the maintenance after improvement, vary considerably; and there are systematic relationships between the relative "success" of improvement, and maintenance quality; and also between these and certain sociological characteristics. Watercourses whose improvement was completed without significant delay or disruptive conflict are generally better maintained than those where the improvement process has been difficult. The better maintained watercourses tend to have all or most of the following characteristics:

1. A large percentage of farmers with landholdings in the 2·5 to 10 ha range. We defined holdings in this range as "small but economically viable" in irrigated Punjab. Watercourses dominated by farmers below this range seem to be very difficult to organize for co-operative programmes, perhaps because they are less committed to farming as a full-time occupation. Larger farmers usually have labourers do their share of the watercourse work, with the result that it is often done carelessly. Large farmers are also more able to violate sanctions, and more involved in conflict.

2. Relatively equal distribution of power and influence among farmers on a watercourse. "Power and influence" was measured by asking sample farmers to rate all the other farmers on the watercourse, and adding the scores. Where influence is more equally distributed, and one or a few farmers do not dominate, farmers seem to co-operate better on collective projects.

3. A large percentage of farmers being perceived by fellow shareholders as having some influence and power. On some watercourses, power and influence scores were uniformly low—no one commanded any respect. Co-operation on such watercourses was much less than on watercourses where the scores were higher across the board—that is, where most shareholders have at least some standing and respect.

4. Concentration of power and influence at the tail or tail and middle of the watercourse. Farmers at the tail of a watercourse usually receive the greatest benefits from improvement, and are thus more highly motivated. If these farmers have comparatively greater influence, they often insure maximum co-operation by others.

5. "Progressiveness" of the community, as measured by the percentage of farmers with a better than primary education, number of institutional services available in the community, and the percentage of farmers who listen to the radio regularly. These three components together were used as a measure of community attitudes toward modernization and change.

6. Previous history of co-operation on community projects, and lack of serious recent conflict. Communities that had successfully co-operated on previous projects, such as building a school, and which had not been divided by serious conflict in recent years, co-operate more effectively on watercourse rehabilitation and maintenance.
7. A small number of shareholders on the watercourse. On the watercourses with the largest number of shareholders—even if they all belonged to one biradari—getting the farmers to work together to rebuild and maintain their watercourse proved very difficult.
8. Membership of most of the shareholders in a single biradari.

In reality the ideal characteristics listed above are not found very often in rural Punjab. None of the watercourses in our sample were well-maintained; but those which were comparatively better maintained share more of these characteristics than those which were in poor condition. None of the watercourses had an effective organizational mechanism to insure that all did their share of the cleaning. On five of the ten watercourses studied, the reconstruction work had not even been completed, because of conflict among the shareholders or between the farmers and government engineers. Our study shows that the quality of improvement and maintenance are closely related to sociological characteristics of the watercourses; but it also shows that present forms of organization are not adequate to insure good maintenance of the system, even on relatively conflict-free watercourses.

Punjabi villages exhibit a considerable variety of structural forms: single, double and multi-biradari villages; villages with strong leaders, and villages with weak leaders; villages with no recent history of serious conflict, and villages where murders occur yearly; and villages of small, medium and large farmers, owners and tenants; a few where landholding distribution is fairly equal, and many where the pattern is highly skewed in favour of a few farmers. Gondalpur's social organization included all of the characteristics shown in the later study to be least conducive to successful co-operation on a watercourse reconstruction project.

However, in contrast with the variation in social organization, there is relatively little variation in cultural values: the concept of izzat discussed here is shared to a large extent by all rural Punjabis, but it leads to the pursuit of different strategies depending on the social context. Both of the studies together show that the organizational and cultural impediments to a co-operative programme such as watercourse reconstruction are serious indeed; and even if the watercourse is successfully rebuilt, the inability of the users to maintain it means the investment in reconstruction may be wasted. However, one can go further than this: these organizational and cultural impediments, together with the ineffectiveness of the overcentralized bureaucratic management structure of the system, are at the root of the low productivity of the system; the minimal payoffs from the huge amount of capital invested in dams, and canals, and SCARP tubewells; and to an undetermined but probably very large extent, the waterlogging and salinization—the processes of desertification.

V. CONCLUSION

For decades, research and development projects on Pakistan's irrigation system have focused solely on the perceived technical problems and on their solution by means of large scale capital-intensive purely technological approaches. The users of the water, the farmers, have been ignored. In the 1970s as a result of the research efforts of a number of American and Pakistani scientists, local-level problems and inefficiencies began to be recognized. However, initially this research too focused solely on technical problems such as watercourse leakage and rehabilitation. Experience with pilot watercourse reconstruction projects soon demonstrated that farmer co-operation was the key to the success of the projects. The focus on farmer co-operation, on which little research had been done, led to an increasing level of collaboration of sociologists and an anthropologist with the engineers, agronomists, and irrigation specialists in an attempt to develop an effective watercourse programme. It was expected that about a dozen experimental "Water User Associations" would be organized under existing laws and their activities monitored. The end product was to be recommendations for forms of organization to be used for establishing associations of irrigators for improving local level water management (see Mirza and Merrey, 1979). However, the cut-off of American aid to Pakistan in 1980, as well as various political developments in that country, made it seem extremely unlikely that this programme will be carried out in the foreseeable future.

This work (including my earliest publications on the subject) was based on the assumption that such "tinkering" with the system could be effective in improving the productivity of the system as a whole, and reversing the process of decline and desertification in the form of declining levels of maintenance at all levels, and waterlogging and salinity.

This assumption now seems highly questionable. The "technical" problems of the system cannot be solved as if they were isolated from the larger social, cultural, and economic context. This point may appear obvious, perhaps, to a social scientist, but it does not generally characterize development policies and programmes, especially of relatively conservative countries such as Pakistan.

The "engineering mentality" has been carried over from the older style development projects such as dam construction to the more people-oriented programmes. For example, the planners of the pilot watercourse reconstruction project in Pakistan believed their own rhetoric that farmers' perceptions of their "self-interest" in watercourse reconstruction would overcome long-standing social and cultural impediments. The social scientists were called in somewhat later and expected to carry out rapid surveys (complete with statistics) to identify the problems and propose solutions to insure the success of the project. Social scientists, can, it is true, often identify social and cultural impediments to seemingly useful projects, and social and cultural factors involved in processes of environmental deterioration; having identified the problems, they can suggest strategies to overcome them. There are undoubtedly many situations where this narrowly conceived role is quite adequate, but Pakistan's irrigation system is not one of them.

In my discussion of the various recommendations and development projects in the Indus Basin, I have drawn attention to the fact that none have dealt with the most fundamental issue: the organization of the system. The study of organization comprehends the nesting of local systems in larger systems, and the complex relationships among social structure, values, technology, and environment. A beginning has been made in this direction with the various proposals to decentralize the management of the system and to organize water users into viable associations,[15] but these are very preliminary and are based on as thin a data base as are many of the technological solutions now being implemented. A great deal more research is needed on the organization of the system at all levels, and especially social constraints and cultural perceptions and motivation. Such research can be used to develop a more comprehensive and realistic model of how the Indus system actually operates. Based on this model, alternative forms of organization can be suggested.

Pakistan's irrigation system—indeed that nation as a whole—is in crisis. Poor management and maintenance of the system at all levels, waterlogging and salinization, low productivity despite capital-intensive inputs (dams, wells, fertilizer, tractors), and the socio-economic inequalities which have increased in recent years as a result of "green revolution" technological changes (Nulty, 1972)—these are all facets of the same fundamental problem: inadequate and inappropriate organization. Capital-intensive technological projects are unlikely to lead to any substantial "development" of the Indus Food Machine unless accompanied by substantial and effective social and economic reorganization.

Endnotes

[1] The findings of the Colorado State University research are reported in many places, including Corey and Clyma (1975); Clyma (1975a,b); Reuss and Kemper (1978); Johnson *et al.* (1977); and Eckert *et al.* (1975).

[2] The research in Gondalpur (a pseudonym) was supported by a Social Science Research Council Foreign Area Fellowship, while the ten watercourse survey as well as the writing of the first version of this paper were supported by Colorado State University's Water Management Research Project in Pakistan, under United States Agency for International Development contract number AID/ta-c-1411. The Mona Experimental Research Project (WAPDA) was also very supportive. I am grateful to H. S. Plunkett, Ashfaq Mirza, John Reuss, and Brian Spooner for helpful comments on earlier versions of this paper. Needless to say, the views expressed are my own and not those of any of the above organizations or persons.

[3] The changes in Gondalpur between 1857 and 1977, both those that preceded and those that followed the introduction of canal irrigation, are the subject of D. Merrey (in preparation).

[4] Much of the material on the watercourse reconstruction project and the implications of the concept of *izzat* have also appeared in Merrey (1979).

[5] *Zāt* is a cognate of the word usually translated as "caste" or "subcaste" in North India, but caste is not a proper translation of zat here since zats as understood in Gondalpur are not endogamous or systematically ranked *vis-à-vis* each other. For a complete discussion of "caste" in Gondalpur (and Pakistani Punjab) see K. Merrey (in preparation).

[6] *Numbardār* is a hereditary position created by the British: he collects the land revenue and irrigation fees for the government, keeping a percentage for himself, and acts as an intermediary between the villagers and government officials.

[7] The arrangement was that I would observe, but not participate in, the process; in fact, people often sought my intervention to influence the engineers and upon occasion I did offer suggestions to the Mona Project personnel—which were rarely followed.

[8] There were significant differences among the branches in the organization and efficiency of the work. Except for a few portions of branch D done collectively, the work on each portion of all the branches was divided among the shareholders proportional to the amount of land they irrigated. The large farmers at the head and tail of branch A had their tenants and servants do the work, while the small farmers in the middle did their own share—and did it more quickly. Most of branch D was done by tenants, servants, and hired labourers, and more time was spent smoking and gossiping than working, significantly slowing the work. All but a few of the branch B shareholders did their own work, and theirs was completed very quickly.

[9] See for example Bilmes (1980), and other references he cites, for discussions of similar villager-official relationships in Thailand.

[10] *Izzat* is the most common and broadest term; there are others but they tend to have more restricted meanings. The term has obvious affinities, conceptual and historical, with the Middle Eastern and Mediterranean concept of "honour" (see Peristany, 1966, and Campbell, 1964 for examples). This discussion does not pretend to be complete: izzat has important ramifications in many areas of Punjabi life.

[11] Religious generosity such as building a mosque earns one "respect" (*adab*) for piety, but is not itself a source of izzat; pious acts score points in a different game.

[12] The numbardar and the cousin with whom he argued over the extent of work on the watercourse are the sons of the two murdered men; their relations are tense in part

because of jealousy and dissatisfaction over the subsequent partitioning of their fathers' land; and in part because each fears the other will gain an advantage. An exchange of sisters would seem to be called for here, but each branch is marrying matrilaterally (outside the village), accentuating the division.

[13] One Pakistani commentator on an earlier draft of this chapter, as well as one of my Gondalpur informants with whom I discussed my conceptualization of izzat, suggested I have confused what my informant calls "false izzat" with "true izzat". "True" izzat refers to the more "positive" characteristics included in the concept, while "false" izzat includes more "negative" behaviours such as undercutting others, and creating fear in others. It is important to note that my informant here is one of the Langah, who are not active participants in the main game of izzat. Other Gondalpur informants, while understanding the distinction, insist nevertheless that obstructionists like the step-nephews do have izzat in most peoples' eyes; men who are feared and referred to as *badmāsh* ("bad character", trouble-maker, bully) are also respected (even admired) and regarded as having izzat; and the *badmāsh* themselves believe they are increasing their izzat by their behaviour.

[14] Some of the material in this section was first presented in D. J. Merrey (1980), but the implications are developed further here.

[15] See Radosevich (1975); Water Management Research Project (1976); Reuss, Skogerboe and Merrey (1979); and D. J. Merrey (1981).

6

Thorns Paired, Sharply Recurved: Cultural Controls and Rangeland Quality in East Africa

Francis P. Conant

Hunter College of the City University of New York
Department of Anthropology
695 Park Avenue, New York, NY 10021, USA

A major premise of this book is that nature as most people conceive it, both popularly and scientifically, is illusory and misleading. The pursuit of objectivity has led to the analytical separation of people from nature, and although more recently it has become *de rigeur* to insist that people are an integral part of the ecosystem, the implications of this new tenet have not been worked out. This case study shows that not only is traditional pastoralism an integral component of the ecosystem but that removal of it can in some cases lead to results that are more undesirable than the effects of overgrazing already diagnosed as incipient desertification. In most cases, of course, where rangeland is deteriorating as a result of overgrazing, removal of the human activity is likely to lead directly to recovery, and may therefore justify the deprivation it causes—but not always. Contrary cases such as the following are known in range ecology, but difficult to find in the literature and tend to be treated as "the exception that proves the rule" rather than as caveats in development planning.   —Ed.

Herding as a traditional subsistence activity is often cited in the literature as among the more prominent causes of environmental degradation and eventual desertification (see, for example, the extensive citations of herding and environmental damage in UNCOD, 1977a). At the same time, either by inference or by reference to actual experiments in "exclosure" (Le Houérou, 1977a) the clear impression is given that only in the absence of all human activity, especially herding, can the full bioclimatic potential of an area be realized.

The present paper questions whether traditional herding practices are all that bad, and whether full bioclimatic recovery is necessarily all that good. By

111

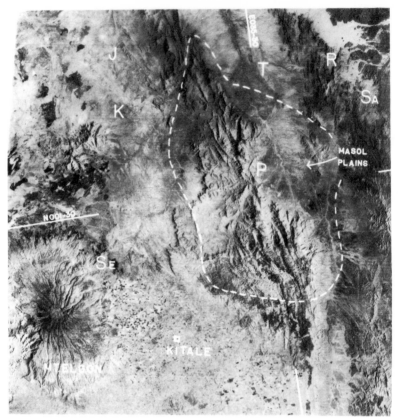

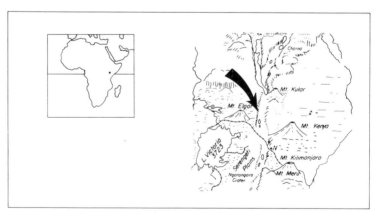

Fig. 1. (above) Landsat 2 image showing approximate area occupied by the Pokot (P) within dashed line (−−−). The grazing area abandoned by the Pokot and which became dominated by *Acacia spp.* is at the head of the arrow indicating the Masol Plains. Other peoples in the region are indicated as follows: Jie (J), Turkana (T),

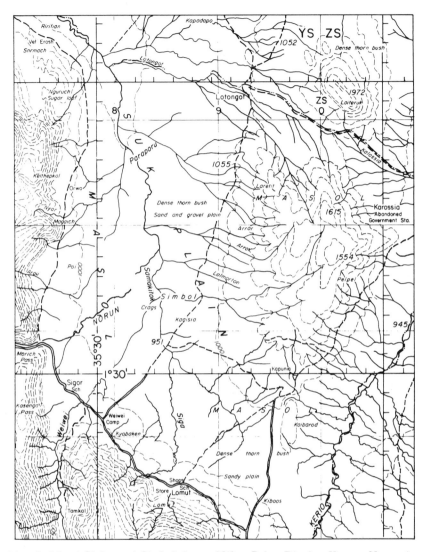

Map I. Masol Plains and Simbol Area of West Pokot District, Kenya. Note: river
courses as shown are only approximate. Scale = 1:250 000. The map base is from
1501, NA 36-12, 1967, Army Map Service, Washington, D.C.

Rendille (R), Samburu (Sa), Sebei (Se) and Karamojong (K). The image centre is
North 01 degrees 30 minutes and East 035 degrees 30 minutes. The scale is nominally
1:1 000 000. The image is geometrically uncorrected.
(below)   Location of study area in Africa (dot) and in East Africa (arrow right).
Based on Lobeck, A. K. (1946) "Physiographic Diagram of Africa". Hammond, New
York.

reference to the Pokot of northwestern Kenya, positive features of traditional herding systems are emphasized, and the negative aspects of biological recovery (which has taken place after forced abandonment of the area) are identified. Preliminary analysis of data from the Landsat satellite system is used to demonstrate the reality of a kind of "green desertification"—the explosive growth of *Acacia* and the retreat of the grassy cover, which has taken place in only five years.

The "Thorns paired, sharply recurved" in the title is a botanical descriptor for several species of *Acacia* (Dale and Greenway, 1961:29–3) which have almost completely taken over a traditional grazing area known as Simbol on the Masol plains of northwestern Kenya (see Fig. 1 and Map 1). The Pokot people (known as "Suk" in the earlier literature) have exploited Masol and the Simbol area for many generations by managing large herds of cattle, goats, some sheep, and sometimes camels. The Pokot were forced to withdraw from the Masol plains in 1974 because of a state of near warfare. The plains were unused until late in 1978, and even now are still largely unoccupied. In 1978 a series of ground observations was made in the area, aerial photography was flown, and Landsat data for the years 1973 and 1978 acquired.[1] These data and the results of their preliminary analysis will be considered after a brief look at Pokot ecology and recent history.

I. POKOT ECOLOGY AND RECENT HISTORY

The Pokot are a farming and herding people of northwestern Kenya who number about 100 000 persons (Central Bureau of Statistics, 1969). They exploit environments ranging from Afro-Alpine highlands of the Cherangani mountains in the south to the semidesert lowlands in the north (Porter, 1965).

This range of ecozones occurs within relatively short north–south and east–west distances. The steep terrain profiles associated with the Cherangani and Sekerr ranges largely account for this compression of ecozones. Elevations utilized by the Pokot range from over 3000 m at Sondang in the Cherangani mountains to 900 m on the Masol plains, less than 20 km to the northeast. The juxtaposition of contrasting ecozones is evident in Fig. 2, which shows a west–east transect of the District at 01° 30′ North Latitude, 035° 30′ to 036° East Longitude.

Most Pokot (about 75 000) exoloit the mid-elevations by shifting culti-vation methods, sometimes combined with an indigenous irrigation technique. Herding Pokot number about 25 000 and are still committed to the traditional open-range system of livestock management. The herding Pokot increasingly depend for their supply of grain on government-licensed trading posts, although traditional exchange relationships with kin in the farming

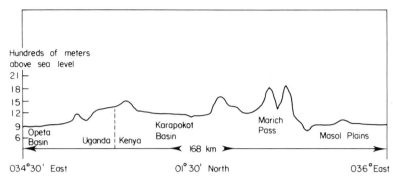

Fig. 2. Terrain profile, West Pokot District. Source: ONC L 5. Scale = 1:1 000 000.

areas also remain active. Milk, meat, blood, and hides are traded for sorghum, maize, and millet, as well as bottle gourds grown in the mountains (Conant, 1965). The herders also maintain alluvial gardens in riverine areas, much as Gulliver has described for the Turkana (1955). During the rainy season the herders congregate on slightly higher ground in temporary nucleated settlements. In the dry season they disperse into smaller units sometimes comprising personnel from two or three households.

Pokot herds consist of Zebu cattle, goats and sheep, some donkeys, and, in northern areas, some camel. The combination of grazers and browsers is a technique for not only exploiting different kinds of plant cover, but also keeping the *Acacia* thicket in check, and by so doing, helping to maintain the grassy cover. In general, men and older boys manage the cattle, although women do the milking. Women and girls, as well as younger boys, manage the goats and sheep. In most household herds the ratio of goats and sheep to cattle is in the order of 10 to 15 to one. This major involvement of women with the small, browsing stock is an important aspect of the Pokot division of labour, and may be true of other peoples as well.

Pokot herding strategies include (1) periodic burning of grass and bushland; (2) varying the composition of the herds in terms of the mix of grazers and browsers as well as a sort between mature and immature animals; (3) seasonal deployment of the stock according to the availability of graze, browse, water, and supplies of grain for human consumption, and (4) dispersal of the herd or fractions thereof over a wide area, thereby minimizing losses in any one area due to disease, drought, or other disasters. One periodic type of disaster is the loss of livestock through raiding by neighbouring peoples.

Pokot livestock management is fairly typical of East African peoples. Similar techniques are found among Karimojong (N. Dyson-Hudson, 1966;

R. Dyson-Hudson, 1972), Turkana and Jie (Gulliver, 1955), Samburu and Rendelle (Spencer, 1973), as well as other farming and herding peoples of Kenya, Uganda, Ethiopia and the Sudan (Huntingford, 1953a,b).

One special historical circumstance true of the Pokot but not widely found elsewhere is the treatment they received during the colonial period. Pokot District (originally known as West Suk District) and its inhabitants were perceived as a buffer between the high-potential areas (the so-called "White Highlands" of Trans-Nzoia District) for commercial farming, ranching and the under-administered and frequently drought- and disease-stricken lowland areas to the northwest, north, and northeast. Movement across the northern boundaries of Pokot District was irregularly policed, while movement across the District's southern boundary with Trans-Nzoia was much more tightly regulated. Generally speaking, non-Pokot could not settle in the District, and the Pokot themselves were given what can only be called mild assistance in developing or changing their area. Until 1967, then, the Pokot (farmers and herders alike) were in a kind of economic holding pattern.

With independence, however, one of President Kenyatta's early actions was to lift travel and settlement restrictions on the so-called "closed districts" and "native reserves" like Pokot District. An influx of outsiders took place with profound effects on Pokot traditional ways. One such effect was the exploitation of the area for its animal protein—intensely in demand in neighbouring Uganda as well as in the urban centres of Kenya itself. Securing livestock for these markets could not be encompassed within traditional raiding patterns. The traditional raids involving Pokot and Turkana, for example, were carried out on foot by some 25 to 50 men, armed with shields, spears, bows, and knives. Acquisition of livestock was the primary objective, and a secondary goal was to earn varying degrees of personal honour by killing or wounding a warrior on the opposite side. The loss of human life in this kind of traditional raiding seems to have been relatively light, with what amounts to dueling between individual attackers and defenders. According to Pokot accounts, a homestead under attack would be left standing with survivors remaining in place. Since personal honour was enhanced mainly through killing an adult male, and since the desired livestock were cattle, the survivors were mainly women and children, goats and sheep. A likely effect of this survivorship, with the livestock skewed strongly toward browsing animals, was to keep the proliferation of *Acacia* in check.

But beginning in the early 1970s the pattern of traditional raiding for cattle took on quite different dimensions. These include the use of automatic weapons, grenades, motor transport, and the slaughter of all possible witnesses to the identities of the raiders. Where before individual warriors craved identification far and wide for their acts of bravery, the new raiders (often referred to as *ngoroko*—a term of uncertain origin) remained anonymous, and increasingly lethal. Livestock taken in this new kind of raiding reportedly was driven into collection centres for sale to buyers from remote urban centres. The periodic slaughter of old and young, male and female, together with the loss of nearly all livestock (cattle, goats, sheep) forced the Pokot to abandon their traditional grazing/browsing areas on the Masol plains in 1974. The Pokot herders on the Masol plains withdrew to four refuge areas at the foothills of the Sekerr and Cherangani ranges. The Masol plains were abandoned. The cultural controls were now absent—no burning off of the old, dead grass or the new *Acacia* seedlings; no cattle to graze, no goats to browse.

II. LANDSAT DATA ON VEGETATION CHANGE ON THE MASOL PLAINS

Preliminary analysis of the Landsat data for the Simbol area in Masol amply confirms what Pokot informants were saying: much grass is gone, and *peli.l* and *panyerit* (*Acacia spp.*) are everywhere. The area is about 7·7 × 8·9 km and the two years of Landsat data which were used are 1973 and 1978. These dates represent the last year of traditional use of the plains, and the first year in which light occupation resumed. The Landsat tapes are both February acquisitions, which means the data were recorded toward the end of the dry season for each year.

Since *Acacia spp.* are sensitive indicators of climatic trends (Langdale-Brown *et al.*, 1964:66) and rangeland conditions (Pratt and Gwynne, 1977:94–99), the available records for rainfall in the region were collected for the years 1973 through 1977. Table I shows that for these years there was a fairly consistent trend toward drier conditions, as compared to the average monthly records for each station in the general area. All five stations show almost half again as many months in 1973–77 below local averages as the number of months above these averages.

As Nicholson has noted (1978:187), rainfall and climatic records for the tropics and subtropics lack much in the way of historical depth, and apparent noise in the series may obscure actual trends or create apparent ones. With this caution in mind, there seems to be no reason to think that a five-year drying trend, 1973–1977, would be a cause of massive change in the plant cover of the Simbol area. A more proximate cause seems likely to be the total withdrawal of the herders from the area, and the removal of cultural constraints on plant cover composition.

A four-way classification of vegetation cover types is presented in Table II.

Table I: Rainfall Trends, 1973–77, Pokot Area and Region, Kenya[a]

Station	Number of Years in Average[b]	Number of Data Months 1973–1977[c]	Above (+) Average Months[d]	Below (−) Average Months[d]	% +	% −
Lodwar	52	60	20	40	33	67
Sigor[e]	18	44	16	28	36	64
Kongelai	23	57	23	34	40	60
Sebit	19	39	16	23	41	59
"Regional"[f]		200	75	125	38	63

[a] Data are from the East African Meteorological Department, Nairobi.
[b] Number of years of recording for which monthly averages exist.
[c] Incomplete records reflect local difficulty in recording and transmitting data.
[d] A simple "above" (+) and "below" (−) was scored for each month 1973–1977 as compared to the equivalent month in the averaged data base.
[e] The station closest (10 miles) to the Simbol grazing area. Sigor however is still within the "rain shadow" of nearby mountains, and according to local account in wetter than on the Masol Plains or at Simbol.
[f] "Regional" only in the sense that the weather stations are in west-northwest-Kenya and within 500 metres of Simbol's elevation of c. 900 m.

Table II. Preliminary Estimates via Landsat of Plant Cover Changes, Simbol Area of the Masol Plains, 1973 to 1978[a]

Vegetation Type	1973 Pixels	1973 % Area[b]	1978 Pixels	1978 % Area	Pixel Change 1973 to 1978
Bushland	3919	23·92	8191	49·99	+4272
Bush-Grassland	6161	37·60	2136	13·04	−4025
Riverine Forest	3787	23·11	3885	23·71	+98
Black Cotton Soil-Bush Grassland	1212	7·40	1151	7·07	−61
Total	15 079	92·03	15 363	93·81	

[a] The Landsat tapes used in this analysis are 1193-07230, 1 February 1973, and 21107-06413, 2 February 1978.
[b] The analysis area in the IMPAC system (Egbert, 1979) used in this research is 128 × 128 picture elements or "pixels". Each pixel is nominally 0·5 hectare, making the entire analysis area about 69 square kilometres.

The classification is based on ground observations combined with inspection of aerial photography for 1956, 1967, and 1978. The vegetation types are (1) Riverine Forest, (2) Bushland, (3) Bush-grassland, and (4) Black Cotton Soil-Bush Grassland. These categories approximate the physiognomic classes of Pratt and Gwynne (1977:44–50). *Riverine Forest* along the Somakitok and Weiwei Rivers bordering Simbol presents a closed and often tiered canopy structure. It gives way abruptly to *Bushland* including some *Commiphora* but mainly *Acacia millifera*, and *nilotica*. The transition from Bushland to *Bush-Grassland* is gradual. Perennial and annual grasses include *Digitaria, Chloris*, and *Dactyloctenium* (Bogdan, 1958). The *Black Cotton Soil Bush-Grassland* grassy mixture is similar except for the background contribution of the soil itself, which is apparent in both the aerial photography and the Landsat data.

Analysis of the Landsat data first involved finding the same ground area in the 1978 aerial photography and in the Landsat data. The *Riverine Forest* category was most useful for this purpose since the dense canopy clearly shows up in the photography as well as in the Landsat data. With registration accomplished, Landsat data values for the different vegetation complexes were extracted; the computer then mapped all locations of these values in the analysis area.

Table II shows little change for *Riverine Forest* and *Black Cotton Soil-Bush Grassland* categories, 1973 to 1978. This is expectable since both riverbeds and black soil distributions are relatively permanent features of the Masol landscape. The major changes are in the *Bushland* and *Grassland* categories. From the last year of regular use of the Masol/Simbol area in 1973 to five years later, when the first Pokot began to return to the plains, the Bushland category increased by more than 4200 data points or "pixels" (picture elements): from 3919 pixels in 1973 to 8191 in 1978. In 1973 the Bushland accounted for almost 24% of classification area, and in 1978 it accounted for almost 50%. Thus, in the absence of the Pokot, *Acacia spp. doubled* its distribution. Grass retreated, from 38% of the classification area in 1973 to 13% in 1978. In fact so little grass remained in the Simbol area in 1978, that the few Pokot who began returning in August avoided the area completely. This avoidance was on account of both the decrease in grass and the increase in the tsetse fly which accompanied the proliferation of dense bush (see Lambrecht, 1972).

The Landsat data strongly indicate that in the five years' absence of the Pokot the Masol Plains in the area of Simbol underwent drastic change in which ligneous growth proliferated and grassy and herbaceous growth declined. All in all, the changes are in the direction of increased biomass, and it is uncertain what the ultimate limits of this increase are, or what the composition of the climax plant community will be. With respect to the notion of "climax", Pratt and Gwynne (1977:27) note that "climax vegetation often exists as a theoretical reconstruction rather than as an observed fact''; also

> It is necessary to avoid the error of regarding the climax as necessarily desirable as a vegetation type. The climax vegetation of East Africa is usually some form of woodland or bushland, whereas the more open country, prized for agriculture, ranching and wildlife viewing is usually not the climax type (ibid.).

III. CULTURAL CONTROLS AND RANGELAND QUALITY

The case of the Pokot shows that a cultural presence is necessary for the maintenance of a benign environment in East Africa. It also shows the Landsat system to be a powerful tool in testing local observations over a much larger area, and with the identification of correct indicators, to monitor the environmental quality of remote areas (Conant, 1978; Otterman, 1974, 1977; Reining, 1978b). The proliferation of *Acacia spp.*, while involving an increase in biological productivity, effectively removes almost the entire area of Simbol from human use. There is no easy remedy. Arboricides are expensive and relatively ineffective against the spread of *Acacia*; of the mechanical controls, chaining is only partly successful; cutting and stumping offers good control, but like other mechanical approaches, is quite costly (see Pratt and Gwynne, 1977:128–138 for a review of the cost effectiveness of different kinds of bush control). While there may be technological solutions to the reclamation of land invaded by *Acacia*, and while these solutions may be imminent, as in the case of portable charcoal-making equipment, the fact remains that the bush has taken over and remains in place for the foreseeable future.

Perhaps faith in quick technological solutions is unwarranted. Instead, indigenous management practices of traditional pastoral populations should be examined anew for the kinds of controls they have exerted over their environments with *positive* results. The literature is rife with negative examples of the misuse of arid and semi-arid environments by pastoralists, sometimes explained by pressures from outside sources, ill-advised administrative policies, and internal conflicts, especially in the context of newly created water points, markets, and transportation networks (see, for example, the case studies in Davies and Skidmore, 1966; Dregne, 1970; and Monod, 1975; Hare *et al.*, 1977).

Among the positive environmental controls available to pastoralists, Pokot included, are size and composition of the household herd, strategies of differential deployment of the herd, and the selective use of fire. With regard to this last, Langdale-Brown and co-authors acknowledge fire as one of the "major factors controlling plant communities" (1964:18) in East Africa. Warren and Maizels (in UNCOD, 1977a:169–260) describe the effects of fire as removing old, dead growth; releasing nutrients from surface litter which would otherwise much more slowly decompose; encouraging the growth of grassy shoots, but discouraging the spread of scrub bush; and as also reducing the prevalence of insect pests (tsetse fly and ticks in particular). Among the possible harmful effects posed by the use of fire is sterilization of nitrogen-fixing bacteria in the soil; also, nutrients released from burned litter may be lost by being blown away, and desirable seedlings and sprouts destroyed.

But for all its important effects on the environment, fire seems to have received slight attention in the ethnographic literature on East African peoples—far less, for example, than the attention paid pastoral strategies of herd deployment, or the growth cycles which the family herds undergo as the household itself changes size and composition. There is, however, one aspect of herd management which also seems understudied, and that is the management of goats. Some time ago Worthington (1958:328) noted that, in spite of being the most numerous domestic animal south of the Sahara, goats have received surprisingly little attention—and this despite their predation of undesirable woody plants and their resistence to many of the diseases transmitted by flies and ticks to cattle. More often than not goats are lumped together with sheep, and together are perceived as a "menace to vegetation and soil" (Kowal and Kassam, 1978:226). Much more rarely are goats credited with a positive contribution to rangeland quality, as by Thomas (1966:55), or, more recently by Pratt and Gwynne (1977) in their consideration of the importance of goats as browsers in bush control.

Despite their numbers and importance, it is not uncommon for goats to be considered along with sheep, donkeys, and camels in a single section or chapter headed "other livestock". Cattle usually receive separate and far more extensive treatment. In the ethnographic literature far less attention is paid the management of small stock than is paid the strategies East African peoples employ for managing cattle. While cattle can and do browse, the grassy cover they require is maintained partly by fire and partly by goats keeping the *Acacia* in check. These positive contributions to the maintenance of East African rangelands seem worth more attention than they have received thus far, especially in the context of studies of desertification processes, and the role of indigenous peoples in these processes.

Endnote

[1] The fieldwork was made possible by a grant from the National Science Foundation for the project "Satellite Analysis of Human Ecosystems in the *Sahel* of East Africa" (BNS77-15622). An emergency grant from the City University of New York (11787E) helped in coping with the situation on the Masol plains, and over the years the Wenner-Gren Foundation for Anthropological Research has helped in the acquisition of satellite data, sponsored a conference on the use of Landsat materials, and assisted in specialized aspects of the fieldwork. An NSF Scientific Equipment Improvement

Grant made possible the acquisition of an IMPAC image analysis system for continuing analysis of the Landsat data. The IMPAC system's designer, Dr Dwight Egbert, also generously made available essential parts of a spectroradiometer in the field. Special appreciation is due to the project's two field associates, Tina Cary, who was on hand for both field sessions, and Barbara Bury, who was so helpful during the second and Nora Conant for her documentary photography and film making.

In addition warm acknowledgment is made for the advice and support of Mr Parmeet Singh and Mr Simon Anzagi, Central Bureau of Statistics, Nairobi; Dr and Mrs Winans of the Ford Foundation, Nairobi; as well as District Commissioner Francis Cherogony, West Pokot District; Mr and Mrs Reynolds, Kapenguria; Messrs David and Festus Koech, Sigor; Mr Jacob Kamkal Kamkele, and Mr and Mrs Komolingolei, formerly of Masol. Opinions expressed in this article, however, are those of the author alone, as are any errors of interpretation or fact.

7

The Design of Rural Resettlement in Dry Lands: Lessons From a Programme in Egypt

Helmi R. Tadros

American University of Cairo
Social Research Center
113 Sharia Kasr el Aini, Cairo, Egypt

Desertification eventually generates the need for resettlement. Most resettlement programmes suffer from the familiar problems of technology transfer: lack of communication between designers and users. Invariably there is no relationship between the settlers and the designers of a new housing development until after the building process has passed beyond the possibility of compromise between the settlers' perceptions and the designers' constraints and ideals. There is no longer any incentive to investigate what the settler would have preferred and why. The following study—in this case from a desert reclamation programme—shows in some detail the economic and social losses that can result. —Ed.

The large scale planning of new settlements in developing countries is a relatively recent phenomenon. It is generally agreed that such planning should be based on adequate pre-investment study of physical and human resources as well as of the social organization and values of the population. Careful planning reduces the possibility of failure. However, careless implementation often results in serious difficulties which later require adjustment—which may be costly.

In Egypt, following the construction of the High Dam, by the end of 1974 about one million feddans (1 feddan = 0·42 ha) had been reclaimed, necessitating the construction of new villages to house settlers. About 500 new villages were constructed, involving the resettlement of about 62 000 families, selected according to specific criteria. Since some of these resettlement projects have now been in operation for a number of years, a considerable amount of administrative experience has been accumulated. Unfortunately, there has been little systematic evaluation of the human and social factors involved, nor have most of the experiences of planners and field administrators been recorded in such a way as to be of benefit in the designing of new projects. In what follows I present the results of a recent study[1] in the North Western Delta area of Egypt, focusing on (1) the suitability of house design, (2) settler reaction, (3) the maintenance record, and (4) the settlers' own design ideas. This information is presented in the hope that it will be of value to planners and administrators of resettlement programmes in drylands elsewhere, as well as to scientists interested more generally in the spatial organization of social life.

I. THE ADMINISTRATIVE CONTEXT

In 1966 the President established the Egyptian General Organization for Land Cultivation and Development (EGOLCD).[2] The purpose of this new agency was to bring reclaimed land to an appropriate level of soil fertility and productivity before distribution to farmers. The associated government resettlement scheme was part of a programme to expand domestic production, promote local industries to meet the increasing demands of the home market and replace imports, and to help relieve population pressures in overcrowded areas, while at the same time raising the living standards of the resettled families.

A significant ingredient in the social organization of these new settlements is the fact that EGOLCD institutes three types of land tenure on its reclaimed land: ownership, tenancy and state farming. In order to become eligible for ownership or tenancy a settler has to meet specific individual and familial qualifications such as nationality, age, occupation, marital status, and family size. Under ownership each family is provided with three to five feddans of land, a house and a cow or water buffalo. The cost of these items is repaid by annual installments over a forty-year period at a low rate of interest. Under tenancy a farmer receives a variety of items including approximately three feddans of land at a very reasonable rent and a house identical to those distributed under ownership—but he is not allowed to make any vertical or horizontal additions to his house, and he is prohibited from cultivating permanent crops such as fruit trees, since as a tenant he can be ejected from both land and house any time he violates his contract with EGOLCD. In state farms the overall responsibility for cultivation lies within the hands of EGOLCD administrative and extension staff, who employ hired labour and use modern farm machinery.

The size of plot to be granted to each settled family is an issue that has received repeated attention in land reclamation schemes throughout the world. Clearly, there is hardly ever an adequate amount of land and water to provide economically sized plots for all those who would like to have them. In Egypt, for example, we are aware that every individual is backed by less than one sixth of a feddan. In other words, it is clear that the average plot is bound to be small—with obvious economic implications.

EGOLCD is at present supervising the cultivation and development of altogether approximately 700 000 feddans in the Nile Valley and the Delta. This area is divided into ten working sectors,[3] each of which covers about 70 000 feddans. Each sector is divided into units called "farms" of about 5000 feddans each. Each farm is composed of central and sub-villages, whose number varies from one farm to another. Central villages bear no names, being identified only by serial numbers that indicate the chronological order of their construction. A central village has between two and fifteen sub-villages associated with it which are also identified by the numbers which they carry. For example, the Fifth Central Village in the Dessoudi Farm, one of the two farms in this study, has six sub-villages which are called 1:5, 2:5, 3:5, 4:5, 5:5, and 6:5 consecutively, while the Sixth Central Village in the same farm has eleven sub-villages. The idea behind the construction of a central

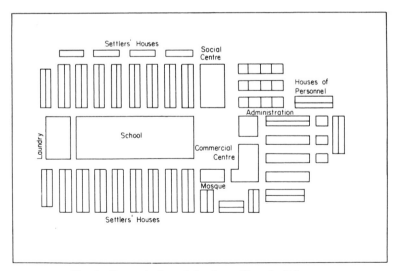

Fig. 1. General plan of the first village in Ibis area.

village and its affiliated sub-villages is to keep the settler as close as possible to his land. According to this system, the central village is set up as a kind of service village, which embraces in its centre almost all of the public facilities needed, such as child daycare centre, school, mosque, hospital, youth club, community development society, commercial centre, agricultural co-operative, veterinary service, work-shop for small industries, post-telegraph-telephone office, police station. The only public facilities provided within the sub-villages are a small mosque, and sometimes small private grocery stores. Other services must be sought at central villages (see Fig. 1).

The present study has been focussed on areas under ownership and tenancy only. Examples from state farming were omitted, mainly because it employs seasonal migrants and is not therefore associated with communities.

II. THE STUDY AREA

The Northwestern Delta Sector covers a cultivated area of about 50 340 feddans, stretching between the Rashid Branch of the Nile Delta on the east, the Mediterranean Sea to the north, the Cairo-Alexandria Desert Road on the west, and the Tahrir Sector to the west and south. It was chosen for this study because it includes adjacent farms which are ecologically similar but sociologically different; they have almost the same type of soil, but are under different land tenure systems—ownership and tenancy. Among the six major component farms of the sector El-Deshoudi Farm was chosen to represent the ownership type, while the adjacent Ibis Extension Farm represents tenancy. Their selection was not, however, intended to represent in any way a cross-section of ownership and tenancy farming as supervised by EGOLCD in its various sectors. Since most of the farms in this sector were originally either beds of Lakes Mariout or Edku, or desert lands, the soil types vary. El-Deshoudi and Ibis Extension Farms, being originally part of Lake Mariout, have saline clayey soil.

III. RESEARCH METHODOLOGY

The material presented here derives mainly from a household survey, although various other techniques such as participant observation, interviews with responsible administrators and other key informants, and a survey of community services and facilities were also carried out in the larger project of which this study formed a part. The household survey covered the total population of the two central villages of El-Deshoudi Farm and their 17 sub-villages (1960 households) and the two central villages and 17 sub-villages of the Ibis Extension Farm (2148 households). The intention was to obtain information about the following:

1. Composition of household at time of resettlement and change therein due to marriage, divorce, migration or death;
2. For every person currently present in each household, details of relationship to head of household, age, sex, religion, place of birth, duration of residence in the new community, marital status, educational status, employment status, occupational history and present occupation.

A socio-economic survey was also administered on a stratified representative sample of 402 landowner household heads in the El-Deshoudi Farm and 390 tenant household heads from the Ibis Extension Farm. Information obtained from this survey included details of residence patterns, intra-family relationships and cooperation among kin and non-kin. The percentage distribution of sample settlers according to the years of settlement revealed that while the majority of owners in the Deshoudi Farm were settled between 1962 and 1964, the majority of tenants were settled between 1967 and 1969, with a mean difference of five years between the two settlements.

IV. THE SETTLEMENT HOUSE AND EGOLCD MODIFICATIONS

It is immediately apparent to an observer that the settlement houses are entirely different from a farmer's house in any traditional Egyptian village. They are built of whitewashed red brick with sloping concrete roofs which allow rain water drainage. Each house is provided with a lavatory and running water. The settlement plan is based on a system of straight, wide streets quite unlike the narrow winding alleys of traditional villages. However, not all houses in the villages of the Northwestern Delta Sector have the same design, since some modifications have been introduced by EGOLCD as a result of (1) experience and observation, and (2) the need to economize in the face of constantly rising costs of construction materials. The following is a description of a settler's house in the first village to be built, and the subsequent modifications introduced by the organization in later villages. In the First Village, as an experiment two types of house were built: 192 houses were built on a one-storey design and 70 on a two-storey design. Each type cost about LE 320 per house (LE 1 = $1·41 approximately). The one-storey house is composed of a living room 4 m by 3 m, two bedrooms each 3 m by 3 m, a bathroom 2 m by 2 m and a kitchen 3 m by 2 m with a pantry 1 m by 1·5 m and a built-in oven. Included in the complex is a roofed stable measuring 6 m by 2·5 m and an open 5 m by 2·5 m backyard. The stable has a separate entrance. A storage area of 1·5 m by 5·5 m, primarily for animal manure, is located next to the stable (see Fig. 2). The two-storey house had the living room, the bathroom, the kitchen with the pantry and the oven on the ground floor, and a staircase leading to two bedrooms with a large balcony on the upper floor. A stable was attached, as with the one storey design. Two-storey houses were discontinued in the first village because it was found that they did not suit the settlers' needs. Bedrooms are preferred on the ground floor since, as will be mentioned later, they were the most favoured storage space, for grain and light agricultural tools.

In the second village the houses were built with an entrance 1·5 m wide opening into a living area. The living area opens on to an unroofed yard 3 m by 6·5 m, containing the lavatory and a 2·5 m by 5 m stable. The house includes two additional rooms each 3 m by 3·5 m. Such a house cost LE 312 in 1956 when the Second Village was constructed (see Fig. 3).

In the third and fourth villages, the most striking change in house design was the disconnection of stables from the houses. Stables were grouped together at the outskirts of the village. Each settler now has his own stable with private entrance and a set of keys (see Fig. 4). The house entrances lead to an unroofed yard 5·5 m by 8 m which may be used for future construction if the settler wants to build an addition to his house. The house includes an unroofed living area 5 m by 2 m, a lavatory and bathroom 1 m by 2 m and two bedrooms 3 m by 6·5 m each.

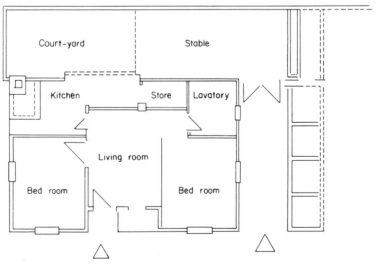

Fig. 2. One-storey peasant house in the first village. Cost in 1954 = LE 320.000.

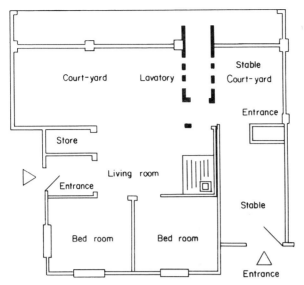

Fig. 3. A peasant house in the second village. Cost in 1956 = LE 312.000.

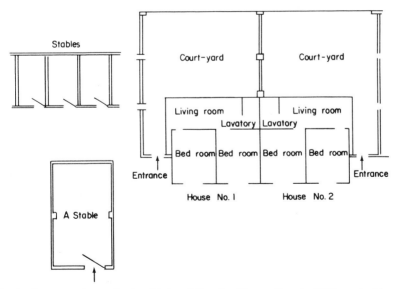

Fig. 4. A peasant house in the third and fourth villages. Cost in 1960: cost of house = LE 250.000; cost of stable = LE 150.000; total cost = LE 400.000.

From the third village onwards, each settler was left to construct the style of oven he preferred. It was realized through experience in the first and second villages that the oven type required by the Egyptian farmer varied according to (1) the type of bread the farmer was accustomed to baking in his place of origin, and (2) the height of the wife which would dictate the height of the opening of the oven so that she might sit comfortably while baking.

In spite of their simplicity in style, these houses were considered quite expensive at LE 250 each after the increase in the price of construction materials. The stable cost an additional LE 150. Thus the total cost per house-stable unit was LE 400 in 1960 at the time the third and fourth villages were constructed.

The houses constructed in the fifth village and those villages following it were of the same design as the third and fourth village houses.

In the first two villages, the EGOLCD undertook the whitewashing of the houses, both inside and out. But in the third and fourth villages whitewashing was left to the settlers, on the grounds that the settlers in the first two villages seemed unconcerned with the appearance and were negligent about the cleanliness of their houses. EGOLCD came to realize, however, that leaving the outside walls bare exposed the bricks to weathering, and they were then obliged to add thicker paint to the houses for protection against the weather in the seventh village onwards, but they still left interior painting to the discretion of the individual settler.

Houses for EGOLCD staff consist of one or two storeys. The one-storey house has three rooms, a kitchen, a hall, a bathroom, a terrace, and a garden. The two-storey house has one room, a hall, a kitchen, a bathroom, and a garden with an open terrace on the ground floor and two rooms and a terrace on the upper floor.

The number of houses in the two farms under study now lags far behind the number of households. Most of the stables have been used as residences; houses are generally shared by more than one family and shacks have also been constructed for additional living space. This inadequacy in the supply of houses has arisen not only as a result of population growth, but because (1) the distribution of land, especially under the tenancy system, did not provide residence because of the financial difficulty which faced the nation after the 1967 war and the readiness of villagers to accept such a situation due to the scarcity of land; (2) decisions on land distribution are sometimes made in response to popular pressure and on occasions of national celebration even before the houses are constructed; (3) it was necessary after the 1967 war to accommodate displaced families from the Suez Canal Zone. Ibis Extension Farm was among the many areas chosen for this purpose.

V. NEIGHBOURHOOD AND THE ALLOCATION OF HOUSES

It was found that only 32% of the owners were given the opportunity to choose their own houses at the time of arrival as compared to 61% in the case of tenants; 10% of the owners got their houses by means of balloting as compared to less than 2% of the tenants; and 59% of the owners did not have any choice and had to accept the house allotted to them as compared to 37% of the tenants. In the first two villages houses were distributed randomly, but from the Third Village onwards settlers originating from the same village were given neighbouring houses where possible. Overall, analysis of neighbourhood in relation to origin indicates that 32% of the owners had all their neighbours from their own place of origin, as compared to 28% of tenants; 30% of owners had some of their original neighbours leaving 28% of the owners who had all their neighbours from places other than their own place of origin. There were two interpretations of this situation:

(1) There is no doubt that it is to the advantage of the settlers that the majority of both owners (68%) and tenants (72%) live with neighbours from areas other than their own places of origin. Such a situation encourages the exchange of experiences and the diffusion of new techniques among settlers in their daily interaction and above all accelerates the process of community assimilation and integration. It might be a good policy for EGOLCD to minimize ethnic segregation in future distribution of houses.

(2) Though 61% of the tenants were given the opportunity to choose their own houses, only 27% of them chose to live with neighbours from their own places of origin. This behaviour can be taken as an indication that tenants with a relatively higher economic background are less traditional than owners, the majority of whom have come from the bottom economic stratum. This phenomenon has also been supported by other data where 48% of the owners stated that they would have preferred to live with neighbours from their own places of origin as compared to 39% of tenants. In the meantime, while 49% of the owners said they would not care about the place of origin of their neighbours if they were good people, tenants still scored a higher rate of 60%.

VI. HOUSEHOLD SIZE AND CROWDEDNESS

About 75% of the owners started their settlement with a household size of between 4 and 7 persons. The total sample mean was 5·7 persons, compared to 65% for the tenants within the same range and with a very close total sample mean of 5·8 persons. To start life in the new communities with a large family size and in many cases as an extended family already, did no doubt have its later repercussions on many aspects of family life and on the community at large, such as the crowdedness of the house, the standard of living of the family, unemployment, population increase and the pressure on the community facilities which would lead to inadequacy of services. By 1973 the average family size had already grown to 7·1 with 2·4 persons per room for the owners, and 6·3 with 2·5 persons per room for the tenants.

The response to questions about the settler's attitude toward housing space in general in relation to household size at the time of arrival indicates that 51% of the owners and 54% of the tenants thought the housing space was suitable, while the remaining 49% of the owners and 46% of the tenants considered it inadequate.

VII. PRESENT REACTION OF SETTLERS TOWARD HOUSING FACILITIES

Present reactions among settlers toward housing facilities were similar. 45% of the owners allowed that the number of rooms was enough as compared to 42% of tenants; 45% of the owners said that the size of rooms was suitable as compared to 42% of tenants; 72% of the owners stated that the size of the courtyard was suitable as compared to 74% of the tenants; 65% of the owners said that they had no storage space problems for their produce as compared to 81% of the tenants; 77% of the owners responded that they had no storage space problem with regard to fuel as compared to 94% of the tenants; 78% of the owners answered that they had no storage problem with regard to agricultural tools as compared to 97% of the tenants.

Arrangements of the rates of owners' dissatisfaction with regard to housing facilities in descending order of magnitude put the number of rooms first, with a rate of 56%; size of rooms second, with a rate of 55%; storage space for farm produce third, at 35%; size of courtyard fourth at 28%; storage space for fuel fifth at 24%; and storage space for agricultural tools last, at 22%. The situation among tenants followed almost the same sequence with the exception of size of courtyard and storage space for farm produce, which were reversed.

Both owners and tenants had reinterpreted the original design. Bedrooms were found to be favoured the most as storage areas for grain with the rather high rate of 50%. Top house grain silos were second at 35% and 32% respectively, and the courtyard third, at only 5% and 3% respectively. The remaining proportions were combinations of these places. It is evident that the need for storage space is felt more acutely than the need for sleeping space.

When settlers were asked where they stored their agricultural tools, it was found that the three most frequently used places were courtyard, at 44% and 67%; and secondly bedrooms at 13%, more or less, for each category.

Concerning storage of fuel, whether in the form of cotton, corn and rice stalks, or dung cakes, both owners and tenants again prefer the house; secondly in front of the house; thirdly the field; and fourthly the courtyard. These preferences have the following disadvantages: (1) the large heaps of cotton, corn, and rice stalks piled on top and in front of houses are apt to catch fire and frequently cause the destruction of whole villages throughout the country; (2) the dung cakes in front, inside, and on top of the house, breed flies and are the source of an unpleasant and pervasive odour; (3) the storage of fuel and agricultural tools in front of the house, in addition to stable waste, blocks streets and alleys; (4) from the aesthetic point of view, the dirtiness of the model village in spite of their wide, straight streets is as bad or worse than traditional villages.

In response to questions about their own design preference the vast majority of both owners and tenants (95% and 85% respectively) prefer the stables attached to the house. Only 6% of the owners said that they would prefer combined or individual stables completely separated from the house, as compared to 15% of the tenants. Again, the two most preferred designs for stables attached to the house were as follows: the stable entered through the main door of the house ranked first, at 52% for the owners and 51% for the tenants; the stable with a separate outside entrance and another internal door into the house, as found in the first village, ranked second at 34% for owners and 30% for tenants.

Reasons given by settlers in support of the stables being attached to the house were: for both owners and tenants, security came first at 86% and 67% respectively; being able to care for the animals, especially when sick or in last days of pregnancy, came second at 32% and 45% respectively; settlers' convenience of feeding and milking, collecting manure and cleaning the stable came third at 9% and 12%; and traditions which discourage women from being seen outside the house came last with rates of around 3%.

Settlers in favour of combined or individual stables completely separated from the house gave reasons that were related to hygiene and cleanliness, with distribution rates of 21% among owners and 33% among tenants.

Though all settlers are charged the same fee by the Agricultural Co-operative for the consumption of house water, only 42% of the owners and 36% of the tenants receive enough water in their homes to meet their daily needs. The remaining 58% of the owners as well as the 64% of the tenants have to seek additional water from other sources like public taps, canals, and mosques.

Observation has also proved that most of the public taps, as well as those installed inside the houses, are missing. Small round pieces of wood which are capable of shutting off the weak water pressure are used in the houses to replace taps when broken or misused. As for public taps, the water runs day and night without control, and consequently every public tap is surrounded by a swamp.

Some of the settlers' houses are lit by electricity. In these cases, current is illegally obtained by throwing two hooked wires on the two main power lines which pass through the village streets to feed EGOLCD offices and other service buildings. Though aware of these violations, EGOLCD turns a blind eye, and looking from a different angle considers the action an indication of social development. The Agricultural Co-operative, on the other hand, is trying hard to charge the settler the sum of 25 piastres (= LE 0·25) per lamp per month.

Settlers were questioned with regard to how they would have designed the house if they had been given the opportunity to do so. The answers show that 16% of the owners would actually have designed the house in the same way as it now exists as compared to 28% of the tenants. Suggested changes in design in a descending order of preference were (1) complete roofing of the courtyard along with the construction of the stable within the house—ranked first at 36%; (2) making the entrance to the house in the middle with one bedroom on each side of the entrance and moving the courtyard to the far end of the house at 25%; (3) the construction of more rooms at 19%. Rates among the tenants in relation to these three changes were very close, at 23% in each case.

Only 3% of the owners and 2% of the tenants said that they would have built the lavatory and the oven downstairs and the bedrooms upstairs. 2% of the owners answered that they would have made more openings in the house by installing extra doors and windows.

VIII. HOUSE IMPROVEMENTS OR ADDITIONS CARRIED OUT BY SETTLERS

A survey of house improvements and additions indicates that 26% of the owners and 27% of the tenants did not carry out any. One percent of the owners and a little over 1% of the tenants had built a second storey. But 50% of the owners and 40% of the tenants had added more rooms by means of horizontal expansion. 38% of the owners compared to 29% of the tenants had partially or completely roofed the courtyard. 29% of the owners and 26% of the tenants had painted their houses inside, while 28% of the owners and 25% of the tenants had painted their houses both inside and outside.

The large proportion of both owners and tenants who added more rooms is evidence of the inadequacy of space. Houses with two storeys and houses painted inside and out belong in most cases to the community leaders who were able to gain prestige through accumulation of wealth in addition to board membership in one of the different community institutions such as the Agricultural Co-operative or the Community Development Society.

Though the foundation of the house was originally designed to support one floor only, EGOLCD does not take the construction of a second storey as a dangerous violation. Instead, it is often seen by responsible administrators, as proof of a settler's economic development. Even for tenants who, according to the contract of the house, are not supposed to make any permanent alteration in the house, the same treatment is given.

IX. CONCLUSION

It becomes clear that a settler's house as presently designed falls far short of meeting many of the farmer's daily needs. It is often said that the poor landless peasant could never have dreamed of a better house. Such an argument, however, is not sound. The Egyptian Government is spending millions on these expensive projects, and careful planning is essential. If the house has become dirty and untidy, the settler is not to be blamed because he was not given any other choice as to where to store necessities. Had the storage space and the right location of the stable been taken into consideration, both the house and the street could have been much better organized. Even health education campaigns carried out by hospital doctors and the Community Development Society could have been more effective.

Combined stables now built on the outskirts of villages and sub-villages in the areas under study have proved to be a complete failure. A settler rarely keeps his livestock in these stables; he brings them home. Combined stables are now used either for the storage of fuel, or in most cases are rented to families with no houses. Since an Egyptian farmer cannot live far from his animals EGOLCD should have insisted on the construction of the stable within the premises of the house, as in the case of the first village where relative cleanliness exists.

Many of these problems could have been avoided if the administration had followed the following steps:

(1) A survey of farmers' houses in the settlers' places of origin would have led to an awareness of the different sub-cultural design preferences. Such a survey would have substantially reduced the trial and error nature of the existing processes.

(2) Administration architects could have consulted with community leaders from the settlers' places of origin concerning the design of a model house.

(3) Architects could have benefitted from the mistakes of the early designs in the first villages and many of the problems of the later houses could have been avoided.

Endnotes

[1] This paper is an extract from a larger research project on the Study and Evaluation of the Rehabilitation Process in the Northwestern Nile Delta (see Tadros, 1976, 1978, 1979), which was carried out by the Social Research Center of the American University in Cairo in close collaboration with the Egyptian General Organisation for Land Cultivation and Development (EGOLCD) of the Ministry of Land Reclamation, and was partly financed by a grant from the U.S. Department of Health, Education and Welfare (HEW). The project was initiated in June 1971, and research activities continued until spring 1975. The author was principal investigator.

[2] This organization has been abolished and the administration of reclaimed areas has been turned over to the Ministry of Land Reclamation.

[3] New sectors have been added recently.

Part II
Regional Programmes

Since the United Nations General Assembly called for an international campaign to combat desertification in 1974, several comprehensive interdisciplinary dryland research programmes have been initiated, and the dryland programmes of existing research institutes have been strengthened and encouraged. The second half of this volume is devoted to selections from the work of two such programmes, from Iran and India. Like other similar programmes they were designed to generate the knowledge necessary to improve the efficiency of dryland resource use, and they have sought to involve and integrate the work of scientists from as many as possible of the disciplines that bear on the interaction of human activity and natural processes. The following selections have been made with the aim of illustrating the distinctiveness and significance of the results obtained in these programmes by their special attention to the problem of incorporating the social dimension.   —Ed.

Section A
The Turan Programme (Iran)

The Turan Programme was initiated by the Department of the Environment in Tehran in 1975, and from the beginning was closely associated with the United Nations Environment Programme and other international agencies involved in the global effort to combat desertification. Its purpose was to investigate the relationship between human activity and natural processes in the more arid parts of the country, and develop scenarios that would assist planners to increase the efficiency of that relationship in the future. The research area was defined to cover a Protected Area, which later became the Turan Biosphere Reserve (under MAB Project No. 8; see Map 1), in north-eastern Iran, plus neighbouring populations—in the districts of Khar and Tauran[1] (Map 2)—that have traditionally used the area, and adjoining land forms that have direct relevance to the Reserve.

This area presents a variety of habitats, including three extensive plains at different altitudes varying from 700 m to 1400 m, a saline river system, three mountain systems rising to a maximum of 2200 m, large areas of broken country, some 200 000 ha of sand with moving dunes, and a vast expanse of barren playa (known in Iran as *kavir*). The 200 mm isohyet passes through the northern part of the area, and the southern plain probably receives less than 100 mm average rainfall per year (Map 4). A light snow cover appears on the higher mountains for two to four months in most years and snow lies on the higher northern plains for short periods. Summers are dry and hot. Only the central salt river flows at least intermittently throughout the year. Rainfall of several millimetres at a time generates sheet run off and wadi (arroyo) flooding. Springs occur along the base of the mountains. Soils are generally light and sandy except for solonchak in the kavir. Vegetation varies according to land form soils and altitudes, and secondarily, according to human activity patterns. It is Irano-Turanian and consists primarily of *Zygophyllum–Salsola, Haloxylon aphyllum, Artemisia–Amygdalus, Calligonum–Stipagrostis* or *Astragalus–Amygdalus* communities. Woody shrubs predominate with ephemerals and annuals growing largely in their protection. Perennial cover over most of the plains varies between 5 and 40%. Flora and mammalian fauna generally show great affinity to the Kara Kum in Soviet Turkmenistan to the north, and include onager, gazelle, ibex, cheetah, and leopard.

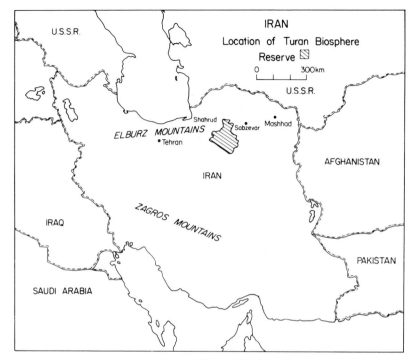

Map 1. Location of Turan Biosphere Reserve.

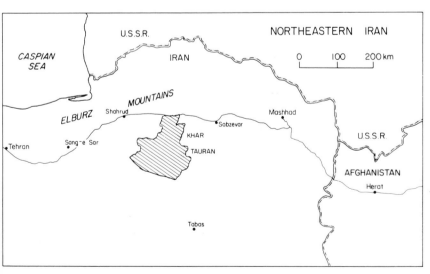

Map 2. Khar and Tauran.

Map 3(a). The Tauran Plain. * Kuh-e Peighamber.

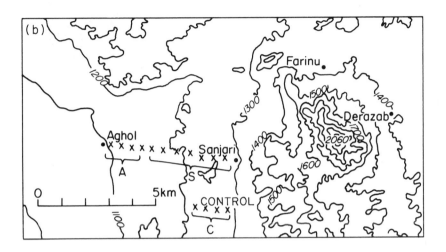

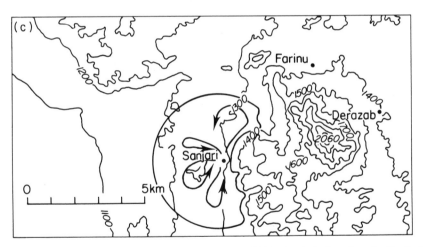

(b) Sanjari—the location of study plots (see p. 241). A: Aghol series; S: Sanjari series; C: Control series.

(c) The foraging area of the herds based on Sanjari is a semicircle extending to a maximum of 3 km from the spring, a distance reached in the overnight grazing period by the goat herd only. Afternoon movements of the goat herd are rotated on the four main routes shown here (see p. 233).

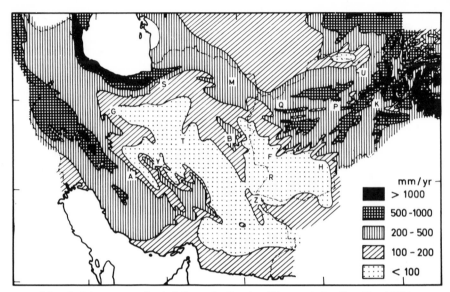

Map 4. Mean annual rainfall on the Iranian Plateau (after Breckle, 1979).
Letters indicate towns. A: Abadeh; B: Birjand; F: Farah; G: Qom; H: Qandahar;
K: Kabul; M: Meshed; P: Panjao; Q: Qadis; R: Zaranj; S: Shahrud, T: Tabas;
U: Kunduz; Y: Yazd; Z: Zahedan.

The predominant economic activity in Turan is pastoralism of various
types, sedentary and transhumant. Some 150 000 sheep and goats winter in
the area from November to May, of which 25 000 belong to the local settled
populations who remain in the area through the summer. Local populations
also keep camels, donkeys and a few cattle. Agriculture is also important
around most villages and is conducted by means of irrigation, from *qānat* (see
below, p. 147), springs, diversion of run off and to a limited extent by direct
rainfall.

The total human population using the area presently is in the region of
3000. The following chapters deal with various aspects of the relationship
between this population and its environment in ecological and historical
perspective. A major theme pervading the section is the significance of
exogenous forces for the explanation of variation in local resource management. Exogenous forces have also—since the autumn of 1978—unfortunately
caused the indefinite discontinuation of the Programme and prevented the
inclusion of work from Iranian colleagues here.[2]

Endnote

[1] Confusion arises unavoidably from the similarity in spelling, pronunciation and
meaning in Turan and Tauran. This confusion is increased by the fact that Turan is
elsewhere often spelt Touran, pronounced as Turan, but with a special, related but

different (botanical) meaning. All three English spellings derive from one Persian spelling (Persian orthography does not distinguish *u* and *au*), but the single Persian spelling represents two words with different pronunciation and meanings, which have nevertheless been confused by Persian usage.

Turan, in Iranian history, denotes what is not Iran—geographically and culturally—in the direction of Central Asia, and especially the neighbouring Turkish areas. Transliterated according to French rules, Turan became Touran, and this form was taken up by botanists, and therefore later also by ecologists generally, throughout Europe in whatever language, to denote the vegetation characteristic of that area.

Tauran is a Persian place name, apparently different etymologically, possibly derived from Tabaran, which is known historically from, among other places, the eastern border of the area which has now become the Turan Biosphere Reserve, where it denotes the group of villages treated in this study and also at earlier times apparently by synecdoche, which is common in Iranian toponymy, the primate settlement on the plain.

When officials looked for a name for the newly created Protected Area (1971) and Biosphere Reserve (1977) they saw on the map Tauran, which they read as Turan and, recognizing in it the Turan of history and botany, which was anyway relatively close geographically, found it ideal.

In the Turan Programme, Turan is used to refer to the whole area of the Biosphere Reserve; Tauran refers to the cluster of villages that are the focus of attention and to the district immediately surrounding them.

[2] The Turan Programme, and therefore also the projects on which this section is based, owes its existence in the first place to Mr Eskandar Firouz who as Director of the Department of the Environment in Teheran both commissioned it and consulted and advised in the details of the research from 1975–1977. At one time or another during that period the Programme called on the majority of the staff of the Department in Tehran, but among them Dr Fred A. Harrington who carried out the first ecological survey of Turan in 1971 and Dr H. Mohammedi who supervised the administration of the Programme, deserve especial mention for their support and help; also, the local employees of the Department—the game wardens in Turan, helped enormously with the physical and logistical problems of fieldwork.

From 1977, when the Programme was transferred to the Ministry of Agriculture, to 1979, we were in the debt of Dr A. H. Borhan, Deputy Minister for Research, for his invaluable support and of Dr I. Nassiri-Toussi, Deputy Director of the Soil Science Institute, for his administrative assistance.

Of the associates of the Turan Programme, the following Chapters have benefitted most from the work of Drs S.-W. Breckle, Robin Dennell, H. Freitag, Peter Moore, H. Rechinger, and Per Wendelbo, and from co-operation with the Soil Science Institute and the Statistical Centre in Tehran. The American Institute of Iranian Studies and the British Institute of Persian Studies in Tehran also rendered valuable assistance. The support of the Secretariat of UNESCO's Programme on Man and the Biosphere was important not only financially but in the opportunities it has afforded for interaction with similar projects in other areas. We are particularly grateful, in this regard, to Dr Francesco di Castri and Mr John Celecia.

Finally, many members of the population that is the subject of this section also participated in the work. Without their co-operation and hospitality it could not have been done. Each of us enjoyed getting to know them and their environment and sharing some of the events of their lives. We hope they will feel that the trouble they went to in assisting us was worthwhile.

8

Conservation at the Local Level: Individual Perceptions and Group Mechanisms

Mary A. Martin

Washington University
Department of Anthropology
St. Louis, MO 63130, U.S.A.

Why do people cause desertification? This question is implicit and unanswered throughout the UNCOD process. Is it possible they do not understand what they are doing? If they understood, how could they continue to do it? The first chapter in this section suggests some answers to these questions for the Tauran case. To begin with, in order to work out the answers, it is necessary to investigate in some detail what the people actually do and why, and then analyse the conditions—social, economic and political—in which they make their decisions. The conclusions suggest that any ascription of intention or fault, if it can be justified at all, must be spread much more widely than the original questions imply.   —Ed.

In much of the ecological literature on the Middle East and Southwest Asia the rural agricultural and pastoral populations stand accused of accelerating the process of desertification by various strategies of pastoral and agricultural production and firewood collection. Both farmers and herdsmen are often portrayed as being unaware of the negative effects of their practices—or if aware, as pursuing them anyway and displaying insensitivity to their own and others' future interests (see, for example, Dasmann, 1976:70–71). Often both ecologists and historians treat all pastoral strategies as the same and equally unresponsive to the environment upon which they depend (see, for example, Pearse, 1971:15; Ashtor, 1976:17). There is, however, a growing literature that demonstrates significant variation in environmental knowledge and perception in the practices of rural populations. There is also evidence of adaptation to ecological change.

Some strategies, especially pastoral strategies, are described as *inherently* destructive, but there are many studies of resource use in drylands which claim that particular strategies are not inherently destructive, though they may result in varying degrees of stress under certain but not all environmental or socio-economic conditions (see above, Sandford, Chapter 4). Such strategies and conditions include, for example, firewood collection, charcoal production for urban markets, grazing in situations of prolonged drought, border closure, and enforced sedentarization.

For a better understanding of desertification we need to examine the social and ecological processes which generate certain incentives and constraints affecting decisions about the use of resources. Of particular concern is the degree to which people are aware of environmental variables, how and to what degree they relate degradation of resources to their agricultural, pastoral or other uses of vegetation and how this relation is translated into individual or collective strategies. Although extraneous political, economic or climatic factors intervene to affect local populations, this chapter concentrates on decision-making at the local level. It is concerned with the decisions of the farmer and herder in the villages—not the bureaucratic and political processes at a provincial or national level, although the information and argument presented here will be relevant at that level. Examination of agricultural and pastoral strategies in Tauran (see Maps 1–3, pp. 139–141) suggests that certain patterns of resource use may contribute to environmental stress and potential desertification, whereas others are responsive to environmental change and serve to prevent or reduce stress.

The 308 households of the Tauran Plain are divided among thirteen villages ranging in size from six to forty-six households. These villages are separated from the nearest provincial centres of Shahrud and Sabzevar (population 44 000 and 54 000 respectively)[2] by natural barriers such as sand, mountains, and a salt river, and the roads that traverse these barriers are often rendered impassable by sheet run off, mud, or loose sand or river-bed gravel. A brief description of the natural conditions in Tauran are given above in the introduction to this section.

These small, isolated settlements are unable to support themselves by either pastoralism or agriculture without exploiting the underground water table. Pastoralism is the primary economic activity in the larger region and over 95% of the Tauran Plain (12 000 ha) is grazed. Irrigated agriculture is practised on a small scale—at present only 160 ha, or not quite 0·5 ha per family. Various types of rain-red farming are also practised but are generally unreliable, although in a good year as much as 240 ha may be ploughed for this purpose.

Even with the use of ground water for irrigation and watering animals, the current population cannot support themselves entirely from the Tauran Plain. Some families always seek additional income both on and off the Plain.

Today this additional income comes from shepherding, truck hauling, and less often, from labouring in urban brick works. In the past the additional income was obtained with greater difficulty through charcoal production, shepherding, hauling by camel, occasional brigandage, and collection and sale of vegetation (for fuel, food, spices, medicine and tanning). What follows is a discussion of the production strategies that are most likely to cause environmental stress by reducing the vegetation cover on the Tauran Plain.

I. AGRICULTURE

Settlement in Tauran has depended on the *qanāt*. Qanats were introduced in Iran during the first millennium BC. They consist of an underground channel that carries groundwater by gravity flow from an aquifer underlying relatively high ground out on to the surface at a point lower in a valley or plain where there is good soil (see Fig. 1 and Spooner, 1982).

In Tauran qanats are relatively short and meagre and the largest area watered by one qanat is only 30 ha. Each village on the Tauran Plain has access to one or more qanats which provide water for all needs, and most families own one or more shares in the flow, which is divided into cycles of 12–14 days. The qanats receive no maintenance, except occasionally in cases of severe damage from run off, and it is not possible to reconstruct the co-operative arrangements involved in their original excavation, which was carried out by an earlier population.

The limited water supply from qanats is manipulated in order to grow a combination of staples such as winter wheat and barley, cash crops such as tobacco and cotton, and fodder crops, such as sorghum, millet and alfalfa. Scarcity of water is a persistent problem. The problem varies according to both direct precipitation and the effect of the previous year's precipitation on the water table. When qanat flow is reduced the villagers change their cropping emphasis in order to maximize productivity by, for example, growing less cotton and more tobacco; when water flow is high sesame or cumin may be planted.

Within living memory the major cropping pattern has changed little with the exception of a change in variety of cotton, and cessation of opium cultivation when it was prohibited in 1955. Figure 2 illustrates the present day cropping pattern on most of the Plain, and Fig. 3 shows how it varies with distance from the water source. The actual amount of land planted in a certain crop varies from year to year depending on water supply and the availability of family labour. There is some variation within Tauran based on differences in microclimate, wind force, soil, water salinity and qanat flow. However, in all villages the amount of land planted in walled gardens remains

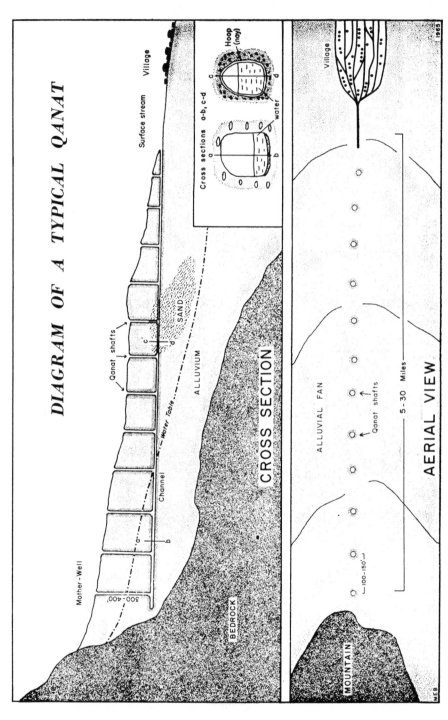

DIAGRAM OF A TYPICAL QANAT

Mother-Well

300-400'

Qanat shafts

Channel

a
b
c
d

Water Table

ALLUVIUM

SAND

Surface stream

Village

BEDROCK

CROSS SECTION

Cross sections a-b, c-d

Hoop
(noy)

a
b
water

c
d
water

MOUNTAIN

ALLUVIAL FAN

Qanat shafts

100-150'

5 - 30 Miles

Village

1965

MEB

AERIAL VIEW

Fig. 1. Permission from University of Wisconsin Press and author (originally published in "City and Village in Iran" by Paul Ward English 1966).

	N	D	J	F	M	A	M	J	J	A	S	O
Open fields— irrigated (no rotation)	Wheat Barley (opium, cumin)							Millet, sorghum Fallow Tobacco				
	Alfalfa											
(rotation)	Cotton		Fallow		Cotton							
	Fallow (alternates with cotton)											
Walled gardens	Grapes, fruit and nut trees, herbs, tomatoes, tobacco seedlings[a] (mulberry trees for silk worms)											
Open fields— dry farmed (no rotation)	Fallow											
	Fallow					Barley Wheat			Fallow			
	N	D	J	F	M	A	M	J	J	A	S	N

[a] Sometimes grown in open fields.
Note: the irrigated agricultural cycle begins in November with the planting of winter wheat. Crops in parentheses are those grown 30 years ago.

Fig. 2. Tauran crop cycle and land use pattern.

constant from year to year since this land is devoted primarily to grape vines, fruit trees, tomatoes and herbs.

Each farmer is responsible for his own small plots and individual farmers pursue various techniques to increase production. These techniques involve different combinations of the application of animal dung and chemical fertilizer, weeding, double cropping and fallowing.

The dung available from livestock stabled in the village is applied as fertilizer—most intensively to those fields which will be double cropped. Chemical fertilizers have been introduced in the last ten years. They have not replaced dung yet (except in a few cases) but are used in conjunction with it.

There are several reasons why fallowing is useful. According to local villagers, good farmers plough their fallow fields and the "lazy" farmer does not. Whereas the unploughed fallow fields may quickly become weed-covered and hence protected from wind or other erosion, this is at the expense of moisture retention (cf. Dasmann, 1976:71; Antoun, 1972:8),[3] and may

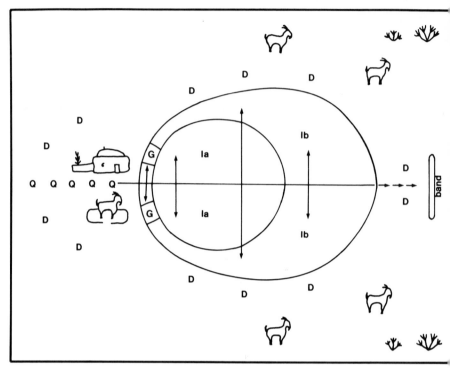

Fig. 3. Land use pattern in relation to qanat opening. D. Dry farming; G. walled gardens; Ia. irrigated fields: double cropped; Ib. irrigated fields: fallow/crops; Q. *qanat* system; ⟶ irrigation channel; grazing (goats and sheep); animal pen; firewood.

aggravate conditions of low nitrogen content (cf. English, 1966:159; Ault, 1972:7).

The practice of ploughing fallowed land could increase erosion. However, the ploughed soil is left in heavy clumps which are not susceptible to transport by wind in contrast to the finely pulverized soil outside cultivated areas regularly trampled by animals. Wind erosion was generally not perceived as a problem among farmers on the Plain, though they felt that the wind itself was a problem for agriculture close to the kavir. Water erosion is not a problem in irrigated fields because of the small size of the individual fields, which average about one tenth of a hectare and are bounded by small ridges to retain water during irrigation (see Fig. 4). Only irrigated fields which border dry river beds are vulnerable to water erosion.

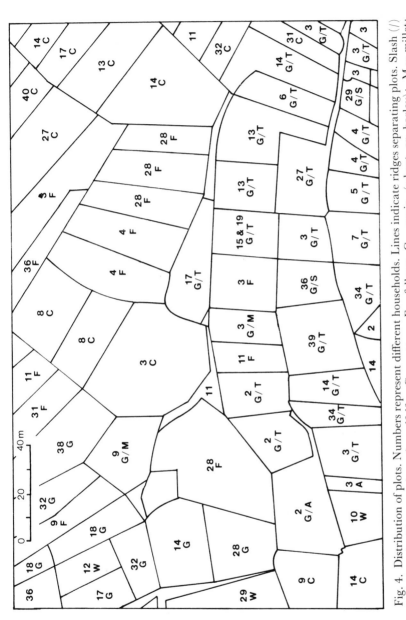

Fig. 4. Distribution of plots. Numbers represent different households. Lines indicate ridges separating plots. Slash (/) indicates double cropping. A = alfalfa; C = cotton; F = fallow; G = grain (wheat or barley); M = millet; S = sorghum; T = tobacco.

Dry farming is important in Tauran for several reasons. It is inefficient to use irrigation water on crops that may not need it. There is also a local preference for dry-farmed wheat for bread-making. In recent years there has been an increased demand for barley to supplement the diets of sheep and goats and dry farming can help to satisfy this demand. Pastoralists feel additional pressure to take advantage of years of good rainfall and plant as much land to barley as the law (which since 1966 prohibits further extension of dry farming) and their labour resources allow. This increased demand is related to the state of the range and the increasing number of animals. In the past the range is said to have supported the animal population except in drought years. As grazing, dry farming and the collection of vegetation for other purposes increased, vegetation cover, particularly on the Tauran Plain appears to have decreased. Since the decrease was in shrubs, flocks became increasingly dependent upon annuals for spring and winter forage. Consequently, in most years—and even more so in drought years—the demand for supplemental feeding increased.

Dry or rain-fed farming (*deim*) includes several strategies of cultivation which have different environmental impacts. The most widely-known form is that which relies on direct rainfall only. This form is currently found in Tauran close to settlements, but it is limited in area because of the irregularity and insufficiency of rainfall.[4]

Other forms of dry farming in Tauran depend less on direct rainfall. During the winter, unwanted qanat water is channelled on to otherwise unirrigated land. In this case the water may not belong to the owner of the land. This winter-watering may then be supplemented by spring rainfall to build up sufficient soil moisture for a crop, such as melons. Another strategy is known also as *band* or check dam farming. Either run-off or unwanted qanat flow is captured behind the band in the winter and spring and is supplemented by direct rainfall (see Fig. 3). The land is farmed by those who contributed to the building of the band.

Bands are constructed throughout the Plain wherever either type of run off can be caught. Both types encourage the accumulation of silt which has been the source of much of the good farmland in Turan (see Dennell, Chapter 9). The availability of tractors with earthmoving attachments have increased the number of check dams constructed on the Plain in the last ten years.

In all forms of dry farming various factors limit potentially negative environmental effects. Generally dry farmed land is not ploughed after harvesting. The grain is harvested primarily by sickle. Stubble is not ploughed under until spring planting. Weeds which have grown up with the crop are not removed until spring ploughing which will only occur if rainfall is promising. The plant *Goebelia pachycarpa*, whose root systems have good soil binding qualities (Iran, 1977:29), are allowed to remain and protect the land against erosion. However, where ploughing is done by tractors brought into Tauran in the autumn and spring, more disruption is caused to soil-stabilizing root systems than is the case with the traditional scratch plough.

In addition to government restriction and irregular and limited rainfall, there is another factor which has limited the extent of dry farming. Small families may not plant grain because of insufficient labour. Depending on the timing of the spring rains the labour demands of dry farming may conflict with those of pastoralism within the family. During the spring animals are taken off the Plain by owners (even they are grazed in the mixed village flock during the rest of the year). It is sometimes necessary, therefore, to choose between dry farming and the increased milk production which will result from the spring graze. The decision will be related to whether or not a man has sons or other male relatives who can help him out. In recent years the availability of tractors for ploughing in the spring has alleviated this problem to some extent.

The individual farmer is in a position to deal with most agricultural problems independently—except those involving the qanat. Individual farmers are responsible for their own land and no one but themselves will suffer if they let the fertility of the land decrease. Agricultural maximization depends on controlling the factors that will continue sustained production such as manuring and fallow. If a man begins to neglect his field (for example, because of opium addiction), only his own yields will be affected. But when the village or qanat needs cleaning or rebuilding, the cultivator must co-operate with other water owners and find some group solution to the problem. However there is no mechanism for group co-operation beyond the individual need for water. There are, for example, cases of villages that were abandoned when problems of reduced qanat flow resulting from moving sand and flooding were not solved.

Dry farming poses a special problem of control and communal organization because unlike irrigation it competes with pastoralism and collecting for the same land. When unirrigated land is not planted, it may be grazed. It is not clear to what degree local populations are aware of the positive or negative effects of dry farming but local control of dry farming would anyway be difficult now because of the current lack of endogenous social mechanisms to facilitate group action.

Because of the variety of factors involved, therefore, agricultural strategies in Tauran cannot be considered inherently destructive or conservative. Qanat-dependent cultivation on individually owned plots is generally aimed at sustained production and current techniques do not contribute to erosion, salinization or loss of fertility. Dry farming includes a number of techniques which in the long term counteract erosion (particularly those associated with bands). The environmental threat posed by dry farming derives: (1) from new technologies such as tractor ploughing and the increased demand for grain as supplemental animal feed, and (2) from the fact that it often constitutes an additional pressure on land that is already being used for grazing and fuel collection. So, some types of dry farming may not only affect current and future crop yields—but also the potential for grazing and fuel collection.

II. FUEL[5]

Removal of shrubs for use as fuel affects the quality and quantity of grazing and can contribute to loss of soil through wind and water erosion. The most important fuel need for the local population is domestic. Villagers classify wood into two types: *hizom*—shrubs, especially *Artemisia*, for bread ovens and space-heaters, and *konda*—thicker branches from species such as *Haloxylon* for longer cooking, for processing of milk, grapes, tomatoes and pomegranates, and for heating water for bathing and washing clothes. Until 1975 village bath houses used wood fuel, but these have all been replaced by new government diesel-burning bath houses.

Since about 1965 kerosene has been available through a co-operative, reducing somewhat the demand for firewood. It is used in samovars for tea making, in cooking stoves, in lamps, and more and more in wealthier houses for kerosene space heaters. In 1977, an energy use survey conducted in Tauran indicated that in one village approximately 1 ton of firewood was consumed per person per year. An additional 21 tons were used for processing the milk of 254 sheep and goats by two families at summer milking stations.

From the point of view of ecological impact the use of vegetation for fuel in Tauran constitutes two separate problems. Domestic fuel use has been one problem, but export of vegetation in the form of charcoal was another. Villagers with insufficient land or animals needed other sources of income if they were to remain in Tauran. In the past charcoal burning was one of the available options for earning additional income. Local use was minimal but urban demand was high—particularly in the period just prior to the introduction of kerosene. The government finally prohibited charcoal production in 1966 throughout Iran.

The collection area for charcoal was different than for domestic fuel. Domestic fuel was collected much closer to villages (Fig. 5), whereas the distribution of charcoal production was independent of village location. In addition there was a different effect on vegetation. For domestic fuel a wide range of species were collected including *Artemisia*, *Haloxylon*, *Calligonum*, *Lactuca*, *Ceratoides*, *Zygophyllum*, *Amygdalus*, and *Ferula*. Charcoal production required the larger shrubs which had pretty much disappeared from the village collecting range, especially *Haloxylon*, *Pistacia*, *Amygdalus*, and

Calligonum. The species which are good for fuel are also preferred for grazing (see Nyerges, Chapter 11). *Artemisia,* in particular, is considered the mark of a good range and its disappearance from the Plain has been a cause for alarm among villagers, who price it for grazing as well as oven fuel. Ten years before the government outlawed both charcoal production and the cutting of live vegetation (1966) residents of Asbkeshan in the southeast of Tauran, who depend primarily on pastoralism, banned charcoal production in their territory. This action was unusual not only because of the nature of the decision but because they were able to enforce it. But this community was small, closely-knit and of tribal origin whose members did not themselves need to supplement their income by producing charcoal. None of the communities on the Plain had these social and economic characteristics.

The 1966 government decree prohibiting the cutting of live vegetation was difficult to enforce because there was not enough dead vegetation to meet the demand despite the increased availability of kerosene. The gendarmerie and later the Department of the Environment game guards, who were charged with enforcement became unpopular in their attempts, especially when sometimes they burned a load of firewood they had intercepted. Their position was made more difficult by their own families' needs. Nevertheless they were effective to the extent that fuel collection ceased on the Tauran Plain. Some residents claim that certain species of shrubby vegetation are beginning to reappear.

Other changes have affected the pattern of fuel collection in recent years. Game guards were transferred off the Plain leaving enforcement in the hands of the gendarmerie alone. Since 1974 the number of vehicles which could be used for hauling firewood have increased. Many of these vehicles were hired by men who were away working as shepherds and needed fuel for their families or by flock owners who needed fuel at summer milking stations. Probably the quantity gathered has not changed because of truck hauling but the pressures may have increased on certain species such as *Haloxylon*— especially close to the main tracks.

There is no mechanism to regulate fuel collection in relation to grazing requirements. When shrubs disappear or are severely reduced, grazing becomes more dependent on weedy plants and annuals. This change in vegetation composition is more important because in the winter or in drought years when annuals are less abundant, shrubs are less available for grazing and flocks require supplemental feeding—particularly the pregnant ewes (and occasionally does). The most common supplemental feed in Tauran is barley because of the absence of other low cost fodders. Larger stock owners may therefore try to expand dry farming operations in years with sufficient rainfall in order to obtain additional fodder as cheaply as possible.

Fuel collection has caused a desertification problem in Tauran primarily as a function of urban demand, especially for charcoal. Although fuel needs

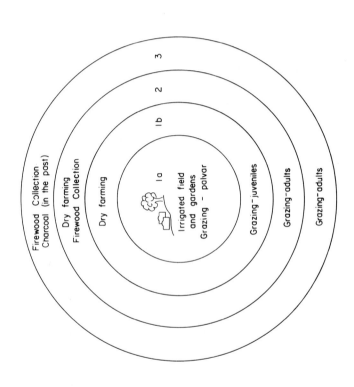

Firewood Collection
Charcoal (in the past)

Dry farming
Firewood Collection

Dry farming

1a

Irrigated field
and gardens
Grazing - palvar

1b

2

3

Grazing-juveniles

Grazing-adults

Grazing-adults

Zone	Estimated biomass	Average stocking rate	Pastoral land use	Agriculture	Firewood gathering
1a	300 kg/ha	1·2 ha/animal	Village flocks Palvar Kids/Lambs Other fattening operations	Qanat irrigation Dry farming	Prohibited
1b	300 kg/ha	1·2 ha/animal	Village flock	Dry farming	Prohibited
2	2–3 tonnes/ha	2 ha/animal	Summer/winter stations of local villagers	—	Allowed (dead vegetation only)
3	2–5 tonnes/ha	up to 3·4 ha/animal	Transhumant stations and occasional village stations	Occasional bands	Allowed (dead vegetation only)

Fig. 5. Distribution of village land use—Tauran Plain.

continue to be a major cause of desertification in many developing countries, this problem has become less severe in Tauran as a result of a combination of factors including the availability of fossil fuels, governmental restriction and general economic change.

III. PASTORALISM

The environmental impact of grazing varies, depending upon which of several management strategies are followed and whether the particular grazing area is also subject to collecting or dry farming.

Most flocks in this part of Iran are transhumant. They enter the area in October–November and leave in April–May and graze away from the main villages (see Table I). The remainder—some 25 000 locally-owned animals— are managed in three ways. Large flocks belonging to individual Tauran owners are maintained away from the villages at year-round stations. An alternative pattern involves wintering in the villages while the summers are spent at milking stations. The third pattern affects most Tauran residents although it represents only a small proportion of the total animal population (see #4—Table I): in each village there are one to three mixed flocks (with animals from several owners) which are grazed on village commons in the Plain by hired shepherds. These flocks return to the village daily to water at the qanat, for milking in the summer, and for barley-straw rations and night shelter in the winter.

Local villagers practising each of the three strategies just described all share in one additional management pattern. All villagers have animals which are kept out of the big flocks at certain times of year and are grazed in the area directly surrounding the village (see Zone 1a, Fig. 5). For example, in the spring young lambs and kids are grazed close to the villages until they are weaned and can be grazed with the larger flocks. A few other animals (known locally as *palvār*) are grazed close to the village and fattened for winter consumption, gift or sale. In the spring each family selects out one to a dozen or so from the main flock. They are grazed daily by an aged or young family member on agricultural stubble and the nearby land (Zone 1a, 1b, in Fig. 5). These zones contain primarily *Goebelia pachycarpa* (eaten only from mid-summer on), a variety of thistles, *Peganum harmala* (rarely eaten even dry), and various annuals. But the palvar also receive a daily supplement of agricultural weeds gathered from the irrigated fields and water channels and a mixture of barley, barley flour and washed straw.

The nature of the pastoralist's right of access to his grazing determines his ability to control his own and others' use of it. There are two basic types of access in Tauran: individual access through permit, ownership or rental; or

Table I. Significant Factors in the Ecology of Pastoral Systems in Turan

Type of Pastoral Group	Access to Grazing	Control of Stocking Rate	Production Emphasis	Shepherd Wages[a] ($ = 70 rials)
I. Sangsari long-range transhumants (winter only in Turan)	Restricted to flocks whose owners have rights to water wells and pens through ownership, rental or permit	Individual owner	Meat—primarily for Teheran market. Milk products and wool for subsistence or sale in summer area. Buy up and fatten Tauran animals for sale in Tehran	Head shepherd: 20 000 rials ($285) per month Assistant: 15 000 rials ($214) per month
II. Village owned flocks which graze year-round at sheep stations	,,	,,	Meat and milk products for sale or subsistence	17 000 rials ($242) per month
III. Villages owned flocks which spend the summer at milking stations and winter in the villages	,,	,,	Meat and milk products for sale and subsistence	—
IV. Mixed village flocks which graze in central Tauran (chekana)	Unrestricted within grazing distance from home village. (Formerly restricted to traditional village territory)	None	Mainly subsistence milk products and meat for owners. Occasional sale of extra animals or products locally	16 000 rials ($229) per month

[a] These figures are for 1978.

access as a function of village residence. Sangsari flocks and the large local flocks have access to areas with shallow wells and sheep pens. Traditionally rights to these areas have been transferred by sale or rental. They are neither surveyed nor fenced; they include an area with an average three kilometre radius (roughly the distance which can be travelled to and from the water source comfortably and allow for grazing during a short winter's day, see Iran, 1977:23). Although there is provision for legal control of grazing ratios in these areas, effective control rests with the shepherd or flock owner. The situation in village territory presents a different management problem. The majority of households had a few animals which were grazed in joint flocks (*chekana*) by hired shepherds (see Table II). Any villager may graze his animals in village territory and there is no provision for control of grazing ratios.

Iran's move to protect its rangeland from overgrazing and fuel collection by nationalizing rangeland is consistent with the actions of other Middle Eastern countries in the last 50 years and with the policies of FAO and other development agencies. In some respects these moves have provided the protection needed by small communities against exploitation of their resources by outsiders—and by themselves. However, in Iran and elsewhere this action has often weakened traditional village or tribal controls. This situation has been reported for nomads in Syria (Draz, 1977) and is the case in Tauran for the sheep stations off the Plain.

The land grazed by village flocks is particularly vulnerable to overgrazing because the number of animals is under no one's control and the same land is also used for dry farming and collecting. The government has tried to enforce the restrictions on the amount of fuel collection and dry farming, but not on the numbers of animals. Oral history suggests that grazing pressure on the Plain has increased over the years. While this increase is difficult to substantiate, it is feasible given the changes which have occurred in the area over the last century. Formerly villages of Tauran were divided into opposing factions which raided and sometimes fought each other. According to oral history, in the 19th century, villagers acted to protect their resources (particularly their animals) against theft by other villages and by outsiders such as nomadic Turkmen from the north and Basseri from the west. Both of these threats—from inside and outside the region—probably served to draw villagers together to protect their resources including the grazing of traditional village territory. With final pacification of the area in the late 1950s, the gendarmerie provided security for grazing, trade (including charcoal), and travel. The large flock owners of Tauran, who had grazing areas off the Plain, provided summer grazing areas for many village families who now graze animals in the Plain year round. These families now speak of spending summers away at milking stations in the 30s. Individual owners are free to keep as many animals as they wish. The constraints are the limits of space for the animals at night, money to pay the hired shepherd who herds them with other village animals, and grain and straw for supplemental winter feeding. The shepherd decides how many animals he will herd, but he usually herds for a number of families who have agreed to co-operate with each other in hiring him, paying him, sending someone to join him nightly in the summer and bring him food.

The situation in the 1970s for the combined village flocks resembles what has been termed the "tragedy of the commons" (Hardin, 1968) or "common pool" situations (Ostrom, 1977:175). However, common pool situations are only problematical when the following conditions are present:

1. Ownership of a resource is held in common.
2. No one owner can control the activities of others.
3. Total use or demand upon the resource exceeds the supply.

When these conditions exist, Ostrom argues, use of the resource by one individual reduces its value to the others. Since Ostron does not imply that a common pool situation will automatically become a "tragedy of the commons", it is worth considering the variety of mechanisms which may serve as constraints on the increase in livestock. For example, in the Sahel, until circumstances led to changed conditions in which carrying capacity was exceeded, a variety of mechanisms such as freedom of movement, timing of movement and control of movement by individual groups existed which served to prevent overgrazing (Hardin, 1977). In Tauran the mechanisms inhibiting overgrazing have varied. They have been designed to enhance milk, meat and wool production for subsistence and market—rather than retaining specific sex ratios or numbers of animals as in Sahel or East Africa where animals have been kept for subsistence and social and economic security. In Tauran all owners are concerned with milk and meat production, although to varying degrees (Table I). Both Sangsari and local large owners produce primarily for markets, although for large local owners subsistence needs are also important. Sangsari concentrate on the raising of sheep breeds with higher meat/fat yields per animal than Tauran sheep which are primarily all-purpose breeds good for wool, meat and milk. Sangsari meat is destined for the Teheran market, whereas Sangsari milk products are taken for family use. All owners see production of milk, meat, and wool as a function of range condition, but it is the large owners who are in a better position to maintain the range in good condition by controlling the number of animals.

One strategy for assuring the best grazing possible (and hence to avoid overgrazing) is to keep outsiders off. The Sangsari, who only use their territories in the winter, often hire persons to patrol their grazing territory—a strategy which is resented by local owners who would like to have access to the

Table II. Household Ownership of Animals in a Tauran Village

	1975	1976	1977	1978
Total number of households	33	33	31	32
Households with no animals	1	1	1	1
Households with palvar only				
(not kept in village flock)	2	2	2	2
Households with animals in				
village flocks				
(a) year-round	23	22	21	23
(b) winter only	3	3	3	3
(c) summer only	1	1	—	—
Households with animals away				
year-round				
(a) With relatives	2	2	2	3
(b) With outside flock for whom				
owner is hired shepherd	1	1	—	—

untouched annuals which are being guarded for Sangsari winter grazing. In addition both Sangsari and large local owners try to reduce grazing pressure by acquiring a number of areas and grazing them in alternate years. Other local owners have one area for winter use and another elsewhere for summer use. In the last ten to fifteen years, however, there has been an overall decrease in the number of owners who have been able to retain empty grazing areas. Several factors may account for this shift. Since 1973, the end of the last drought period, herds have been expanding. In addition this period has been one of great prosperity in Iran during which several large Sangsari and local owners bought up sheep stations and hired more local shepherds. If stocking rates have increased, it has primarily been a response to a series of good years. This pattern of increased stocking rates in good times may also have occurred in the past as security and weather conditions changed.

The villagers who are obliged to share grazing on the Plain are aware that they are getting less milk, meat, and wool per animal than they would if their animals grazed elsewhere (Table III). On the one hand their options for grazing elsewhere are limited. For example, at least two Tauran owners summer off the Plain but must winter in the village because there are no available sheep stations to buy or rent year-round. On the other hand, for the villager with few animals there are pressures to keep animals in the village flock—which means accepting lowered production. If the animals are grazed away from the village in the summer, then the family is deprived of fresh milk[6]—or must leave the village for the summer and live by the animals and process the products. Although some women enjoy this life away from the village, most consider it a hardship. In addition agricultural labour needs

Table III. Grazing Pattern and Annual Production of Meat, Milk and Wool

Product	Year-round on Tauran Plain	Winter on Tauran Plain Summer off the Plain	Year-round off the Plain
Wool	Low	Medium	High
Lamb weights[7] (6–8 mo)	Gross weight: 20–21 kg	—	Gross weight: 25–30 kg Meat yield: 12 kg
Clarified butter (per animal)	300–400 g	750 g	750–1500 g

must be co-ordinated with pastoral labour demands—which is easier in the village.

The villager is constantly seeking a solution to the problem of finding better grazing off the Plain. In a few cases villagers who have relatives with large flocks away at sheep stations for the summer or year will persuade or cajole their relatives into taking some or all of their animals. It is not just the male owner or shepherd who must be convinced: an arrangement must be made with his wife and daughters who will have to do the milking and process the milk. This type of co-operation between families is rare, but may be exchanged for help with weeding and harvesting. Other options for grazing off the Plain occur in the winter when milking is not a problem. At this time those village men who work as shepherds for Sangsari or other large flock owners may take their own animals with them. This action reduces the need for supplemental feed because they are grazing better range. They are also reducing the grazing pressure on the Plain although this is not their goal. They wish to find better grazing and raise production, save shepherding and feed costs, and relieve the family for the additional responsibility of looking after the animals while he is gone.

Another constraint on the increase of livestock in Tauran and hence a check on overgrazing is what has been termed the managerial cost (Brokensha *et al.* 1977:12)—which refers to labour availability. It was noted earlier that for Tauran families without enough land and animals to support them, shepherding for the Sangsari has enabled them to meet expenses and even to save money to buy land without migrating to the cities. Sangsari wages are higher than those paid locally (see Table I) and there is the added advantage that shepherds can limit their work to only part of the year (winter only or summer only) which frees the shepherd for family and agricultural obligations in Tauran. However, it is often difficult to find competent shepherds for village flocks. Village shepherds not only receive lower wages but must deal with several owners. The men available for the village positions are those who have chosen lower wages for various reasons—or who do not have the choice to be higher paid shepherds (some are mentally retarded, or have other psychological problems; others are too young). Some local shepherds are therefore less attentive to the sheep than the top Sangsari shepherds who receive higher wages for being trustworthy and having the knowledge and skill that will serve to protect an owner's animal investment (for example, matching up mothers and offspring in the spring).[8] For this reason, it is likely that village flocks suffer higher mortality rates because of inadequate shepherding, although Sangsari flocks may have higher mortality due to wolves and snakes. Mortality due to disease is a significant factor in Tauran. Both individual and mixed flocks are subject to higher disease mortality than are flocks in centrally located regions because of lack of veterinary care. Presently local owners have inoculated their own animals on the basis of their understanding of animal diseases—but this is an individual strategy dependent upon money for drugs and individual understanding of the proper drugs for various diseases.

Other labour factors which may affect animal numbers in village flocks are a result of the obligations of the owners to assist the village shepherd. For example, as noted earlier, village animal owners have an obligation in the summer to go (or send a substitute) to take food and spend the night with the shepherd out on the range. In the spring, they (or their substitute) must take turns grazing kids and lambs separately until the two herds are merged. The amount of time they spend is figured on the basis of the number of animals owned. Also in the spring some owners take their animals out of the village flock and move away from the Plain for about a month in the spring when the annual vegetation is most lush. This is difficult since it conflicts with ploughing and planting, but essential if there is to be any hope for good milk production. The more animals a village owner has, the more essential it is that he invest more time in their maintenance—particularly in spring and summer when demands on his labour are high. If a man has few family members to help him out, he will have to choose that year between various pastoral and agricultural tasks—and may decide for the time being not to expand his animal holdings.

Various mechanisms have been operating to ensure access to good grazing and the consequent higher yields. These strategies are: (1) for large owners— individual control over grazing areas ensuring exclusive access to these areas, acquiring a number of areas and grazing them in alternate years; and (2) for villagers to graze some animals off the Plain with relatives taking animals out of the village herd for the spring graze. However, in the common grazing land should animal numbers increase to levels exceeding carrying capacity, the problem of control arises.

IV. THE PROBLEM OF CONTROL

In this discussion, three strategies of production—dry farming, collecting, and village pastoralism—all primarily for local consumption—have been singled out as potentially destructive towards vegetation on the Tauran Plain, particularly where they compete for the same land. With regard to collecting, during the period of research, dealing with governmental control was perceived as a more acute problem for local residents than were the effects of the practice. With grazing, however, the situation is more complicated. Villagers are aware that by grazing on the Plain they must settle for lowered milk, meat and wool production. To raise production they will try to move animals off the Plain whenever possible or supplement the diet with barley and straw. Although they are aware that the vegetation quality is poor, it is not known to what degree they see dry farming, fuel collection, and increased numbers of animals as contributing factors. It is no doubt also true that perception of the problem varies, and those who use the range most— particularly older residents and those who go out daily (such as shepherds)— are more aware of the state of the range. Assuming that villagers had recognized the negative impact of these activities on pastoral production, why do they not organize to protect communal grazing and fuel resources? What differentiates them from, for example, northeastern Afghanistan (Edelberg and Jones, 1979; Shahrani, 1979), mediaeval England (Ault, 1972), Greece (Koster, 1977), the Swiss Alps (Netting, 1976) and Mexico (Lewis, 1959), where local communities have established control over the common resources?

The lack of communal organization has been explained in a variety of ways. One explanation refers to national character which, for example, would suggest that there is a "strongly individualistic, indeed anarchical, strain in Iranian character which so often prevents sustained co-operation ..." (O'Donnell, 1980:viii). Another suggestion is that this behaviour is the result of people following an unwritten cultural "rule". For example, Banfield (1958) in his discussion of southern Italian society suggests that the people there behave as if they were following a "rule" which he calls "amoral familism" in which one will "maximize the material, short run advantage of the nuclear family; assume that all others will do likewise ..." (Banfield, 1958:83).[9] A similar explanation of the behaviour of peasants, in Tzintzuntzan, Mexico claims that they act on the basis of a model termed the "image of limited good". According to this view, there is a common perception in peasant societies that ... all the desired things in life such as land, wealth, health ... exist in finite quality and are always in short supply, as far as the peasant is concerned ... there is no way directly within peasant power to increase the available quantities ... an individual or a family can improve a position only at the expense of others . . . mutual suspicion seriously limits co-operative approaches to village problems (Foster, 1965:296–7, 308). These examples illustrate various attempts to explain the lack of co-operative effort on the basis of national character or cultural rules or "ethos".[10] But none explains why co-operative behaviour *does* occur in some situations for control of areas of common grazing, fuel collection or agriculture.

What then are the conditions under which co-operation occurs which will help us explain co-operation or the lack of it in a peasant society such as in Tauran—or for that matter in any type of society? In the *Logic of Collective Action*, Olson suggests that though all of the members of a group (large or small) "have a common interest in obtaining collective benefit, they have no common interest in paying the cost of that collective good" (1965:21). In Tauran one of those costs is setting up a formal village or Plain-wide organization for enforcing grazing, fuel cutting, and dry farming. Olson's contention that "the cost of establishing an organization entails that the first unit of a collective good obtained will be relatively expensive" would find agreement among Tauran villagers. They have in the past only co-operated in smaller groups—primarily as shareholders in the qanat water, co-operating to clean the storage pool or arrange for repair in case of serious damage interrupting the essential flow of water.

Another important factor is that "the behaviour of the individual in the group is due partly to the fact that each individual in a group may place different value upon the collective value wanted by his group" (Olson, 1965:22) and that willing participation can arise only if the resource in question works the same way for all users . . . (Doherty and Jodha, 1977:7). When we examine the Tauran communities we see that in fact resources are not used in the same way at all. Fuel is used differently by different families depending upon the size of family, the number of animals owned, and differential ownership of garden lots (which affects fuel needs for making pomegranate and tomato paste and grape syrup).

Charcoal production also affected communities differently. In Tauran only one case was found of a village which protected its territory against outside exploitation—the village of Asbkeshan cited above. The inhabitants of this village were closely related and led by a strong relative at a time when Iranian villages were armed and could enforce such a ban. They also had no extensive agriculture and were solely dependent upon their pastures which were threatened by charcoal production. On the other hand, on the Plain the social context of charcoal burning was different. It supported local men rather than outsiders. In addition, it may not have threatened grazing territories of large flock owners as much as it did in Asbkeshan. It may have affected small village owners but they would not have the power or organization to control it.

The degree of use of the Plain for grazing by a household is determined by the size of its holding, their access to a summer or year-round station, access to relatives who can take animals off the Plain, their ability to supply supplemental barley and straw, and to meet the labour requirements of milk processing, and assistance to the shepherd. Dry farming depends on the amount of family or hired labour available for ploughing and harvesting which is linked to the demand for labour for the household's irrigated farming and pastoral interests. In the case of each family, these factors combine in a variety of ways to give each a different interest in the use of the Plain. Even if there may be a general consensus that the range of the Plain is worse than off it, the next step—to seeing this as due to overgrazing, dry farming and collecting (singly or together), and a further step of organization for control— is lacking. The "individual benefits of participation are not equated with group benefits . . . and group pressure for participation in an organization for control is not very strong" (Bennett, 1979:10).

These explanations of lack of collective action have been phrased in terms of the individual decision-maker and the benefits he would or would not receive by paying the costs associated with co-operative action. It is important to consider these individuals in their socio-political context. For example, the groups listed above which effectively exercise communal control of a common resource range from tribally organized groups in Afghanistan to feudal society and autonomous small villages in Europe and Mexico and the nature and extent of their group control varied considerably. In the present context the tribally organized groups are the most interesting. The presence or absence of tribal organization within a society is important because it can serve as the basis for resource control in the absence of a political superstructure such as might be provided by the central government.

The recent history of the Middle East is filled with examples of the disintegration of tribally organized society and its replacement by the social and administrative framework generated by the rise of the nation state. This change has occurred at a time when population growth and other changes have increased both local pressure and general demand on natural resources. In Tauran the tribal structure was changing by the turn of the century for a number of reasons which were difficult to reconstruct because of the limited time depth of the oral histories and travellers' reports upon which we are primarily dependent. For example, residents' and travellers' reports refer to the power of khans whose position seems to have derived from a tribal base in the area. Until the 50s villages were endogamous communities interrelated by intermarriage among elite tribal families. With the social and economic change that began to accelerate in the larger society in the late 50s, and the unprecedented rise in the power of the central government and its extension into previously isolated districts, the political and economic base of this tribal elite disintegrated, and most survivors migrated to the towns. In addition, in the 60s, the central government nationalized rangeland and water resources—which negated many of the traditional rights of villagers over their resources. Village territories were greatly reduced and villagers do not have legal recourse to government agents such as the gendarmerie to protect the larger area they exploit. Currently we are left with a situation in which the only control mechanisms on the Plain are imposed from outside such as those on dry farming, charcoal burning and firewood collection. The government has not yet imposed grazing restrictions on the Plain but individual villages can no longer control their own territory even if they should organize to do so. As a result villagers attempt to avoid grazing on the Plain whenever possible within the limits of their family capabilities.

This chapter has argued that in Tauran traditional forms of resource use are not inherently destructive, common use is not always a tragedy of the commons and central government policies may overlook or compromise valid local strategies. Although desertification undoubtedly occurred to some extent before the modern period, there appears to have been a social mechanism—in the form of tribal organization—that may have facilitated communal control and conservation. In the last two decades or so that internal mechanism has disappeared as a function of change in the larger society, and the associated administrative changes have not been sufficient to replace it. It is not too late to re-evaluate the problems of control in regions like the Middle East where tribal organization may have offered some solutions.

Endnotes

[1] The data from Tauran derive from ethnographic field research conducted over a period of 17 months between November 1974 and December 1978. I wish to acknowledge my appreciation to the MAB Secretariat (UNESCO, Paris) and Sigma XI for financial support and to the residents of Tauran, the director, staff and local game guards of the Department of the Environment (Tehran), the research division of the Ministry of Agriculture (Tehran), and the botanists who worked with the Tauran Programme—S.-W. Breckle, H. Freitag, K. H. Rechinger, H. Runemark, and P. Wendelbo—for their invaluable assistance. I am also grateful to Brian Spooner, John Bennett, Lee Horne, A. Endre Nyerges, and Stephen Sandford for suggestions which were helpful in analysing and interpreting the material presented here, and to Lee Horne for drawing Figs 3 and 4. The final responsibility for both facts and arguments is of course my own.

[2] All figures are taken from the National Spatial Strategy Plan—First Stage Report (Iran, 1976) and an unpublished Agricultural Census conducted by the Statistics Centre (Iran, 1973).

[3] Antoun (1972) conducted anthropological research in an Arab village in Jordan. His discussion of the effects of ploughing fallow land is based on interviews with villagers and references from Shihabi (1935) and Keen (1946).

[4] Rainfall and dry farming records for the area are incomplete but are supplemented from oral history and a dendrochronological study of a dominant shrub. 1958–1963 and 1968–1973 were bad years, with 1971 especially bad. During the earlier period the Chubdari, a group of nomads, settled at the northeast corner of the area. An analysis of *Zygophyllum* growth rings suggests that over an 88 year period bad years occurred every 2–3 years (see Bhadresa, in Iran, 1977:69). A bad year for *Zygophyllum* may not necessarily be a bad year for flocks or agriculture, depending upon the timing of the rains. For grazing even a low yearly rainfall will not be disastrous if spring rainfall occurs at the right time and the crop of annuals flourishes.

[5] A more comprehensive discussion of fuel is given by Horne, below, Chapter 10. This section is concerned only with aspects that bear on the problem of conservation and control.

[6] A few families with many children do have milch cows—despite the problem of finding adequate and inexpensive fodder. Alfalfa is grown in small amounts only, because it needs at least weekly watering.

[7] For *palvar* which receive supplemental feeding, lamb gross weights can be 30–35 kg and meat yield is 18 kg. These animals do not graze with range flocks.

[8] There have been discussions about the problems of hired shepherds elsewhere in Iran—mainly among nomadic groups such as the Basseri (Barth, 1961, 1964), Komachi (Bradburd, 1980), and Qashqai (Beck, 1980).

[9] Banfield's thesis is criticized by Silverman (1968) who argues that the social system is the basis not the result of the ethos of "amoral familism" and that social structural features of a society have their basis in the agricultural system. She compares two areas in central and southern Italy, and attempts to show how the agricultural organization is related to several features of the social structure.

[10] See Goodwin for other explanations of this type concerning peasant uncooperativeness (1979).

9

Dryland Agriculture and Soil Conservation: An Archaeological Study of Check-dam Farming and Wadi Siltation

R. W. Dennell

University of Sheffield
Department of Prehistory and Archaeology
Sheffield S10 2TN, U.K.

Check-dam farming (see the discussion of current strategies in Chapter 8) has received little attention from either historians, or planners. The reason is not difficult to find. Check-dams represent minimal investment and are most important in economically marginal and isolated areas. They were probably, however, one of the earliest forms of land use in this region and since they represent a technology that has survived up to the present, and they involve the redirection of run off and the redistribution of soil, they are likely to have been responsible for a considerable degree of ecological change. This chapter shows how in one case—the Tauran Plain—they have not only changed the topography and drainage, but efficiently husbanded and possibly increased total soil deposits. Check-dam farming, therefore, though it may never have been the only farming strategy in Tauran must have been at least an important supplementary strategy over considerable periods, and responsible for increased population levels and greater selfsufficiency. In the context of this volume the correlation of selfsufficiency and conservation is significant.   —Ed.

Water management plays an important role in the present-day economy of the villages in Tauran. One method that is commonly found in this area is also widespread throughout the desert regions of Iran, and other arid areas such as the Negev (Evenari *et al.*, 1971) and Pakistani Baluchistan (Raikes, 1965). It is known by a variety of names, of which check-dam, flood channel and bund agriculture are probably the commonest. Its underlying principles are very simple: in areas where river flow falls to a minimum during the summer, the length of time that water can be utilized for crop cultivation can be extended by building a crude barrage across the stream channel. As the flood level drops, water behind the check-dam can be diverted through channels on to prepared land. In some cases an important subsidiary purpose is the collection of silt to improve meagre and poor quality soils. The check-dam may be made of wood, rushes, mud or stone, and often has to be rebuilt each year. From the latter part of the 1st millennium BC onwards, more elaborate versions were sometimes constructed of cut stone and mortar across many of the major river channels in Iran (see Iran 1977:3). Check-dams silt up and lead to a superficial change of land form which is not always immediately recognizable as such. Consequently, this type of water management has been frequently misunderstood by archaeologists and hydrologists working in the Middle East. The possibility for confusion is increased by the fact that the same term bund, which is Persian originally and is used in the vernacular from India to Senegal, may also denote other kinds of barrier. Sometimes, the remains of check-dams have been mistaken for dams that were used to store water in permanent reservoirs under moister conditions than the present (Raikes, 1965).

The study of the relationship between this type of water-management and historical and environmental change has presented two major problems. The first is one of dating. The check-dams themselves rarely include material that can be used to date their construction, even if they are sufficiently well-preserved to be studied. Attempts to date them indirectly by reference to sherd scatters found in their vicinity are often suspect, since there is no reason why the production of the sherds should be related to the use, let alone the construction, of the check-dam. This problem has tempted some scholars into circular arguments. For example, Evenari et al. (1971:97) suggested that Middle Bronze Age (MBA) agriculture in the Negev was based upon check-dam farming on the grounds that MBA settlements were near the remains of check-dams; yet these were dated to this period because they were near MBA sherd scatters. In this context, it is noteworthy that Stager (1976), working in the Judaean desert, directly dated one such system to the Iron Age (c. 600 BC) from sherds found in silts trapped behind a check-dam, and saw no reason to assign an earlier date to similar remains in the Negev. We should note that in this study, Stager was considerably aided by the fact that the original check-dams had been built of stone, and were still visible.

The second problem is that the life-history of check-dam systems is far from clear. Little attempt has so far been made to relate them to their ecological and economic contexts by establishing their rate of sedimentation, the length of time that they were used, or their effect upon the landscape. The reasons for this neglect are not difficult to suggest. At first sight, they are usually unpromising topics for detailed study. Often, they have simply been dismissed as "primitive" and thus unworthy of further study, particularly if more grandiose remains exist in the area.

In cases like Tauran, however, their history deserves careful consideration. First, they are essential for year-to-year crop cultivation (see Martin, Chapter 8). Secondly, present-day irrigated agriculture in Tauran occurs on extensive areas of silts, which could not have formed naturally under present or recent climatic conditions, but appear to have been deposited behind check-dams that were built several centuries ago. Detailed investigations of one stream channel showed that up to 6 m of silts were trapped in this way, probably within a period of two to three hundred years. This system subsequently fell out of use, and the silts were eroded. Today, irrigated agriculture is practised either on those silts that are still in place, or on those that have been re-trapped behind modern check-dams. Thus, the present inhabitants of Tauran are adapting their agriculture both to contemporary climatic and economic conditions, and—unknowingly—to an earlier system of land use that extensively modified the landscape. The following investigation of this case of long term agricultural adaptation begins with a discussion of the present potential of the natural resources for crop cultivation in Tauran and the ways in which that potential is being realized and then moves into a reconstruction of selected historical components of the system on the basis of archaeological research.

I. AGRICULTURE IN TAURAN

A. The Potential for Crop Cultivation

If we discuss this issue in terms of the two basic physiological requirements for crop cultivation—water and soil—it is obvious that without the use of technologies for conserving water and accumulating soil the area has an extremely low potential.

1. Water: precipitation and drainage

Detailed climatic data are not available, and the nearest relevant information is from Shahrud and Sabzevar, some 200 and 100 km to the north-west and north-east respectively, and Tabas, 200 km to the south. On the basis of this information (see Iran, 1977:72), rainfall is likely to average 120–150 mm per year, to fluctuate considerably from year to year, and to be negligible during the summer. Average annual totals of precipitation are less important than the intensity, duration and spacing of rainfall. Although quantitative data on these aspects are lacking from Tauran, its rainfall may be further characterized as erratic, intense, ephemeral and localized. Consequently, dry farming is a high risk enterprise, since rain is likely to fall in the wrong place, at the wrong time, too infrequently or too violently to guarantee an economically worthwhile crop. As mean temperatures during the summer average 25°C, and maximum air temperatures can exceed 55°C in the sand (personal observation), evapotranspiration rates are high and emphasize the need for irrigation.

The principal drainage is the Hojjaj river which flows—only after rain— westwards along the southern edge of the sand sea, turns north through its southwestern corner and eventually joins the Kal-i-Shur and the Great Kavir. The Hojjaj is fed by numerous ephemeral tributaries (wadis) which run northwards from Mt. Peighambar through the adjoining foothills and across the Tauran Plain. Gradients vary considerably. On the lower slopes of Mt. Peighambar, they are almost vertical, but they slacken to less than 2% in the northern part of the area.

The deposits associated with modern stream flow without check-dams tend to be poorly-sorted, coarse sands and gravels along most of the channel profile. Peak discharge along the uppermost sections of wadis can be violent and transport boulders up to a metre in diameter. The texture of deposits in wide channels, notably that of the Hojjaj, varies greatly across the bed, and ranges from clay to gravel over a short distance.

Stream behaviour over the last few millennia is likely to have been similar to today's. Sections along the Hojjaj, a large tributary which flows past Barm, and another downstream of Shakhbaz show two super-imposed deposits of gravels. The earlier is very heavily cemented and coarse; those overlying are finer and only slightly cemented. Neither can at present be dated. However, the unsorted nature of these gravels is consistent with the view that past stream flow has been, like the present, erratic, intense and ephemeral.

2. Soil

Soil development in Tauran has been negligible, and the present surface consists of transported material or locally weathered bed-rock. The southern part of the area is defined by the limestone of Mt. Peighambar, c. 2000 m high. Against this, there rests a low range of heavily dissected hills of metamorphic rocks, mainly conglomerates and shales. These hills also dip from the west and so are lower and narrower in the eastern part of Tauran. North of these is a low and generally flat plain of compacted sands and gravels. A section through a qanat (for definition, see above, p. 167) shaft near Kariz indicates that the gravels are at least 15 m thick in the northern part of the plain. This part of the area is now covered by a thin mantle of drifting sand derived from the sand desert that defines the northern border of Tauran.

From an agricultural viewpoint, the sands and gravels of the plain are more suitable for cultivation than the bed-rock of the foot-hills since they offer less resistance to tillage and root development. On the other hand, they contain little in the way of organic nutrients and are very porous.

It is evident from this brief account that Tauran lacks both water-retentive soils and regular supplies of water, particularly during the summer. Check-dam farming presents a satisfactory solution to the former problem by creating pockets of alluvium along stream channels, and partially solves the latter problem by reducing the loss of stream-water.

3. Potential for check-dam technology

The conditions for check-dam farming vary considerably according to the type of wadi and the part of its course. A large stream, such as the Hojjaj, carries a large volume of water at peak discharge and is thus difficult to obstruct. Upstream sections of tributary streams are also unsuitable for damming, since the channel is narrow and steep and peak flow too violent, for check-dam technology or any accumulation of sediment. Most of the stream-load in these conditions is transported rather than suspended, and the resulting sediments are coarse and retain little moisture. Finally, the catchment areas of upstream sections of these tributary wadis are small and do not produce much material, even though rates per unit area may be greater than in large catchments (see Evenari *et al.*, 1971:145–7).

Downstream wadi sections also tend to be unsuitable for flood-channel agriculture. In the first place, because of the width of the channel the dam would have to be impractically long. Secondly, as the stream gradient is slight and the area behind the dam is large, the silts deposited after each flood would be thinly spread, and a long period of sedimentation would be needed before the silts were deep enough to retain sufficient moisture to facilitate crop growth. A third consideration is that much of the sediment would have been deposited before the water reached the dam. Finally, there is the danger that no water would reach the dam, since it may peter out upstream after a minor flood or change its channel after an unusually violent one.

Because of these factors, the mid-stream sections of wadis are most suitable for check-dam technology. In no case does the width of channel necessitate a dam longer than 50 m at most, and its gradients are sufficiently gentle to ensure that only the finer components of the sediment will be transported, yet steep enough that a reasonable depth of sediment will be deposited. The stream course is also strongly influenced by the surrounding topography and the stream flow is thus more predictable. As a final point, a large amount of material is likely to be contributed by run-off from the surrounding slopes.

In Tauran, supplementary irrigation water is in many cases provided from qanats, which ensure a continuous supply; typical deliveries are in the order of one to two cusecs (see Iran, 1977:9). Although they considerably improve crop yields, qanats are not essential to check-dam farming, and many of the systems in Tauran function without a supplement. Unlike check-dams, qanats require specialized skills and considerable outlays of capital for construction and maintenance. They also impose a different series of constraints upon agricultural settlements. The cost of construction is obviously a major factor, and one which rises considerably with length. The location of a qanat exit is also an important consideration, since it may be difficult to make this coincide with areas of good arable soil. Qanat costs also rise considerably if the shafts and tunnel have to be dug through very hard or very soft deposits. In the latter case, the channel has to be lined with baked clay hoops to prevent collapse during construction and use. Check-dams, therefore, offer considerable economic advantages in areas such as Tauran where the size of holding and degree of capitalization and of integration into the larger economy has always been relatively small.

B. Modern Agriculture in Tauran

Crop cultivation in Tauran is now based upon four major technologies which utilize different situations and are marked by differing levels of productivity and labour requirements.

The simplest, and most extensive in terms of acreage, is dry farming. This is undertaken largely on the plain, where the surface is coarse and porous. As might be expected from the discussion so far, this method of farming is hazardous and opportunistic: yields are low, the dangers of failure high, and overall it forms an ancillary role to pastoralism and irrigation agriculture.

The next simplest technology is check-dam farming. Many of the dams in current use are recent and made with the use of bulldozers; some in the northern part of the plain are threatened by drifting sand and may be abandoned in the next few years. In view of the argument in the previous section, it is perhaps surprising that most modern instances of check-dam farming are found in the down-stream rather than mid-stream sections of wadis. The reasons for this, which derive largely from factors of previous land uses, will be presented in greater detail later. Briefly, the sediments trapped behind many modern check-dams on the plain are derived from an earlier system of check-dams further upstream, where considerable volume of silt have been stored over the last millennium.

In the upstream sections of wadis, particularly at Takht, near the foot of Mt. Peighambar, slope-terracing is found, and constitutes the third type of agricultural land system. Material transported down-slope by run-off is

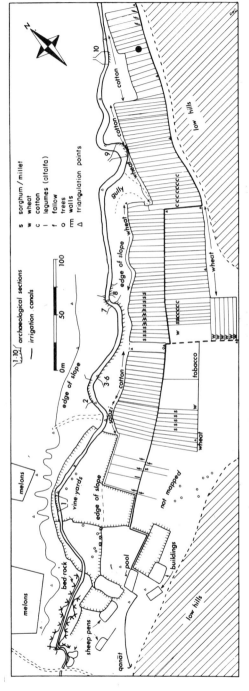

Fig. 1. Modern land use and drainage of the Kalata basin.

retained behind earth, and often stone, walls parallel to the long axis of the wadi. As the stream-bed is narrow and steep, revetements are often needed to prevent the erosion of the banks and fields during peak flooding. This technique of accumulating soil is supplemented by check-dams across part of the wadi channel. At Takht, irrigation water is provided from a large spring at the foot of Mt. Peighambar.

The fourth and most complex agricultural system used in Tauran is qanat irrigation on flat or gently sloping land. A convenient example is the field system at Kalata (Fig. 1): this comprises 15 ha of land which slopes gently from south to north and is divided into over a hundred strip fields, each c. 40 m long and c. 6 m wide, irrigated from a qanat at the south-east end of the field system. In 1977, wheat, cotton and sorghum/millet were the main crops; some tobacco and pulses were also grown. Soil condition was maintained with dung from local sheep and goat pens. Terracing has been undertaken here both to facilitate irrigation and to prevent loss of water through seepage into neighbouring fields or the present stream bed. Similar examples can be found on the west and north side of Baghestan and at or near most villages in the area. This type of agriculture is the most productive of those practised in Tauran today, but also the most labour intensive.

The agricultural technologies currently used in Tauran represent types of farming which become progressively more complex, labour-intensive and costly. The contrasts between dry-farming and qanat irrigation are evident; the former is simple, needs little initial investment, and almost no attention between planting and harvesting, whereas the latter's high productivity is feasible only with considerable initial and recurrent inputs of labour and cost. It is highly improbable that the present population of Tauran could survive in permanent settlements without qanats for irrigation.

In any analysis of the total agricultural system today, it is essential to consider that the areas of flat, easily-tilled and water-retentive soils that are not cultivated and irrigated by qanats were created by an earlier agricultural system based upon check-dam farming along the mid-sections of stream channels. The evidence for this argument can now be considered.

II. THE HISTORY OF CHECK-DAM FARMING IN TAURAN

A. Kalata

We should note at once that, unlike areas such as the Negev or Baluchistan, Tauran has no archaeological example of check-dams. The evidence for earlier check-dam systems is indirect and consists largely of the geomorphological consequences of check-dam farming. In this argument, the sediments at Kalata play a key role.

The Kalata basin lies 200 m east of Baghestan, at the junction of the foot-hills and plain. It is approximately 600 m long, 200 m wide at its northern end and tapers to an apex near the road from Baghestan to Nauva. Its sides are defined by low ridges of conglomerates which enclose an area of ap-proximately 15 ha. The basin is drained by a channel which carries water from Mt. Peighambar towards the Hojjaj. Today, the western part of the basin is uncultivated, apart from two melon fields, and is largely used for grazing flocks from Baghestan. In contrast, the eastern side is irrigated by qanat and intensively cultivated (see Fig. 1). According to local information, some dwellings and a vineyard were constructed about 70 years ago on the south-east part of the basin. Since then and especially over the last 30 years, irrigation farming has been extended northwards. The present stream is eroding land near the western edge of these fields and redepositing the material behind a check-dam at Deh-e Nau, 1 km downstream.

1. The sediments at Kalata

In 1976, when this research was begun, the present stream had down-cut through 6 m of deposits and exposed numerous sections. These sections were investigated in greater detail in the following year and were found to contain three types of deposit.

(a) Uniform silts. These were the most readily apparent type of deposit exposed along the present channel. In view of the present and past stream behaviour in Tauran, it seems most improbable that these silts were deposited under conditions of unimpeded stream flow. Their homogeneity, texture and depth also indicate that they were not formed by the downward transport of material weathered from the surrounding slopes. On the other hand, they are too coarse to have been deposited by aeolian action. Samples of air-borne dust were collected by the writer in the Kalata basin in 1977, and comprised particles which were far smaller than those in the silts.

The most likely explanation is that they represent the deposition of sediment under conditions of gently flowing water as a result of human modification of the stream's behaviour. This could most easily have been effected by the construction of a check-dam across the wadi channel, causing stream-flow to become both less ephemeral and less intense. Under these conditions, silts would have been deposited and could subsequently have been used for cultivation.

(b) Varve-like deposits. One of the most remarkable discoveries in the 1977 field season was a series of varve-like deposits at the base of the sediments cut by the present stream. These deposits were masked by a crust of slope-wash and a thick mat of thorny vegetation, and discovered only after clearing and straightening large sections along the present gully. Altogether, over 70 m of varved sections along 350 m of the present wadi were drawn; their location is shown in Fig. 1.

Although these deposits are not, strictly speaking, varves in the sense that they were formed by the annual deposition of detritus from the melt-water of a glacier, I have used this term (as none other is readily available), in order to emphasize their morphology and sedimentary characteristics. Each is *c*. 2–10 cm thick, and comprises a basal band of sandy silt which then grades into silt and finally a clay skin. An important detail of these varves is that the clay skin was intact and showed no signs of disturbance by cultivation or trampling, nor distortion and fissuring by prolonged exposure to direct sunlight. Most of them taper slightly upstream. This detail may be explained as a result of the gravity flow of material already deposited, and by the fact that the largest amount of suspended sediment would also be deposited downstream where the depth of water was greatest. As Fig. 2 shows, few varves show a perfectly smooth slope, and several are uneven. These irregularities may result from unevennesses in the underlying surface, minor differences in the velocity and turbulence of stream flow and post-depositional compression.

At least 10 and up to 60 varves were noticed in each section. Their length is difficult to estimate, largely because most are cut by recent side gullies. As a result, most of the undisturbed varved sections were only 5–10 m long, although two were traced over 20 m.

Details of the gradient and thickness of varves in each major section are summarized in Figs 3 and 4. Although the varves within each section exhibit considerable variation in respect to their slope and thickness, these variables appear to differ markedly between section. Thus the section furthest upstream (Fig. 5) contains varves which are generally thick and gently sloping; those furthest downstream are thin and steep. Varves in the intermediate sections gradually decrease in thickness as well as slope. An attempt to interpret this pattern in terms of the way sedimentation may have proceeded in the Kalata basin will be made later in this chapter.

The uppermost varves in each section grade gently and imperceptibly into the uniform silts which have already been described. Both ends of varves in four sections also grade into this type of silt, so that varved deposits and the lower part of the silts should be regarded as contemporaneous. In other words, the uniform silts probably represent varves, which were subsequently disturbed by exposure to air and cultivation.

The textural gradation within each varve from silt to clay provides a clear indication that this deposition occurred under gently flowing, or even standing, water. Their alluvial origins are also evidenced by the deposits at the base of the section furthest upstream (Fig. 5). In this section, the basal

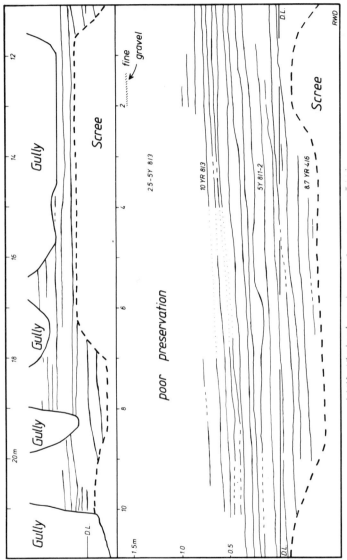

Fig. 2. Mid-stream varved section at Kalata.

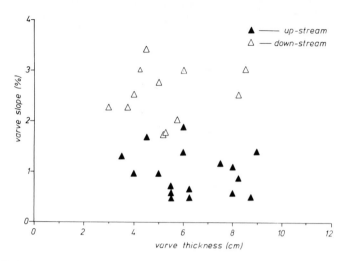

Fig. 3. Relationship between varve thickness and slope in up-stream and down-stream sections at Kalata.

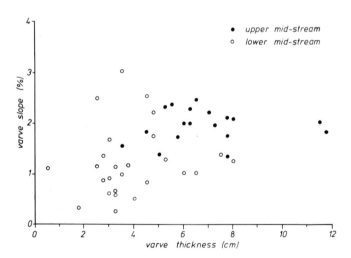

Fig. 4. Relationship between varve thickness and slope in mid-stream sections of Kalata.

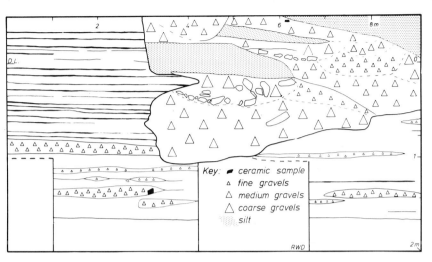

Fig. 5. Up-stream varved section at Kalata.

deposits consisted of bands of gravels alternating with sands and becoming increasingly fine until they merged into varves like those just described.

The dating of the sections by direct stratigraphic evidence presented major difficulties. The fineness of each varve precluded the possibility that large objects such as archaeological artefacts would be transported downstream whilst sedimentation proceeded. If, as suggested, these sediments were deposited in standing, or at most, gently flowing water, it is unlikely that people would throw in diagnostic material for the sake of a future archaeologist. In Tauran today, the writer saw only one object of potential archaeological value which had been thrown into water behind a modern check-dam, and this was an old tin can discarded by children. The problem of obtaining direct dating evidence was solved fortuitously by the observation that the basal varve of the section which lay furthest upstream contained a few flecks of gravel. A small sounding was therefore excavated alongside the section, and further gravel lenses exposed, the fifth and lowest of which was coarser than the others and contained in section an abraded fragment of a baked clay brick (Fig. 5). This was subsequently dated by thermoluminescence to AD 975 ± 120. Whilst one date is scarcely a secure chronological framework, this evidence does suggest that sedimentation is unlikely to have been *before* the 9th–11th centuries AD.

One further discovery associated with these varved sediments is shown in Fig. 6 and consists of a shallow, V-shaped cut, incised into a series of varves. A reasonable interpretation of this feature is that it was an irrigation ditch. It appears to have been recut on several occasions—perhaps as many as ten—so that its base rose in pace with the deposition of varves. Eventually, it filled with a fine, mottled clay and was probably abandoned when it could no longer be kept clear or was no longer required.

(c) Gravels. The uniform silts and varves along the Kalata channel eventually attained a thickness of over 6 m without apparent interruption. At some stage, however, they were down-cut by stream action which also deposited thick bands of coarsely sorted sands and gravels.

The situation is summarized by Fig. 5 which shows a section of evenly-laid varves cut by a thick layer of coarse and unevenly-bedded gravels with lenses of sands and silts. A large composite section over 30 m long and containing similar gravels was found further downstream and could be traced until it cut the long section of varves shown in Fig. 2. There is every reason to regard these gravels as indicative of a return by the stream to a pattern of unhindered, short-lived and violent flow.

2. The sedimentation of the Kalata channel

Although the varved sections at Kalata are readily explained as the result of sedimentation under gently-flowing water, their preservation presents major difficulties. Modern instances of check-dam farming in Tauran provide no immediately apparent and convincing explanation of the phenomenon. Sediments that have been trapped are normally cultivated soon after they have been exposed, and thus their surface is disturbed by tillage, plant growth and trampling. None of the modern check-dams in Tauran observed by the writer retains water throughout the summer, let alone from year to year. Thus in the summer of 1976, only two check-dams—one immediately west of Baghestan, and the other, 1 km north-east of Kalata—retained standing water. In the following year, both were dry: the former was under cultivation, and the latter was unused and its surface had developed into a deeply fissured mud pavement.

We are thus forced to consider how the varves at Kalata remained moist or under water not only throughout the year, but also from year to year. Three explanations can be considered.

The first is that sediments were deposited under water that was deep enough to cover them throughout the summer, even after losses through seepage and evaporation. This explanation seems most improbable. The overall distance covered by the sections is 200 m; and as the mean gradient for the varves in the upstream section is *c*. 1·15 cm per m, the drop in height along a varve of this length would have been 253 cm. In other words, water trapped behind a dam would have had to be at least 2·5 m deep to ensure that the sediments at the end furthest away from the dam were covered with water and did not dry out. Evaporation rates in Tauran are unknown, but likely to be high. In the Negev, they range from 2155 to 4867 mm per year (Evenari *et al.*, 1971:34); similar rates are recorded from other arid regions (Walton, 1969:31). Summer evaporation rates are less amply documented, but likely to constitute half the annual total over a four month period. If the varves at Kalata were formed under a permanent body of water which was not replenished during the summer, it must have been at least 1 and possibly 2·5 m deep over its entire surface. Loss through seepage is difficult to quantify but likely to have been considerable. Because of these factors, the depth of water behind the check-dam would probably have been 2·5 m deep to allow for the gradient, and as much again to compensate for evaporation. As the varved section furthest upstream is 2 m thick, the face of the dam would also have had to be raised by roughly this amount to maintain the depth of water at its previous level.

It seems unlikely that the sediments at Kalata were laid under water that was as much as 5–7 m deep. In the first place, a dam which was ultimately 7 m high would have been substantial and likely to leave some trace, especially if constructed of stone. Secondly, even though an earth dam would leave little evidence, one of this height would encounter serious problems for seepage and settlement (see Hudson, 1969:89–116). In any case, a body of water 7 m deep would have inundated the entire wadi and prevented the cultivation of any crops. Thirdly, the presence of an irrigation ditch in the downstream part of a varved section strongly implies that the depth of water at this point was not great. The argument that the ditch could have been dug and cleaned at the end of the summer, when the water level would have been at its lowest, is easily countered by the fact that there would have been nothing to irrigate at the end of the growing season and shortly before the advent of autumn rains. Moreover, it seems doubtful whether rainfall in Tauran would have provided sufficient run-off from the catchment area at the Kalata basin to result in a body of water of this size. Finally, this explanation might account for the preservation of the varves, but not of the uniform silts, whose featureless appearance is most readily explained as a result of cultivation and prolonged exposure to air.

If the silts at Kalata could not have remained under a continuous body of standing water throughout the year for reasons already presented, we should consider the possibility that they were exposed to air, but remained undisturbed. One model for the preservation of varved deposits might be that they were laid under flood-waters which slowly receded and gradually exposed sediments. These were cultivated and their surface disturbed. Some still remained under water and were not exposed until after crops had been planted. Fissures which developed in their surfaces once exposed to air could then have been refilled with sediment when the process of deposition was repeated.

This hypothesis has further advantages in that it does not necessitate the construction of a high dam, and explains the preservation both of the uniform silts as a result of tillage and exposure to air, and of the varves. It also allows for the annual cultivation of crops on most of the land behind a check-dam. Nevertheless, it still remains unacceptable. Fissures as deep as those which develop in modern sediment surfaces in Tauran would have penetrated several varves and are almost certain to have left some trace. Although almost three-quarters of a kilometre of varved surfaces were drawn by the writer, none showed any evidence of fissuring and subsequent infilling.

As it seems highly unlikely that the varved deposits at Kalata were preserved under a body of standing water, we should examine the possibility that the water in which silts were deposited after run-off and stream-flow was supplemented by additional water from a spring or qanat. This model would envisage the deposition of varved sediments under a shallow body of flood water; as it receded, these deposits would have been cultivated, and their surfaces disturbed. Irrigation water from a spring or qanat would have flowed along the lowest part of the channel and ditches such as the one shown in Fig. 6 and ensured the preservation of each varve.

Springs seem an unlikely way of supplementing the standing water retained behind a check-dam at Kalata. Only two are known in the vicinity today, and both are small. One is located 50 m upstream of the south end of the area shown in Fig. 1, has a negligible flow, and may be simply seepage from a nearby qanat. The other is better described as a permanent puddle and lies in the bed of the present stream a few metres north of the wall enclosing the present vineyard. As the land above this spring is intensely irrigated, it too could represent seepage.

On the other hand, a qanat is a more likely source of supplementary water. They have been used in Iran for at least two and possibly three thousand years, and there are modern examples in the vicinity of Kalata. As the upper sections of the stream-bed drop almost 7 m through bed-rock (see Fig. 10), the most convenient point for the exit of a qanat would have been at the head of the channel. Water could then have fed downstream to irrigate crops growing on silts behind a check-dam.

The deposition and preservation of the varves along the Kalata channel are most plausibly explained as the result of two interacting processes and technologies. The first is a simple, low dam across the channel to trap sediments which could be cultivated after water had receded; the second is a qanat to provide additional irrigation water. The uniform silts thus represent sediments which have been cultivated, whilst the varves would have remained submerged under a long, thin strip of permanent water behind one, or several dams. This model is shown diagrammatically in Fig. 7.

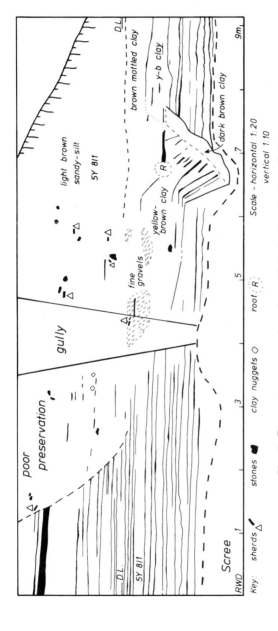

Fig. 6. Down-stream varved section with irrigation ditch at Kalata.

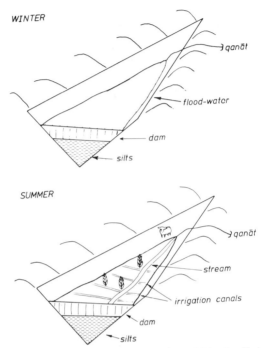

Fig. 7. Suggested model for sedimentation within the Kalata basin.

3. Sedimentation rates within the Kalata basin

The rate at which the Kalata basin filled with silt would have depended upon several factors which fall into three main groups. The first is the run-off rate. This is affected by numerous variables, notably the amount, distribution and intensity of rainfall, and the size, slope and cover of the catchment area. Sediment yield is the second major parameter influencing sediment rates, and this can be viewed as dependent upon run-off rates and the type of bed-rock and soil/surface cover. Thirdly, the sedimentation rate would have depended upon variations of profile within the channel and the sediment yield per year.

None of these variables is documented in detail in Tauran and a detailed model of the sedimentation rate at Kalata is not feasible. Nevertheless, it seems useful to suggest a simple model of the way sedimentation may have proceeded at Kalata, if only to highlight major areas of uncertainty and to suggest the course of future research.

We can begin by assuming that the annual run-off and sediment yield remained constant, and that the Kalata basin can be approximated to the shape and size shown in Fig. 9. It is clear from this diagram that the surface

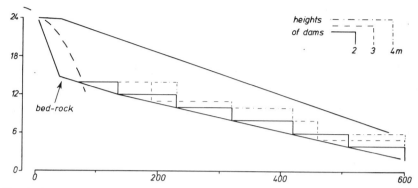

Fig. 8. Suggested numbers and heights of dams needed to cause siltation along the Kalata channel.

area of water—and hence sediment—retained behind a check-dam would greatly increase as siltation proceeded. Thus an area of water whose surface was one metre above the lowest point in the water would enclose only 3125 m², but 50 000 m² when four metres above this point. A tentative estimate can be made of the relationship between the area and thickness of sediment deposited at various stages of in-filling. In simple terms, a given volume of sediment would be deposited over an increasingly large area through time, and become correspondingly thinner. Consequently, the rate of sedimentation would decrease with time. An attempt to quantify this process is shown in Table I, in which it is supposed that a layer of sediment 10 cm thick was laid down when the surface area of the water was 1 m above the base of the channel and covered 3125 m². For the sake of simplicity, it is assumed that the area of sediment would have been the same. The volume of this sediment can be easily calculated as 3125 m³. As the data in Table I indicates, this volume of sediment would result in a layer only 0·6 cm thick by the time its surface had increased to 50 000 m².

This information enables us to calculate a crude retardation factor to express the rate at which sedimentation would have proceeded. As Table I shows, the final rate of sedimentation would have been only one-seventeenth as fast as during initial siltation. In other words, seventeen times as many depositional events would have been needed to produce a given depth of deposit as during the earliest stages of siltation.

If we are to attempt an application of this data to the varves at Kalata, we need to know the thickness of varves laid down during the initial process of sedimentation. As they vary in thickness both within and between sections, only an approximate value of mean thickness can be obtained by estimating

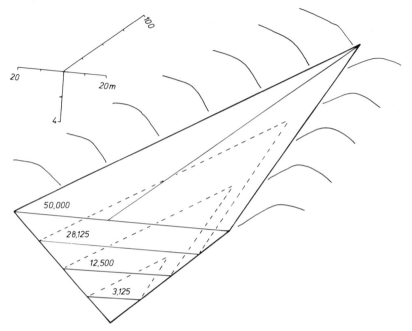

Fig. 9. Suggested dimensions and surface area of Kalata basin at various stages of infilling.

the number of varves within a section one metre thick at various points along the stream channel. This information is contained in Table II. By applying the retardation factors estimated from the data in Table I, we can estimate the numbers of depositional events—represented as varves—required to fill the channel to a depth of 4 m, and the mean thickness of sediments laid down by depositional events at various stages of siltation. The results of these tabulations are given in Tables II and III, and indicate that between 460 and 800 episodes of deposition, each comprising the same volume of sediment, would have been required to effect the infilling of the Kalata basin.

How long need this process have taken? The crux of this question is the time represented by each varve. A reliable answer necessitates the compilation of detailed local climatic and hydrological data on topics such as the amount, frequency and intensity of rainfall, rates of infiltration and run-off on different types of ground cover, slope and catchment area. Until this information is compiled, it is impossible to estimate the numbers of days during the year and from year to year, on which sufficient rain would have fallen to initiate stream-flow and run-off.

Table I. Suggested relationship between varve thickness and surface area in Kalata basin at various stages of infilling. The volume of sediment per depositional event is assumed to remain constant.

Height of sediment surface above channel base (m)	Dimensions of basin			Varve vol. volume (m³)	Varve dimensions	
	Length (m)	Breadth (m)	Area (L × ½B) (m²)		Varve th. thickness (cm)	Retardation factor
1	125	50	3125	312·5	10	1
2	250	100	12 500	312·5	2·5	4
3	375	150	28 125	312·5	1·1	9·1
4	500	200	50 000	312·5	0·6	16·7

Table II. Suggested numbers of depositional events, represented by varves, to result in the infilling of the Kalata basin.

Location of section	Upstream	Upper mid-stream	Lower mid-stream	Downstream	Surface area (m²)	Retardation factor
Observed number of varves per 1 m of deposit	16	15	26	19	3125	1
Suggested number of depositional events						
between 1–2 m above channel base	64	60	104	76	12 500	4
between 2–3 m above channel base	147	137	237	173	28 125	9·1
between 3–4 m above channel base	273	251	435	318	50 000	16·7
Suggested number of depositional events from 0–4 m above base	500	463	802	586		

Table III. Suggested thickness of deposits laid down in each depositional event at various stages during the infilling of the Kalata basin. The volume of sediment per event is assumed to remain constant.

Location of section	Retardation factor	Upstream	Upper mid-stream	Lower mid-stream	Downstream
Number of varves observed between					
0–1 m above channel base		16	15	26	19
Mean thickness (cm)		6·25	6·67	2·60	5·25
Expected mean thickness between					
1–2 m above channel base	(0·25)	1·57	1·67	0·65	1·31
2–3 m above channel base	1/9·1 (0·11)	0·69	0·73	0·29	0·58
3–4 m above channel base	1/16·7 (0·06)	0·37	0·40	0·16	0·31

These factors emphasize the cautious and tentative nature of this attempt to gauge the length of time before sedimentation ceased. For the sake of simplicity and expediency, I have assumed that on average and over a ten-year cycle, water would have flowed along the Kalata channel at least once, and possibly as many as four times, during the year. In other words, each varve is regarded as sediment deposited in at least three, and possibly as many as 12 months. In Table IV, an attempt is made to convert the number of depositional events, estimated from Table II, into years.

These estimates suggest that the basin at Kalata could have been filled with silts after, as a minimum, 115 years, or at most, 800 years. Two factors argue against a slow mean rate of deposition over a long period. First, if rain fell with sufficient intensity and duration to initiate significant run-off and stream-flow only, on average, once a year, then it follows that in some years, no sedimentation took place. If that had been the case, one would expect some varves to show evidence of erosion and possibly fissuring; these phenomena were not observed. Secondly, there appears to be an inverse relationship, in arid areas elsewhere, between the frequency and intensity of rainfall (Evenari *et al.*, 1971:32). If rainfall had been sufficiently intense to cause sediment transport only, on average, once a year, it seems likely that violent flooding would have occurred, and be represented by periodic deposits of coarse material. As these are not found, a more frequent but less intense pattern of rainfall may be implied by the regularity and homogeneity of the varves.

Data from the Negev indicates that on average, there are three days during the year when more than 10 mm of rain falls; at Beersheva (average annual precipitation, *c.* 200 mm), the average number of days with this amount of

Table IV. Suggested number of years represented by the Kalata silts. The expected number of depositional events is based on data in Table II.

Locale of section	Upstream	Upper mid-stream	Lower mid-stream	Downstream
Number of expected depositional events from 0–4m above channel base	500	463	802	586
Length of time in years represented by sediments if each depositional event occurred:				
once a year	500	463	802	586
twice a year	250	232	401	293
three times a year	167	154	267	196
four times a year	125	116	200	147

rain is six. Although extrapolations from one arid region to another are dangerous, an average of three days per year, with sufficient rain to cause run-off and subsequent sedimentation does not seem unreasonable. In this case, the silts at Kalata could represent 150–250 years of deposition (see Table IV).

As a final point, it is worth considering how many catchment dams would have been needed to cause the siltation along the Kalata channel. Modern analogies from Tauran suggest that several dams are likely to have been used at any time, since a series of dams allows siltation to cover a larger area each year than if only one is used. Figure 8 shows a schematic view of the longitudinal profile of the wadi; as can be seen, the stream bed falls 12 m over a horizontal distance of 600 m. In order for the whole of the basin floor to be covered with silt each year, it would have been necessary to build six, four or three dams, each 2, 3 or 4 m high respectively. As siltation proceeded, the dams could have been raised or built elsewhere.

As emphasized above, this model is highly simplistic, and intended only to provide a rough approximation of the time and methods needed to fill the wadi with silt. Its limitations are obvious, and underline the need for detailed data on present and past climatic and hydrological conditions in the area. Some of these factors can be discussed at greater length.

First, the annual sediment yield is likely to have fluctuated considerably from year to year, and possibly on a long term basis. Each varved section showed considerable variation in varve thickness which is unlikely to have been caused entirely by irregularities in the channel profile. On a short-term basis, these fluctuations are an indirect reflection of the duration and intensity of the rain which results in run-off and stream-flow. Moore's dendrochronological studies of *Zygophyllum* (Iran, 1977:67–69) provides some indication that rainfall in Tauran this century has been subject to short-term deviations from the mean on a three- to five-year basis. As there is no reason at present to postulate a significantly different pattern of precipitation a thousand years ago, sediment yields at Kalata could also have varied considerably over five year periods. On a longer-term basis, human modification of the slope-cover within the catchment area could also have affected erosion rates. Increases of, for example, grazing pressure on the slopes around the Kalata basin might have increased the rate of surface erosion by trampling and reducing the effect of vegetation in conserving soil cover. Sedimentation rates might not, therefore, have decelerated as much as indicated by Tables II and III from time to time.

The amount of run-off at Kalata is obviously one of the most important factors affecting the siltation of the wadi, but one that is at present totally unknown. As Cooke and Warren (1973:19–20) point out, "data on desert run-off are even more scarce than data on climatic conditions; and it is difficult to estimate run-off from climatic records because it is a precipitation residual, affected by rates of precipitation, infiltration capacities, water deficiencies in surface material, and local topography, all of which are usually unknown variables." As a result, extrapolation of run-off rates from other dryland regions is rash. For example, data from the Negev (Evenari *et al.*, 1971) indicate predictable run-off rates if the annual rainfall, and size, slope and surface cover of the catchment area are known. Recourse to this type of data would suggest that annual run-off at Kalata, enclosing *c.* 15 ha of uncleared slopes with an average gradient of 5–10% would be *c.* 140 m³ ha if annual rainfall averaged 130 mm. However, run-off rates in the Negev are strongly influenced by the prevalence of loessic soils which form an impervious crust after showers, and thus little water is lost through infiltration. In Tauran, loessic soils are absent and much rain-water is probably absorbed into the ground surface. Data from the U.S. Soil Conservation Service suggests that only 75 mm of run-off would result from 125 mm of rain if the ground surface had medium permeability and storms occurred 2–4 days apart (see Hudson, 1969:47–8): however, these are crude estimates that exclude local factors of topography and climate. Raikes (1965) estimated the total discharge from a catchment area in Baluchistan that cover 2 square miles and was associated with a bund system as 60 acre/ft. This amount can be converted to 14·2 cubic metres per hectare. As a comparison, a catchment area in the Negev enclosing 345 ha produced an average annual run off of 24 cubic metres per hectare (Evenari *et al.*, 1971:145); the difference between these two estimates may reflect the low infiltration capacity of the Negev ground surface.

4. *The erosion of the silts at Kalata*

Although evidence for the erosion of the silts in the Kalata basin is clear and widespread, little can be said about its date and causes until more evidence is available. In particular, ceramic samples from the top of the silts and from the gravels cutting them should be dated by typological and chronometric methods.

If we accept for the present that sedimentation had begun by the 10th century AD and continued for at least two centuries, erosion is unlikely to have taken place before the 13th century. According to local information, it had ceased when the vineyard was constructed on the east slopes of the southern end of the basin in the early years of this century.

The reasons why the silts were eroded and the check-dams abandoned are likely to have been complex, and may have involved a combination of geomorphological, climatic, social and economic factors which operated both within and outside Tauran. It would be exceedingly rash to adduce a single cause for this erosion at the expense of other factors. Even if the down-cutting of the silts were found to be contemporaneous with the Mongol invasions, it would be incautious to attribute it solely to them. Local and less dramatic factors may have been responsible. One is that the process of sedimentation became increasingly unstable; in other words, the control of flood-water and the maintenance of check-dams may have become increasingly difficult. A suggestion that this may have happened is provided by the gradients of the original channel and its silts. As Fig. 10 shows, the original gradient along most of the wadi bed is likely to have been, as today, $c.$ 2%; by the time sedimentation was complete, the gradient along the top of the silts had risen by two-thirds to 3·3%. Consequently, flood waters were flowing along steeper gradients than before and across areas of easily eroded silts. At some point long-established methods of water control and soil conservation proved ineffective, and led to the collapse and abandonment of this agricultural system. Exogenous environmental and socio-economic factors may therefore have played only an ancillary role.

B. The Significance of Sedimentation Elsewhere in Tauran

Kalata is the only stream system in Tauran which has been extensively examined from an archaeological-geomorphological viewpoint so far. However, similar indications of wadi siltation were observed by the writer elsewhere in Tauran in 1976–77. South of Shakhbaz, a small wadi contained almost two metres of sands and silts which overlay gravels and were cut by the present channel into which it feeds. The area of silts is about half a hectare and does not lie near any obvious source of spring-water or a qanat exit. In view of these factors, it may have been used for cultivating trees rather than crops (S. Breckle, personal communication). More dramatic evidence of siltation can be seen to the south and west of Baghestan. Between the village and its qanat, there is a section 8 m deep where the villagers quarry to make mud-brick. Conglomerate bed-rock is exposed on the east and west sides of this cut. In 1976, it was noticed that a silt-filled channel was exposed in this section and had cut into underlying silts. Varves, similar to those found at Kalata, were seen at the base of the lower silts in the following year (see Fig. 11). The original channel downstream and to the north of this section is currently occupied by an extensive system of strip fields and gardens that are irrigated by qanat. Traces of the original sediment could be seen against conglomerate bluffs bordering these fields. To the south of Baghestan and upstream of this channel, silts attain a considerable thickness. In 1977, the writer examined the qanat shafts south of Baghestan with the aid of climbing tackle, and found that they had been cut through up to 10 m of uniform, light brown silts similar to those at Kalata. One shaft had been cut through loosely compacted water-lain sands and gravels.

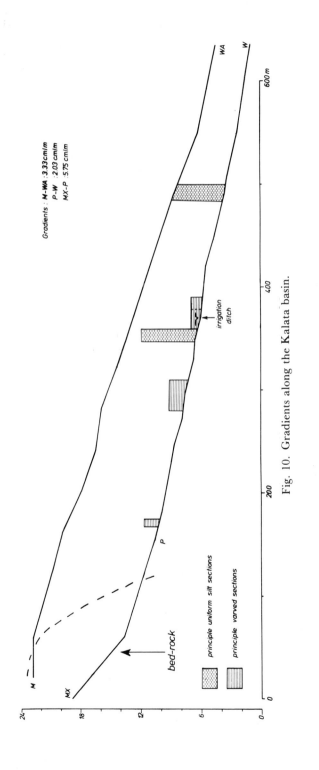

Fig. 10. Gradients along the Kalata basin.

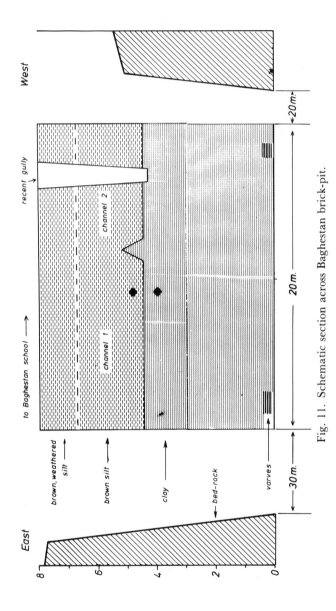

Fig. 11. Schematic section across Baghestan brick-pit.

III. CONCLUSIONS

The fieldwork discussed in this chapter was designed for the investigation of two problems. The first one was archaeological: an elucidation of the history and development of check-dam farming systems in dryland regions such as Tauran. The second, more directly relevant to contemporary dryland research, was the relationship of past patterns of land use to present-day agricultural systems in Tauran. Although the data and interpretation in this chapter are only preliminary, some conclusions can be reached.

The historical study of check-dam farming is not as intractable or profitless as has sometimes been supposed when—as in Tauran—there is no direct historical or archaeological evidence. Indeed, too much attention has been paid in the past to direct evidence in the form of remains of dams, field boundaries and channels and not enough to indirect evidence discernible with the aid of hypotheses about their environmental consequences and life-histories. If we treat the fieldwork at Kalata as a methodological exercise, it is apparent that much can be learned from what initially seemed an unpromising 100 000 cubic metres of silts. Given sufficient patience and favourable conditions of preservation, it is possible to model the historical development of check-dam systems in some detail. In the case of Kalata, the estimates given in this chapter could be checked by dating the sixty-odd samples that were collected from the uppermost silts; the date at which erosion was initiated could also be established in the same way by dating sherds contained in the gravels that cut the silts. Once time-scales are allocated to the process involved, they could be compared with modern analogous conditions of run-off and siltation.

This type of investigation underlines the necessity of an integrated, inter-disciplinary approach that utilizes and shares the expertise of several natural and social sciences. The sediments, and especially the varves, at Kalata represent a unique record not only of a past agricultural system but of its environmental context. Much could be learnt if the varved sediments prove to contain pollen, diatoms or other organic remains. They might also contain aeolian dust which could be used to monitor past fluctuations in sand mobility in the sand sea, 5 km to the north. This study was severely hampered by the absence of comparative data on present climatic and hydrological conditions in Tauran, the type and erosive rate of its surface cover, and its geo-morphology. On the other hand, a study of the modern hydrology of Tauran could easily reach misleading conclusions if the fact was overlooked that few of the stream channels had been modified by siltation.

So far as contemporary dryland studies are concerned, the investigation in and around Kalata would suggest that past farming systems have had profound effects upon the present-day agriculture of the area. Indeed, the check-dam and qanat systems used today have been as much adapted to previous agricultural usage of the area as to prevailing climatic and economic conditions. Modern qanat irrigation is primarily concerned with distributing water on to arable land; the evidence from Kalata and Baghestan indicates that most of this arable land was created by a previous system of flood-channel farming. Much of the sediment behind many of the check-dams now in use is redeposited material that had previously been trapped behind similar dams further upstream. The overall effect of these activities in the past has been to store large amounts of sediments which would otherwise have been lost, and to convert them into areas of high productivity. In view of many recent attempts to raise the agricultural productivity of drylands by using complex and expensive technologies from industrialized countries, it is also worth pointing out that in Tauran the benefits from the previous use of check-dams have lasted for several centuries, and were achieved by using an extremely simple technology.

10
The Demand for Fuel:
Ecological Implications of Socio-Economic Change

Lee Horne

University of Pennsylvania
Department of Anthropology
Philadelphia, PA 19104, U.S.A.

Blame for desertification is too often cast not only on non-industrialized populations but on traditional technologies of production, without careful analysis of the total range of factors involved. This chapter picks up the topic of one of the sections of Chapter 8 and pursues it further. It is devoted to the problem of fuel needs and supplies—well known to have a serious environmental impact but rarely studied—and shows that the degree of impact, as in the case of so many other factors, is a function of the larger economy. —Ed.

Dryland ecosystems are characterized by sparse vegetation with a low and widely variable annual productivity. Nevertheless, throughout the Middle East this sparse vegetation is an essential resource upon which depend the livelihood and very biological survival of the local populations. Pastoralism is a major means by which this natural resource is exploited and has been the focus of much technological and anthropological research. An equally essential aspect of the relationship between settlement and environment is the collection of woody vegetation for fuel—equally essential to the populations involved but, in the Middle East at least, hardly studied at all. This chapter explores the requirements and consequences of energy supply and consumption in Khar and Tauran and aims to show that behind average annual fuel consumption figures lies a multiplicity of factors influencing levels of demand and intensities of exploitation.

I. THE CONTEXT OF FUEL USE IN KHAR
AND TAURAN[1]

Although this study focuses on the settlements of the Tauran Plain, continuous reference is made to the entire Khar and Tauran area (about 3300 km²) which is the more relevant territorial unit for the study of the production and use of fuel.

About half the population of 2500 live in mud-brick villages on the Tauran Plain, and about a quarter live in Khar in three villages on the western side of the sand sea. Most of the rest are scattered to the east and south of the central plain. In 1966, the most recent year for which complete census data are available, there were 15 year-round settlements on the Tauran Plain, including the village of Baghestan where most of the data presented in this chapter were gathered. These villages ranged in size from 11 to 178 persons (mean 80·4) and all but one were less than 2 ha in extent. Mean household size was 3·9 persons. The Plain as a whole held 10 persons per square kilometre; 1206 persons living in 307 households. The population has increased since 1966 but probably to no more than 1350 (11 persons per square kilometre) at most.

Several villages maintain summer milking stations on the lower parts of the Plain where pasture and firewood are most plentiful. There is only a single winter station on the Plain: during the colder months flocks are either kept at the villages or taken away from the Plain entirely. In summer at least 20 and perhaps 30 households move to summer stations on the Plain and the population becomes more dispersed. Those who leave the Plain entirely move 20 km or more away to where they have rented or bought rights to rangeland which is more productive than the Plain. All these families have been counted in with their home villages in the figures given above.

In addition to those on the central Plain, there are three relatively large villages (mean size 228 persons) in Khar to the west of the sand sea, and a scatter of small outlying settlements throughout the rest of the district. In 1966 outliers ranged between 1 and 17 households with a maximum of 82 persons at any single one, and with a mean size of 31·1 persons. These villages are proportionately smaller in area than the central ones. Khar and Tauran as a whole, including the outliers, contained 0·8 persons per square kilometre in 1966 (2530 persons over 3300 km²).

Most of the impact on the environment around these rural settlements comes not from the amount or type of land given over to residential occupation or irrigated fields but from the use of the natural vegetation as a productive resource.[2] On the Plain, for example, no more than 25 ha are covered by the structures and spaces of the villages. If abandoned sites with standing ruins and the few seasonal shelters and stations are included, then perhaps another 5 to 10 ha should be added to this figure. The total is no more than 0·3% of the Plain's area. Irrigated fields account for more: about 250 ha, which still is only 2% of the Plain. Even dry farming, the most extensive form of agriculture, probably never exceeds 8% of the Plain's surface (although at one time or another a much higher percentage appears to have been ploughed and sown). Settlements and field systems off the Plain are ever sparser. Thus, even though residential land and fields are completely cleared of natural vegetation, they comprise relatively little of the area. That leaves well over 90% of all land in Khar and Tauran available for exploitation as pasture, fuel and construction material. The productivity of this resource is the concern of the research reported here.

II. FUEL USE ON THE TAURAN PLAIN[3]

Firewood, paraffin and farming by-products (such as dung and cotton stalks) are all used for fuel in Tauran. Of these, firewood is the most important in terms of the total number of calories. According to estimates made during the period from 1975 to 1978, firewood is burned at an annual rate of 5·3 tons per household. Paraffin is of relatively recent introduction. Its use is widespread and significant, but limited to certain activities—meal preparation, space heating and lighting in particular. Dung and agricultural by-products are used only incidentally for fuel; dung is used primarily to fertilize and condition irrigated fields. This section details the use of these three fuels in Tauran today as a background for assessing the impact of the local vegetation and as a baseline for understanding how both energy demands and environmental impact have fluctuated in the past.

Fuel is burned in Tauran today for two main purposes; winter heating and food preparation, whether year round cooking or seasonal processing. It is also used in lesser quantities to heat water and to burn gypsum plaster. The equipment used varies with the activity and the fuel. Wood-burning space heaters (*bokhāri farangi*) are bought in the city and are made from metal drums, with doors cut into the sides for loading and draught. They are set in the centre of the living room and plastered to the floor, usually in conjunction with a shallow hearth for coals from the heater which are used for cooking and tea-making in the winter. In the summer they are often removed and stored in another room. A narrow stovepipe vents the smoke through a hole in the domed roof. Only the main living-sleeping room is heated, a space about 35 to 40 m³ on the average. *Korsi*, frames supporting an arrangement of quilts heated from underneath by charcoal or dung, are common in colder areas of western and northern Iran, but are not found in Tauran. Almost no one today uses wall fire-places (*bokhāri geli*) though some can still be seen in older houses.

Paraffin camp stoves are also used for heating, but supplement rather than replace wood stoves. Milking stations occupied only in the summer need no provision for heating; winter stations rely on interior multi-purpose open hearths which burn firewood—they are too isolated to depend on paraffin supplies.

Although clothes are sometimes washed in the cold water of an irrigation channel or storage pond (the villages have no piped water), bathing and clothes washing are often carried out at the same time, and large cauldrons of water are set up over brushwood fires at the water sources. Hot water is dipped out and poured over the body, after which the clothes may be soaked directly in the cauldron or scrubbed in pans. In the village of Baghestan, only two courtyards had private showers; the water was heated from a recess on the outside of the shower building in one case and from a hearth on the roof in the other. Water was hauled up to storage drums on the shower roofs and piped down by gravity into the shower room. The traditional village bath house (*hammām*) in the area is a communal structure, semi-subterranean and of the style sometimes called Turkish, with seating platforms and niches, arched vaults and several chambers of water. Ordinarily it is used only in the winter, when men are hired to collect firewood. The one in Baghestan has fallen into disrepair in anticipation of a modern, government subsidized bath house already under construction in 1978; it will use water lifted from the irrigation channel by a diesel pump and heated by paraffin.

The other major fuel need is for cooking, both for immediate consumption and for storage. Storable products are important in this climate because of the pronounced seasonality of crops and milk and because of the lack of refrigeration facilities. In order to carry almost any food except grain through the winter some kind of drying (often aided by heating) or cooking is necessary. In this way surpluses beyond what is eaten in season can be used to supplement wheat products during the rest of the year. Furthermore, preserving food in transportable, imperishable form facilitates exchange with other communities or sale of surplus in distant markets.

Cooking with firewood is done mainly over open hearths (*kalgāh*) and in bread ovens (*tanur*). Outdoor hearths and bread ovens are mud structures built in unroofed courtyards, sometimes hooded by a small domed roof. Hearths are variable in diameter and tailored to the size of the copper pots used. In summer stations these pots hold 120 to 180 kg of milk, and are plastered to their hearths, increasing the efficiency of the heating process. In villages and at summer station hearths where ordinary meals are cooked, hearths are usually built side by side in pairs, one large and one small. Cooking involves frying, boiling and stewing; roasting is never done, even for community feasts.

Whilst most villagers can build hearths, bread ovens are more complex and require the aid of part-time specialists. Not every village has an oven-builder, but one can be brought in from nearby. In villages bread ovens are above-ground structures about a metre high, with enough room for the baker to squat on top and reach in to slap flat sheets of bread dough against the hot walls of the clay chamber, which is lined with a fine local clay of ceramic quality. There were 6 to 8 ovens for the 33 households in Baghestan; each housewife would probably like to have her own, but they are not high on the list of domestic priorities. Sharing in fact saves fuel since women try to bake while the oven is still hot from the previous baker's firing. There is no particular pattern to the social relationships in sharing, though usually the women are close through kinship or marriage. In summer stations, bread ovens are usually sunk into the ground, simplifying construction where specialists are not available.

Grape syrup (*shira*) for domestic consumption and for sale is processed from surplus grapes in the autumn. Special oversize hearths are built in the grape gardens to boil down the grape juice—a process taking hours of heating. Not every household grows enough grapes to make this syrup, nor can it be made when the grape harvest fails. With improving transportation facilities to the city, it is likely that surplus grapes will be sold directly rather than being processed into syrup. Tomatoes are preserved by boiling down into a paste (*robb*).

Firewood collection is primarily a domestic, non-commercial activity carried out at irregular intervals without records of any kind. Observed and reported collecting activities were counted in terms of donkey loads, but were not actually weighed. Informants claimed that an average load of dry firewood weighs 75 kg (25 *man* in the local system of weights) but they do not ordinarily weigh them. They also say a donkey load of grain in sacks weighs between 75 and 90 kg, and unlike firewood, grain *is* weighed out on standard steelyard scales. Bundles of firewood are unwieldy and less compact than sacks of grain and probably do weigh somewhat less per load. A figure of 75 kg was therefore used in these calculations. In somewhat the same fashion a camel load was figured at 200 kg. Both figures are within the range given for these animals in other sources, most of which are as much estimates as ours.

Brushwood for fuel is, whenever possible, collected by men: women who have no adult male household members in residence hire men to collect for them. The round trip may take up to half a day and requires travelling several kilometres outside the settled areas. Since a household averages six collection trips a month (less in summer, more in winter) roughly 36 days per year are required to collect firewood. Usually, whole bushes are uprooted with an *adze*, bound into large bundles and balanced on the back of a donkey for the return trip. Women carry firewood on their heads. Generally, people collect wood where they, their kin or their village have grazing rights. There are no other locally imposed management practices.

Firewood is normally stored in roofed storerooms or unused stables, protected from rain, animal browsing and, probably, official eyes, since much of the firewood is green and therefore illegally collected. Usually no more than a donkey load is kept on hand at any one time—two bundles roughly 1 to 1·5 m³ each. At milking stations firewood is used as fast as it is collected and women help with the collecting.

Firewood is collected without cost except for the equipment used (an adze, rope, and a donkey). Paraffin, however, must be bought from the local distribution centre in the village of Zamanabad. From Baghestan the trip by donkey is only 3 km and takes an hour there and back, but for villages at the back of the plain paraffin is much less convenient to obtain. In 1976–78 paraffin cost 2·5 rials a litre ($US.035) in Zamanabad.

The usual social unit for collecting and using firewood is the household, which in Tauran corresponds to a nuclear family, sometimes extended by one or more unmarried adult relatives. As mentioned above, firewood consumption for an average village household is estimated at 5·3 t per year. In Baghestan (1976–78) average household size was 5·1 persons; the 1966 Census figure for the entire Plain was 3·9 persons per household. Projecting Baghestan's mean annual firewood consumption rate per household to the entire Plain (310 households) produces an estimate of 1650 metric tons per year (dry weight). Per capita figures, sometimes needed for comparative purposes, would be one ton per person if the figure of 5·1 persons per household is used, and 1·4 t per person if the figure of 3·9 persons per household is used. Using household rather than individual statistics, however, is preferable and more directly representative of actual behaviour. Furthermore, the way firewood is used in Baghestan and elsewhere (see for example Earl, 1975; Fleuret and Fleuret, 1978) suggests that in villages, economies of scale operate at the household level. That is, in similar situations, fuel use varies less between households of different sizes than average per capita figures would suggest.

Because pastoral products (such as yoghurt, clarified butter, and concentrated buttermilk) require so much heating and boiling, summer milking stations use significantly greater quantities of firewood per household than do villages. The milking station included in the sample had about 250 milkers (sheep and goats) and used 21 t of firewood per summer. The residential population was composed of members from several Baghestan households who pooled their animals. Taking 2½ as the average number of households at that station, consumption was about 8·4 t per household for those five months of spring and summer. When winter firewood use back in the village is included, each of these transhumant households used 12·1 t per year—a total of 6·8 t more per year than for a sedentary household from the same village. If 25 households spend their summers at milking stations on the Plain, and if they use similar quantities, then 170 t of firewood should be added to the total given above for the entire Tauran Plain.

There are several alternatives to firewood. As mentioned above, dung is sometimes used, but its primary value is as fertilizer or in construction. Other agricultural by-products are used in minor quantities. For the last twenty years or so, however, paraffin has been an important substitute for firewood in the area. An average household in Baghestan uses 500 to 800 kg per year, depending on the severity of the winter and the need for space heating. Other things being equal, households would burn 25% to 40% more firewood to replace the calories now supplied by paraffin. Of course, some of this paraffin is used for lighting, replacing lamp oil rather than firewood.

Diesel fuel is used in the one modern bath house and several mills, and petrol for an increasing number of motorcycles and other vehicles. Except for the bath house supply, these uses do not replace traditional fuels and were not included in this study.

III. THE ECOLOGICAL IMPACT OF FIREWOOD CUTTING IN TAURAN

A description and preliminary mapping of plant species and communities in the Turan Biosphere Reserve appears in the case study report of the Turan Programme (Iran, 1977). Species observed or reported to be used as fuel include *Amygdalus* spp., *Artemisia herba-alba*, *Astragalus squarrosus*, *Calligonum* spp., *Ceratoides latens*, *Convolvulus*, *Cousinia* spp., *Ferula foetida*, *Goebelia pachycarpa*, *Haloxylon* spp., *Lactuca orientalis*, *Lycium depressum*, *Salsola* spp., and *Zygophyllum eurypterum*. Most of these do not die back in the winter, but need their above-ground parts for viability and production. Some of them, especially *Artemisia*, *Amygdalus*, *Calligonum*, *Lactuca* and *Salsola*, are also eaten by goats or sheep. (The degree of competition for vegetation between people and animals is considered in Spooner *et al.*, 1980.)

Measured biomass and annual production figures are not available for the areas which provide firewood to Khar and Tauran, but some estimates based on harvesting and weighing were made for two tracts near Delbar, 75 km to the west. In one, *Zygophyllum eurypterum* shrubs were estimated at 2·6 t per ha (above ground, air dry biomass); in the other, *Artemisia herba-alba* shrubs were calculated at 1·17 t per ha (Moore and Bhadresa, Chapter 13). Different kinds of *Zygophyllum* and *Artemisia* communities are widespread throughout Khar and Tauran, constituting perhaps one half the total area. Visual estimates for *all* major communities in Khar and Tauran, including these ranged from 1·0 to 6·0 t per ha, *Zygophyllum–Salsola* communities averaged 3·0 t per ha and *Artemisia–Amygdalus* communities 1·6 t per ha (Breckle, Chapter 14). Coupled with extrapolations based on plant cover and densities in transects run on the Tauran Plain itself (Nyerges, 1980), these figures suggest an average biomass of 2 to 3 t per ha for plains and stable sand vegetation, with the higher weights found in the sand. Variation is high, however, and some areas are virtually devoid of vegetation, while others may have biomass as high as 8·0 t per ha (Helmut Freitag, personal communication). Comparative estimates for similar conditions elsewhere in the Middle East and Afghanistan also support a range of 2 to 3 t per ha.

Productivity is much more difficult to estimate. For these shrubby communities, average annual production is likely to be on the order of 200 to 800 kg per ha, based on average annual rainfall and comparative data where productivity was measured directly (for example, McArthur and Harrington, 1978; Thalen, 1979; and Casimir *et al.*, 1980).

At present, firewood for the Tauran Plain is collected from an area somewhere between 120 km² (roughly coincident with the Plain itself) and 200 km², based on collection and travel times and *in situ* observations of cutting. Although nearly all firewood is collected from within this area, cutting is not uniformly distributed. Most collecting takes place at the edges of and beyond the Plain itself for two reasons: (1) the Plain, especially the central zone, offers very little brush suitable for firewood and (2) the risk of discovery and confiscation is far greater next to settlements. During periods in which control is enforced and paraffin is available for cooking and tea-making, collection takes place mostly in a fan-shaped area including moving and stable sand that lies to the north of the Plain, where shrubs are plentiful and the chances of being caught slight. Only when paraffin supplies run out at the local distribution centre is it worth collecting on the Plain proper. Then, however, the entire area is scoured for whatever will burn.

The villages on the plain annually remove 1820 t of brush over these 200 km², or 91 kg per ha. Even though per capita use of fuel is highest at summer stations, those settlements are much more dispersed, and the number of households at each one much less than at the central villages so that the intensity of collection is correspondingly less. Winter sheep pens use the least amount of fuel per capita and are suitably dispersed rather than clustered. The villages on the Plain put the heaviest burden on plant productivity.

If these 200 square kilometres have an average above-ground biomass of 2·5 t per ha, roughly 4% of all plant material is cut annually for fuel. Or, to put it another way, a biomass equivalent to that of 723 ha is removed each year. If shrubs of the type used for fuel take five to ten years to grow to suitable size for cutting (Thalen, 1979:274) removing 4% of the biomass annually might be tolerable. But several qualifying factors must be taken into consideration:

(1) For domestic consumption, brush is not uniformly or randomly cut over the 200 km² from which it is collected. The area nearest settlement tends to be totally stripped of suitable shrubs, the farther reaches more lightly used. For a very small settlement, less than 100 ha might be severely affected

(probably the case at small summer stations) and recovery is still possible should the settlement be abandoned or moved. For an area the size of the centre of the Tauran Plain the consequences for regeneration are of a much more serious scale.

(2) Plants are differentially selected. Where there is a choice, some shrubs are preferred over others for their burning qualities, odour, smokelessness and so forth. Certain species may become scarce or absent because of their preferred status as fuel. In any case, even where the biomass is mostly shrubby, it is not entirely so, and the rate of removal of shrubs alone will be higher than 4%.

(3) Firewood cutting removes the vegetation in a particularly destructive way by uprooting rather than lopping, thereby effectively preventing regeneration (McArthur and Harrington, 1978:598; Thalen, 1979).

(4) Firewood cutting takes place in concert with other activities which also affect the vegetation, the most important one in Tauran being sheep and goat herding. Because fuel cutting and browsing exploit different elements of the plant community in different ways. it is difficult to assess their cumulative effects.

In the absence of detailed field studies designed to isolate the effects of firewood collection from other types of land use in Khar and Tauran, it is difficult to gauge its overall impact on the environment. There are no rules of thumb establishing tolerable shrub cutting rates the way there are for tree plantations or pastoral stocking capacities. Certainly a marked zone of deterioration from a combination of cutting and browsing in the immediate vicinity of settlement is obvious. From a local point of view, however, the situation is not as bad as it might be. Villagers are not forced to depend on dung for fuel but continue to manure their fields in order to maintain and improve yields. They do not travel long distances to collect wood nor must they buy it from specialists. Moreover, they claim the range at the edge of the Plain has actually improved since the prohibition of charcoal production in the 1960s. Nevertheless both they and outside observers know the range is not as good as it should be. With a growing population the potential for increasing rates of deterioration of vegetation and soils also grows. Because multiple factors, of which population size is only one, determine how much and what kind of fuel will be used at any particular time, the following section considers fluctuations in rural energy demands.

IV. HOW RURAL ENERGY DEMANDS VARY

Although social differentiation in terms of wealth or status is not great in Tauran, fuel consumption does vary from household to household and from village to village. For example, some households own more sheep and goats than others and have greater quantities of milk products to process. Some have larger grape gardens and process more syrup. The wealthier or more religious are more likely to give community feasts. The wealthier can better afford paraffin and paraffin-burning equipment. Moreover the location and degree of winter insulation (such as wall thickness, tightness of doors and windows, or orientation of houses and settlements) affects fuel consumption for winter heating.

In addition, yearly fluctuations occur in both weather and market conditions which in turn affect local economic strategies and therefore the total quantity of fuel used in Tauran in any particular year. Such fluctuations have so far been observed for only a very short period of time, too short a time to assess how they might affect the vegetation.

Variation over the long term and major changes in the way fuel is used are best studied from historical (both oral and written) and archaeological evidence. In the past, for example, fossil fuels were unavailable and local vegetation was the only fuel source used. Moreover, local industries existed which have since disappeared. Like other productive activity in the area, these industries were dependent on the vegetation, either for fuel (brick baking, pottery firing and copper smelting) or as a resource to be processed (charcoal production). There is no local memory of any of the former activities; they are known from their archaeological remains alone. Charcoal production, on the other hand, was still widely practised until its prohibition by the central government in 1966. It provides an excellent example of a traditional productive activity from the past whose effects are still apparent today.

Just before its prohibition, charcoal production had become so widespread in Khar and Tauran that according to local accounts, "there were pits every ten metres" and the sands were "ablaze" with charcoal burners' fires. How long such intense activity had been going on is uncertain. Gabriel (1935) noted that in the early 1930s charcoal and pastoral products were traded by Tauran villagers for city goods such as tea, sugar and cloth. He either saw or heard of charcoal being made at Torud, Khar, Tauran and Sanjari—that is, from one end of the local area to the other. During our fieldwork, one villager, now in his sixties, said that in his father's time there had been little burning. As he grew up, however, production increased until 100 to 150 camel loads a day (possibly as much as 30 t) came out of the area. It does seem that such intense production would be self-limiting, in that production at the rates reported for that time could not be sustained for any appreciable length of time without seriously reducing if not completely destroying the preferred woody resources: wild almond, saxaul, pistachio and *Calligonum*.

If charcoal had been used for domestic purposes only, production could be estimated from population size and postulated fuel needs. Since it was essentially a commercial product, however, intended for shipment out of the area rather than for local consumption, it may never be possible to measure how much was produced over any particular period of time. Charcoal pits are not very visible archaeologically because they tend to be located away from settlements, are not associated with particular features such as water or terrain, and may today lie under blown sand. Only a few were actually discovered during the study period, and none were excavated.

A more feasible approach to the impact on the environment and the scale of destruction to the vegetation is through the technology of production. Traditional methods of producing charcoal in Iran include both above-ground stacking or kilns typical of the forested regions of the Caspian and Northern mountains, and underground pits typical of the more arid parts of the country such as the areas around Shiraz (Uhart, 1952), Tabas (Mojtahedi, 1955) and Tauran. The underground pits recorded by Uhart near Shiraz were small (1 to 1·5 m³) and produced only 60 kg of charcoal. In Tabas, 200 km to the south of Tauran, Mojtahedi recorded pits that were larger (up to 10 m³) and that yielded 500 to 700 kg of charcoal, closer to the size of those described to us in Tauran.

In Khar and Tauran wild almond (*Amygdalus*), saxaul (*Haloxylon*), pistachio (*Pistacia*) and *Calligonum* were preferred species, supplemented by several kinds of brushwood. Pits were located in relation to these resources. A basic charge at the bottom of the pit was lit so that the fire would spread as the pit was filled, the top was sealed to carbonize the wood and slowly smother the fire, and the charcoal was allowed to cool before being loaded in sacks to be taken by camel to Sabzevar, 150 km northeast.

According to local villagers, a man could burn two camel loads or about 500 kg in a single firing. Today rangeland has been nationalized, but at that time access to grazing territory was controlled by local individuals and groups. Presumably anyone with grazing rights in an area could cut brush and wood as well as graze animals. We do know that at Asbkeshan, one of the Tauran outliers, the local group in control of the range did not allow charcoal production on their pasture.

A Tauran villager told us that at one point he was making bimonthly trips to Sabzevar to sell his charcoal. At that rate he must have been working very nearly full time—a round trip to Sabzevar by camel takes 8 to 10 days, to which must be added time for wood collecting, pit construction if necessary, and the burning and cooling process. It is likely that like other special activities, production was scheduled to take place during slack periods in the agricultural round, and was therefore seasonal rather than continuous. Probably those with insufficient land and animals were more likely to have taken up production. In Khar and Tauran it appears that wood in general was free, at least to those with access to range, and production and transport were undertaken by the same worker, though he might have to hire a camel at 50 to 100 rials per load. We do not know whether a charcoal tax was levied in Sabzevar as it was in Shiraz (Uhart, 1952:10). We were told that during the past 40 years or so charcoal in Khar and Tauran rose from 400 rials a ton to 2200 rials a ton, and just before its prohibition would bring as much as 3200 rials. In 1966 there were 71·5 rials to the U.S. dollar.

Charcoal has a calorific value of about 7100 kcal per kg, air dry wood a calorific value of 3500 kcal per kg (Earl, 1975:24). Charcoal's higher calorific value makes it more compact than wood and easier to store and transport. Transport costs, according to Earl, are a good predictor of whether an area will produce charcoal on firewood for commercial purposes (1975:74); it is a five day trip by camel or donkey from Tauran to Sabzevar. In Khar and Tauran charcoal production, unlike agriculture or pastoralism, was not subject to short-term fluctuations in rainfall, and in fact probably increased during periods of drought when there was less work for shepherds, failure of surplus and cash crops, and rising prices in urban areas whose meagre local supplies were reduced by lack of rain (see also Gronhaug, 1978 for similar conditions near Herat). For these and other reasons Khar and Tauran exported fuel as charcoal rather than wood in spite of the labour involved in production.

Because primitive methods do not convert wood to charcoal very efficiently, a preference for charcoal has important environmental implications. In Shiraz, Uhart (1952) conducted experiments which yielded only 8% to 10% (from green wood by weight) using the traditional pit method (11% to 17% if the wood had been airdry). At those rates, a single camel load of charcoal (200 kg) used perhaps 1200 to 1800 kg of wood. A kilogramme of charcoal may have twice the calories of a kilogramme of dry wood, but it takes at least four kilogrammes of wood to produce one kilogramme of charcoal. Charcoal production rapidly removes and converts vegetation to cash or exchange products; per calorie, it is at least twice as destructive to the environment as is woodcutting.

Unlike firewood collection, charcoal production was not settlement based. A charcoal burner could travel to his site and sleep there during the process if he needed to. The finished product could be sent directly to the city. The commercial demand for charcoal even drew charcoal burners from the towns out into Khar and Tauran to exploit the stands of vegetation which were more abundant there than they were around the towns. Operations were spread throughout the rangeland where only seasonal stations and occasional outliers are found. The density of Khar and Tauran as a whole is only 0·8 persons per km², a density low enough that brush for firewood and even for charcoal should be plentiful if it were actually collected throughout the area. If Khar and Tauran are considered as part of the Sabzevar hinterland, as they appear to have been throughout most of their history, the real intensity of fuel use comes into perspective. In 1956, when charcoal production was at a peak, the population density of the census district of Sabzevar was 8·9 persons per square kilometre according to the National Census of that year. The census district is an area roughly 82 km in radius which, for administrative reasons stops just short of Khar and Tauran. If Khar and Tauran are included in the district, the density drops to 7·8 persons per square kilometre (192 050 persons over 24 588 square kilometres). Not only is this density figure closer to the real size of the population who depended on the area for fuel, but the form of the fuel, as we have seen, was charcoal and therefore more demanding on the vegetation than firewood would have been. Furthermore, in the area around Sabzevar, charcoal was used for more than just cooking and heating; it was essential to some industrial activities such as metallurgy because of its reducing properties and high temperatures. Other industrial activities, though not requiring charcoal, were nevertheless heavy users of fuel: ceramic production, lime and gypsum production and brick baking especially. Today these industries use fossil fuels, but 30 years ago they still relied on vegetation throughout much of Iran (Wulff, 1966; Centlivres-Demont, 1971).

Between 1900 and the mid-1950s the population of Sabzevar and its hinterlands probably doubled; between 1956 and 1966 it grew another 5% (Bharier, 1972; Iran, 1961, 1969). The combination of population growth, increasing urbanization and increasing per capita energy need for industries not yet converted to fossil fuels made the pressure on the environment in Khar and Tauran during the 1940s, 1950s and 1960s greater than it had probably ever been. The drought in 1958–63 and the unsettled political conditions of that period intensified the situation, forcing men who could not find work to turn to charcoal burning. Everyone in Khar and Tauran today agrees that the effects of charcoal production during that period were disastrous, and that since its prohibition both firewood and pasture have increased dramatically. The increased use in the central Sabzevar area of alternative energy sources such as electricity and fossil fuels has also had a relieving effect on the hinterland.

V. CONCLUSION

When people depend upon the natural vegetation for a livelihood, their settlement pattern and economic strategies are susceptible to changes in rainfall and the condition of the vegetation and soils. Rural economies such as those of Khar and Tauran are also susceptible to changes in market conditions and in the demand for rural products in urban centres. In some cases these products only indirectly involve the local vegetation as in the case of milk products or smelted metals. In other cases the link is direct, as in the traditional dependence of the city on its hinterland for firewood and charcoal. Thus, in Khar and Tauran, while the technology of the production and use of energy has changed little over a thousand years and more, demands for energy have fluctuated. These fluctuations are both spatial and temporal. Some of the variation may be attributed to changes in demography while others are the result of shifts in per capita consumption. Less obvious, perhaps, are changes in subsistence strategies such as switches from camel herding (which does not in Iran include milk processing) to sheep and goat herding (which does). The condition of the vegetation at any particular time is a product of all these events.

Domestic and industrial fuels, whether firewood, charcoal or modern alternatives, have long linked rural and urban economies. In the past urban areas were dependent on their hinterlands for fuel supply, and the demand for firewood and charcoal helped support a dispersed population who might otherwise have migrated to the city. Today the roles are reversing as many rural areas increasingly depend on cities to supply fossil fuels and equipment. Although in Khar and Tauran domestic consumption still relies primarily on locally cut firewood, the transition to dependence on sources from outside has certainly begun. The intent of this discussion has been not to weigh up the advantages and disadvantages of alternative fuel technologies, but to show that levels of demand and environmental impact are tied to events and conditions outside Khar and Tauran, and in ways that may not be immediately apparent. An understanding of the social and economic contexts of the exploitation of rural resources is essential to the study of processes of desertification or to the assessment of potential outcomes of management decisions.

Endnotes

[1] Information on fuel use in Khar and Tauran was collected initially in co-operation with a project sponsored jointly by the Materials and Energy Research Centre of the Arya-Mehr University of Technology (MERC), Tehran, and the Centre National de la Recherche Scientifique (CNRS), Paris. In 1977 Vincent Woollam of MERC visited Tauran to study the feasibility of solar energy devices in Iran's rural settlements as part of the MERC-CNRS project; the results, based on data supplied by Mary Martin, Brian Spooner and myself, appear in a report filed in 1977 (Woollam, 1977) which was revised in 1980. For this chapter I have drawn on Woollam's report, my own field data and that supplied by Siegmar Breckle, Helmut Freitag, Mary Martin, Brian Spooner and many other associates of the Turan Programme. I am grateful to all of these. I owe a special debt, however, to Martin Martin, who not only provided the bulk of the fuel use data, but has read and made helpful comments on earlier drafts of this chapter. I alone am responsible for errors in fact and interpretation.

[2] The amount of land under each type of land use was estimated from aerial photographs and on-the-ground observation. The area of the plain itself is about 120 km².

[3] Most of the quantified data appearing in this chapter were provided by six households from Baghestan, a village on the Tauran Plain. Interviews covered frequency of firewood collection trips, weights of donkey loads, kinds of wood preferred and collected, and quantities of fuel required in various activities. One household supplied similar data on paraffin use. Two of the households move to a milking station for the spring and summer; the rest are permanent residents of the village. These same six households had already been the focus of intensive study for other aspects of the Programme, and the context of their responses and reliability of their information were well known. They made estimates not only for their own households, but for "typical" domestic consumption as well. The head man of a neighbouring village was also interviewed. In addition to this period of intense data gathering in 1977, observations and discussions of fuel use continued over the course of four years of field work by a number of Turan Programme associates, especially Mary Martin.

11

Pastoralists, Flocks and Vegetation: Processes of Co-adaptation[1]

A. Endre Nyerges

University of Pennsylvania
Department of Anthropology
Philadelphia, PA 19104, U.S.A.

By synthesizing the approaches of anthropology and evolutionary ecology this chapter develops an important new argument for caution in pastoral development and for careful re-evaluation of the assumptions on which development has so far been based. Underlying the apparent similarity between arid rangelands in different parts of the world is a difference in the history of man–animal–plant interaction. Disturbances in the course of that interaction can lead to unexpected results (as for example in Conant, Chapter 6). Substantiation of these processes of co-adaptation requires considerable research. A beginning was made in Turan and is reported here. —Ed.

Much current academic and political interest has focused on the problems of environmental deterioration in the arid rangelands of Africa and the Middle East. These problems are widely interpreted as the result of mismanagement of livestock and vegetation by traditional pastoralists. In this view, rangeland degradation occurs in an accelerating spiral of overuse, leading to lowered productivity, which leads in turn to further overuse. Typically, development projects respond to these problems by attempting to institute radical management changes. These changes may include sedentarizing nomads, curing cattle diseases, forbidding firewood cutting, replanting ranges, digging boreholes, excluding goats, and introducing new techniques of animal husbandry, grazing rotation, and veterinary care.

Pastoral development projects, however, are not always successful. Problems have been encountered and, in some cases, the development changes brought about in Asian and African rangelands over the last century have themselves contributed to deterioration. A number of authors (for example Baker, 1975; Ormerod, 1976; Sobania, n.d.) argue that the recent drought in East Africa and the Sahel had a disastrous impact on nomad populations because uninformed manipulation of these societies had produced unpredicted and ramifying changes both in the environment and in the nomads' capacity to exploit it. In particular, both colonial and national policies of sedentarization restricted nomad movements, resulting in severe local degradation around settlements. At the same time, development planners succeeded only in bringing about other innovations, such as boreholes and veterinary services, that upset previous ecological constraints on traditional husbandry and further added to the processes of overstocking and overgrazing. In retrospect, massive intervention based on technology and principles of range management developed for ranching in the American West and Australia may not provide the optimal solutions to problems of Asian and African pastoralism, because of significant differences in the ecological and cultural histories of Western and non-Western rangeland ecosystems.

Rangelands in America and Australia have been ranched by capitalist entrepreneurs for a few hundred years at most, but traditional pastoralism is an ancient subsistence adaptation to the dry lands of Asia and Africa. The various components of rangeland ecosystems under traditional management, including human societies, domesticated animals, and range vegetation, have been subject to selective pressures resulting from pastoral production over substantial, if varying, periods of time. Consequently, the ecological characteristics of all organisms in these ecosystems are the result of long-term and complex interactions, both among species and between species and the inorganic environment (see Lundholm, 1976). These interactions among species in rangelands constitute a series of co-adaptations, in which plants have adapted to deter the pressures of herbivory, and people and animals have adapted to optimize their exploitation of the depleted and herbivore-adapted plant populations. Development projects have typically ignored these adaptive processes, because in many cases ranges have deteriorated, and human and livestock populations have increased to such an extent that previous adaptations no longer appear effective. Yet the most pragmatic solutions to the modern problems of pastoralism may consist in readjusting systems to conform to and make use of traditional management strategies which have proved adaptive in the past, alongside the introduction of new technologies designed to supplement the established systems, rather than in remaking them completely.

In dealing with the complex problems stemming from drought and overstocking in arid ranges, planners need to understand both the cultural and ecological contexts in which these problems occur. They also need to understand the basic ecological relationships that have evolved among people, animals, and plants in these ranges. Specifically, until recently little information has been available on the relationship between herds of domesticated animals and resources of water and vegetation for any location in Africa or Asia (see, for example, Field, 1979; Thalen, 1979). Information on the relationships between foraging herds and the management strategies and production goals of shepherds and herd owners in traditional societies remains limited (Dahl and Hjort, 1976). Domesticated animals, however, are the basic means by which pastoralists exploit their environment. Information on their ecology and "ethno"-management is essential to pinpoint the shortcomings of a pastoral system of production, and to suggest development alternatives that take advantage of established ecological relationships.

Most studies of rangelands that are conducted as part of development projects use measurements of productivity that emphasize questions of range management and economics. Such studies often indicate that pastoralism substantially degrades range vegetation, and that traditional management fails to optimize herd productivity in market terms (see Rapp, 1974; Warren and Maizels, 1977). Studies of productivity and biomass, however, tend to neglect the many cultural factors which govern resource exploitation under traditional management, and also ignore the adaptive processes that modify plant–animal relationships in rangelands.

Planners and range ecologists have increasingly come to recognize that traditional patterns of exploitation pose significant constraints on development, because they may be difficult to change without incurring negative social and ecological consequences. Productivity studies of plants and animals alone, therefore, may not provide an adequate basis from which to generate management proposals (Harrington, 1980). As one response to this problem, social scientists are sometimes now included in development projects, but so far the overall contribution of anthropology to the solution of range management problems has been negligible.

Generally, anthropological research on pastoralism has used the method of participant observation to analyse pastoral societies in terms of the ecological, economic, and political pressures affecting them (for example, Barth, 1961; Dyson-Hudson, 1966; see Spooner, 1973, for a synthesis). In such analyses, features of pastoralist social organization are explained as influenced by and responsive to a difficult environment and the herding technology. For example, age-grade restrictions among the Rendille, a tribe of camel nomads in northern Kenya, result in delayed marriage or emigration of as many as one-third of Rendille women. Spencer (1973) and Rainy (1976) interpret these restrictions as a mechanism of population regulation in an area where the carrying capacity for camel husbandry has been reached.

Some anthropologists, then, have considered various traits of pastoral societies as ecological adaptations. At the same time, however, they have implicitly assumed that the animals, the technology, and the environment are unchanging givens in the system. As a result, these factors, which do in fact vary and are crucial to an understanding of pastoralism, have not been adequately considered in research, although a significant exception is the simulation study by Dahl and Hjort (1976) which models the relationship between traditional management, drought, and herd population dynamics in Africa.

Finally, some anthropological studies of pastoralism, such as that recently completed in Tunisia (Bedoian, 1978), have been based on the ecosystem concept in an effort to integrate information on human behaviour with information on rangeland ecology. Ecosystems analysis can be used to outline patterns of energy flow and homeostasis in rangelands under traditional management, and to provide information on economics, human population stability, and the quality of successive years in which rainfall varies. Such information can be readily incorporated in range management projects (Novikoff et al., 1973–1977). Ecosystems analysis in anthropology, however, relegates cultural variables to the position of regulators of calorie flow, and, as is also the case with studies of range ecology, this approach cannot adequately explain either the origin of adaptive traits or their importance to individual organisms in the system.

The various disciplines concerned with pastoralism each provide insights into the ecology of rangelands under traditional management. In particular, ecosystems analysis in both anthropology and range ecology can be used to describe interactions among trophic levels in these systems, thereby avoiding the error of separating the human and environmental components. But while ecosystems analysis can outline the complex feedback interactions that occur in rangelands, it does not provide a coherent approach to the study of co-adaptive processes in rangelands in which human, animal, and plant populations have all adapted in response to selective pressures generated by their interactions with one another. These co-adaptations between plant populations on the one hand, and the people and animals that exploit these populations on the other, can be studied through a set of basic interrelated questions on cultural adaptations and the adaptations of range plants and domesticated animals, which are the subject of this chapter. Briefly, these questions concern:

1. The herding and husbanding strategies used by traditional pastoralists to optimize foraging and to maintain production in the face of the severe social and ecological constraints on life in dry lands.

2. The foraging adaptations of domesticated animals which enable them to survive and flourish in depleted rangelands.

3. The impact of sustained pastoral activity on range plants, and the competitive and defensive strategies that these plants have evolved to meet the pressures of herbivory and nutrient limitations.

In what follows, I discuss these aspects of the ecology of pastoralism in a conceptual framework based on anthropology and evolutionary biology that is designed as a guide for pre-development field studies. I then summarize results of preliminary field research on the ecology of domesticated sheep and goats carried out in 1976 and 1978 in Tauran. This research in the rangelands bordering the central Iranian desert forms the basis for a long-term case study of the interrelationships among traditional management, domesticated animals, and vegetation in dry lands.

I. TRADITIONAL PASTORALISM: AN EVOLUTIONARY APPROACH

A rangeland ecosystem exploited by pastoralists can be simply described as a plant–animal–human food chain. The crucial human components in the set of relations that make up this chain include the cultural value of stock, the pattern of settlement, the economics of husbandry, the production goals of flock owners, and traditional strategies of herd management and migration. The animal components include production parameters, and the behaviour, foraging adaptations, and population dynamics of livestock. They also include the limits which the scarcity and spatio-temporal variability of water and vegetation place on herd size and animal niche characteristics. Finally, the plant components include anti-herbivore defences and the reproductive and other life-history strategies of range species, as well as their abundance and distribution in space and time, especially as a function of grazing pressure. This list is scarcely exhaustive, but data on these topics can indicate, in outline form at least, the essential patterns of ecological relations between a traditional pastoral society and its dryland environment.

The relationships among range plants, domesticated animals, and human societies have been characterized by the development of co-adaptations. These adaptive patterns, of course, have been disturbed by recent economic and social changes that have, in many instances, pushed herd sizes far above the earlier levels and resulted in severe overstocking and overgrazing. But in many ways, the co-adaptations that have their origins in long-term evolutionary processes are still significant features of rangeland ecology and traditional pastoral management. To understand them requires a brief analysis of the ecological history of pastoralism.

Before the advent of pastoralism, rangeland ecosystems had existed for millennia in Asia and Africa (see, for example, Niklewski and van Zeist, 1970), although they expanded subsequently under human use (Brice, 1978). During the long pre-pastoral period, wild herbivores were more diversified and mobile than domesticated herds are today, and range vegetation was adapted to a different and less concentrated pattern of grazing. The adaptive patterns which evolved among plants and animals during that period may have been similar to those now found in arid regions which were until recently unoccupied by pastoralists, such as in the Americas, Australia, and the Kalahari (Orians and Solbrig, 1977; W. Low, 1977; Zummer-Linder, 1976).

Archaeological evidence shows that the earliest domestication of sheep and goats occurred in various locations in the Middle East at about 10 000 BP (Cohen, 1977; Flannery, 1965). Originally a component of mixed-farming Neolithic village subsistence, and only later an independent production system, this technology spread through vast regions of African and Asian grazing lands, although the date of its onset in any particular location has not been determined.

Whether rapid or gradual, the introduction of domesticated animals to a rangeland ecosystem constituted an invasion with major ecological consequences, including vegetation change and degradation. Similar processes of degradation have ensued, in more recent history, with the introduction of starlings to North America, feral goats on Pacific islands, and sheep and rabbits on Australian ranges (Elton, 1958; Mueller-Dombois and Spatz, 1975; Crisp and Lange, 1976). As regular patterns of grazing were eventually instituted around villages and along nomad migration routes, however, it is probable that ramifying adaptive and co-adaptive responses by surviving species in the ecosystem emerged and continued developing to the present where, over vast areas, *they are still maintained by traditional systems of management.*

I therefore hypothesize that, with the advent of domesticated herds, all species in rangeland ecosystems adapted in new directions in response to the ecological stresses and interactions generated by the permanent establishment of pastoral production systems. Except where recent disruption has occurred, these adaptations are reflected in the current ecological structure of rangelands.

Investigation of these adaptive relationships in rangelands demands an explicitly Darwinian approach. Darwin's theory of evolution holds that the present attributes and interactions of organisms are the result of processes of natural selection operating on the genetic characteristics of species over evolutionary time. Individuals with advantageous variations tend to survive and produce more offspring (have greater fitness) than many of their conspecifics in the face of often intense competition for limited resources of food and mates. To the extent that an advantageous variation has a heritable component, it may be passed on to future generations and it may increase in the population (Darwin, 1859). Recently, some have extended evolutionary analysis to encompass cultural as well as genetic adaptations (see, for example, Irons and Chagnon, 1979).

Investigations based on the evolutionary approach can be used to supplement ecosystems analysis (Orians and Solbrig, 1977:2–5). In evolutionary ecology the goal of research is to understand the adaptive strategies which organisms have produced in response to selective pressures. In this approach, *strategies* are defined as the ways organisms allocate resources that enable them to survive and reproduce optimally under resource constraints and pressures of predation or herbivory. In research on the ecology of rangeland ecosystems under traditional management, evolutionary analysis can contribute an understanding of the herding and husbanding strategies of pastoralists, the foraging strategies of domesticated animals, and the anti-herbivore and life-history strategies of range plants. In these ranges, many of the strategies of plants, animals, and people have evolved as co-adaptations to one another, in the sense that as people and animals have evolved strategies to exploit plant resources more effectively, plants have produced anti-herbivore adaptations in response to these pressures. In turn, pastoralists and animals have been forced to intensify their exploitation strategies, creating further selective pressures on plant populations. The analysis of these strategies and of the ramifying processes of co-adaptation is necessary in pastoral development because planners need to understand why existing ecological patterns occur in rangelands before they attempt to change or manipulate them.

In the face of ever-increasing pressures from desertification, political action, development, and the expansion of agriculture, traditional pastoralism has become less widespread, but it still persists in large regions of the Middle East and Africa where, until very recently, it appears to have approached a long-term optimum from the points of view of both environment and society (Barth, 1960). In particular, traditional management strategies evolved as an effective means of exploiting the available resources of vegetation and water, and of compensating for progressive range degradation as forage plants and domesticated animals co-evolved with one another. It is now increasingly apparent that the introduction of modern technology and management practices in arid ranges may be costly, difficult, and even disruptive. The most successful development projects, consequently, may be those that either incorporate, or only disrupt minimally, these past patterns of co-adaptation among plants, animals, and people. In what follows, I offer a framework for research on these components of the evolutionary ecology of rangelands.

II. ADAPTIVE PATTERNS IN RANGELAND ECOSYSTEMS

A. The Responses of Vegetation to Grazing

Changes in a plant community brought on by intensified herbivory eventually result in a grazing climax that may be described as "anti-pastoral". Arid regions are stressful, nutrient-poor environments for most plants, and loss of foliage and seeds due to herbivory is a problem that is likely to be serious, especially for the perennial browse species. Plants may minimize these foraging pressures through escape in space and time, including the strategy of predator "satiation", which is the synchronous production of leaves and seeds with the result that available herbivores cannot consume them all (see Janzen, 1976). They may also grow thorns. The most important defence, however, is the production of toxins and other digestibility-reducing substances, which may be differentially distributed in the plant parts most valuable to herbivores (Harper, 1977; Janzen, 1977, 1979; McKey, 1978; Rosenthal and Janzen, 1979).

Where pastoralists have only recently occupied a rangeland, such as in southwestern North America and Australia, some of the spiny and toxic native range plants have been able to survive the new pressures and increase relative to the favoured forage species (W. Low, 1977). As Kingsbury (1978) points out, these adaptations characteristic of some species of American and Australian range plants have originated under the pressures of foraging by wild herbivores. They do not reflect evolutionary responses of range plants to cattle and sheep ranching, which has affected only the current distribution and abundance of these species. This relationship, however, does not necessarily apply to the anti-herbivore adaptations of plants in ranges exploited by the oldest pastoral systems, such as in the Middle East and perhaps Africa. In these systems, toxic defences in plants may have co-evolved specifically with the foraging adaptations and heavy foraging pressures of domesticated animals. The evolutionary relationship between plants and animals in these ranges may be intense and characterized by co-adaptations, with animals able to consume some amounts of plants and their toxins safely, but usually unable to consume enough of a plant or its seeds to preclude the plant's survival or reproduction.

In general, toxicity is favoured in the plant community under longstanding patterns of pastoral exploitation. In traditional systems, where this co-evolution may be extensive, a sustained level of foraging may be possible while overgrazing and further degradation are prevented by toxicity in plants. This evolutionary interaction, however, is not necessarily stable, and the relationship may fluctuate in a cycle. At first, an increasing animal population overgrazes the nontoxic species until only toxic species remain, animal starvation and poisoning ensue, and a crash in the population of grazers and the eventual return of nontoxic species to the rangeland follows (see, for example, Freeland, 1974). Legesse (1973:15–17) points out, however, that where pastoral populations are highly mobile, pasture degraded by cattle herders may be subsequently occupied by camel herders and never have an opportunity to recover to the point where it can again be used as cattle range. Such progressive deterioration appears to be typical of some East African ranges where pastoral production is relatively recent (*c.* 3500 BP), and plant–animal co-evolution has perhaps not had sufficient time to occur. The amount of time or number of generations required for such complex interactions to evolve in rangelands, however, has not yet been determined.

The relationship between a pastoral society and the environment, then, is partly some function of the time depth of production in an area. In addition, the impact of pastoral production on range vegetation is strongly patterned in space by the location of water sources, which affects the location of settlements and more temporary pastoral camps. Although animals may adapt to survive long periods away from water (Schmidt-Nielsen, 1964), the pressure of grazing is nonetheless heavier near such central water sources than farther out on the range, and overuse depletes or alters the vegetation near settlements (Barker and Lange, 1969; Hamilton and Watt, 1970; Lange, 1969; Spencer, 1973). Thus the frequency and distribution of settlements and permanent water sources are major factors in the impact of grazing on range vegetation. Pastoral ecology, therefore, becomes a study of the spatial distribution of populations and resources and the consequences of this distribution for the adaptive strategies of people, plants, and animals.

B. The Ecology of Domesticated Animals

The basic productivity of a pastoral system, and the role of stock in converting plant biomass into milk, meat, and other products, depends on animal adaptations for foraging on the available vegetation. These foraging adaptations may take several forms. For example, under conditions in which flock sizes are allowed to approach carrying capacity, competition results in partitioning of resources among the different species herded and among age and sex categories, which alleviates competition. These quantitative and qualitative differences in foraging can be further exploited by shepherds in their strategies.

Range vegetation available to domesticated herds is generally depleted, of poor nutritional quality, and partly toxic. In addition to the niche differentiation already mentioned, animals may adapt to these conditions through foraging strategies and physiological mechanisms in order to optimize nutrient intake (Westoby, 1974), to survive periods of scarcity (Ferguson, 1971), and to minimize or avoid consumption of toxins (Freeland and Janzen, 1974; Montgomery, 1978). Animal adaptations to plant poisons include detoxification of secondary compounds in the gut and defensive foraging strategies specific to the local flora, many of which may be learned by individuals, especially in infancy. In general, both foraging by domesticated animals and the control of foraging by shepherds and herd owners are geared to optimize consumption of available vegetation, but these strategies must be understood in the context of plant adaptations to escape or minimize herbivory.

C. Management of Herds

Finally, pastoralists themselves have evolved management strategies to optimize both the plant–animal relationship and the subsistence or market value of herd products. Herd management has two general aspects: the owner's function of *husbandry* and the shepherd's function of *herding* (both of which may be carried out by one person or divided among two or more persons—see Paine, 1972:79, and Asad, 1970:51). Husbandry involves making decisions about herd size, species composition, sex ratios, age structure, and migration patterns in response to spatio-temporal variations in vegetation and water resources and variations in subsistence or market demands on productivity. In a market economy, for example, the owner attempts to optimize the cash value of his herd, emphasizing growth of capital and formation of a profit. In a subsistence economy, the owner attempts to optimize the ratio between the population supported and the consumable herd products. This difference in economic goals is often reflected in meat versus milk production or in emphasis on secondary animal products such as wool or skins, with concomitant differences in herd structure.

Sandford (Chapter 4) defines two basic types of husbanding strategies. *Opportunist* strategies involve maximizing herd reproduction and mobility in order to track and exploit major regional, seasonal, and year-to-year fluctuations in resources. Under this type of management, herd sizes may drop only through disease, starvation, or theft (see Sweet, 1965). Obviously, this type of management can be extremely disruptive if, as has often happened, the normal ecological constraints on herd size are disrupted by veterinary programmes, improved water sources, or increased security. In contrast, *conservative* strategies, generally favoured by range scientists, involve maintaining a relatively constant herd size below carrying capacity and, for the sake of future benefit, using less than the maximum possible amount of forage.

A pastoralist's choice of which strategy to follow, either opportunist or conservative, depends on his perception of how to optimize his success under given ecological and social conditions. Mobility, for example, is a crucial adaptation of large herbivores to the scarcity and spatio-temporal variability of resources in dry lands. A pastoralist's freedom to move flocks continually from depleted areas to fresh pastures, however, is generally limited by the restricted availability of water. In addition, market location, government policy, competition from neighbouring groups, the location of farmlands, and the availability of labour, equipment, and fuel for milk processing may further constrain movements. Consequently, pastoral movement patterns are compromise strategies that optimize herd access to the best pastures, taking into account constraints of water availability and local socio-economic conditions that may vary considerably in space and time (see, for example, Western and Dunne, 1979). Pastoral settlements reflect these strategies and may take the form of transient nomadic encampments along established routes of migration, seasonal camps, transhumant villages, and perennial villages in agricultural areas.

Conservative husbanding strategies that avoid overstocking and ensure the future quality of forage are favoured by pastoralists where movement is restricted, and when future generations can be expected to occupy the same area (see Janzen, 1978). Conservatism is also favoured under conditions of relative environmental stability. Opportunism, on the other hand, is a strategy favoured where resources fluctuate greatly. Opportunist strategies are further favoured where occupation of a site is likely to be transient due to insecurity or resource variability, or where the opportunity exists to convert excess stock rapidly into other capital (Koster, 1977:82).

If management is divisible into husbandry and herding components, the overriding husbandry decisions on herd movement and population dynamics are made by the owner, while herding, or day-to-day management, is the shepherd's function. Herding involves optimizing animal foraging through local strategies of habitat division and the dispersal of grazing pressure. Herding strategies take advantage of animal differences in diet and mobility, whether due to species, age, or gender, in order to disperse and vary grazing pressure on the range through space and time, allowing pastoralists to exploit local resource variations. Such strategies are also closely linked to conservationist husbanding—the dispersal of animals limits the pressure of grazing on any one pasture at any one time, thereby preserving range quality.

Husbanding and herding are essentially complementary. In order to optimize production in a highly fluctuating environment, however, the opportunist husbanding strategies of maximizing herd reproduction and mobility and taking advantage of good years to increase herd size may be stressed (a form of r-selection).[2] By contrast, in more stable environments husbandry may be conservative and animal populations may remain near carrying capacity. Under these conditions, the shepherd's strategies of optimizing foraging in a particular location may be relatively more important (*K*-selection).

Rangelands under traditional management are characterized by the evolution of co-adaptations among plants, animals, and pastoralists. These evolutionary relationships may vary from place to place and over time, as a function of a host of ecological and social factors. The anti-herbivore adaptations of range plants, for example, have evolved under pressures of herbivory that vary in space because the location of water and settlements determines the relative accessibility of different pastures to foraging herds. Foraging pressures also vary in time because the numbers of animals that are herded on a range may fluctuate seasonally or as part of a cycle of drought and overgrazing. The total time depth of pastoral production in different rangeland areas also varies substantially. As a result of these variations in foraging pressures, animals encounter different forage characteristics. In order to survive, they must be adapted both physiologically and behaviourally to the nutritional characteristics, including the anti-herbivore adaptations, of the locally available forage.

While plants and animals in rangelands have co-adapted to one another, pastoralists have attempted to optimize the relationship between their herds and the range vegetation through strategies of husbandry and herding. In choosing among possible optimization strategies, however, pastoralists are constrained by water availability, settlement patterns, and other social and economic variables. For example, in areas where forage and water are scarce and variably distributed, pastoralists may attempt to exploit regional resource variations opportunistically by overgrazing one pasture and then moving on to undepleted ranges. Alternatively, pastoralists may attempt to employ conservative strategies that avoid local overuse and optimize the long-term exploitation of a limited range area. But they are likely to do so only under conditions of security, restriction on movements, land tenure rights, and a stability of resources that may be the product of an extensive history of chemical co-evolution between plants and animals. In any particular case, then, the interactions among plants, animals, and pastoralists are complex and can only be understood sufficiently for the purpose of modelling and eventual manipulation in development projects if intensive case studies of individual pastoral systems are conducted first.

III. THE ECOLOGY OF DOMESTICATED ANIMALS IN TAURAN

This framework for the study of traditional pastoralism was applied in preliminary fieldwork, focused on the ecology of domesticated sheep and goats, that I conducted July 7 to August 15, 1976, and July 25 to August 21, 1978, at Sanjari in southern Tauran. Sanjari is an ideal location for research on traditional pastoralism not only because pastoral technology is of great antiquity throughout Iran, but also because the current ecological and social conditions of this area, including resource scarcity, professional shepherding, repeated use by the same group, and overlapping use by different groups, provide a crucial combination of factors relevant to developing a framework of ideas about traditional management and co-adaptive processes in rangelands.

A. The Study Site

Sanjari (Map 3, p. 140) is an isolated summer milking station next to a mountainside spring at 1300 m. The sparse rainfall in the area, approximately 100 to 200 mm, falls unimodally in winter, and this spring is the only locally available water source in summer. To the west and south, Sanjari faces out onto rangeland, cut by frequent dry streambeds, that slopes down to the kavir (an internal-drainage basin of saline mud flats) some 20 km away at 700 m altitude. Characteristically, the vegetation on this range is a low shrubland of 20% cover, dominated by *Artemisia herba-alba*, *Zygophyllum eurypterum*, and *Salsola aurantiaca*. *Pteropyrum aucheri* and *Amygdalus lycioides* are a common shrub association in dry streambeds (Freitag, personal communication, and see the Preliminary Vegetation Map in Iran, 1977; Rechinger, 1977).

The area around Sanjari seems now to be rarely visited by the non-domesticated large herbivores that are found elsewhere in Turan—jebeer and goitered gazelle (*Gazella dorcas* and *G. subgutturosa*), onager (*Equus hemionus*), urial (*Ovis orientalis*), and feral camels. Of these, only occasional gazelle were observed or reported in the study area. Apparently, the non-domesticated large herbivores do not now compete with the sheep and goats at Sanjari for the same forage plants.

A more significant source of competition for water and pasture resources in this area, however, is the use of the range by overlapping and neighbouring pastoral groups and the agricultural use of some areas. The entire range area in Tauran is heavily used by pastoralists, especially in winter, and the ecology of the flocks at Sanjari needs to be understood in this context. Because of the scarcity of resources, and competition for these resources from other groups, the pastoralists at Sanjari must attempt to optimize their use of a particular range area, over both the short and long terms, through development of locally-adapted strategies of herd management.

From April to October, Sanjari is the base for approximately 1100 sheep and goats divided between three flocks. Most of the animals are owned by a wealthy resident of Salehabad, 20 km to the north in the cluster of villages in the Tauran Plain. Station personnel, numbering eight to fourteen persons, are women from the owner's family, who process milk, together with hired shepherds and their families. Activity at the station, aside from maintenance of a small garden, focuses exclusively on pastoralism. Flocks are tended, lactating animals are milked, milk is processed into clarified butter and other products, and firewood is cut as fuel for use in milk processing. During the winter, the settlement serves as an occasional watering point only. The animals, culled to about 600 sheep and goats and separated in two flocks, are sheltered at night-time in a covered pen (*aghol*), next to a well 5 km from Sanjari. They are then grazed only during the day on pastures that are partially differentiated from summer grazing areas. At this time, the animals are tended by a few shepherds, while the rest of the human population returns to Tauran villages.

In summer 1976 at Sanjari, goats outnumbered sheep by four to one. This ratio reflects the greater value of goats, which may be milked from shortly after the birth season in April until the September rut, while sheep lactation ceases in mid-July. Sheep are valued for their tail fat, wool, and meat. The animals are separated into the following three flocks, each herded by a different shepherd:

1. The *goat herd* consisting of 300 milch goats and over 200 gelding and non-lactating does, accompanied by approximately 20 rams.
2. The *sheep herd* consisting of 220 ewes and lambs, accompanied by approximately 30 buck-goats.
3. The *kid herd* consisting of more than 300 kids under one year old.

Note that bucks and rams are not herded with females of their species until autumn, presumably to ensure springtime births.

B. Methods

During a total of 210 hours' contact with animals during 35 separate grazing periods, I made observations on the distance and direction of herd movement, and on shepherds' activities in relation to their herds. I also conducted a single 24-hour behaviour study of the goat herd based on the "instantaneous scan" method (after Altmann, 1974). This study involved noting the behaviour of ten animals in the herd every five minutes (interpolating 2:00 to 4:00 a.m.). Behaviour was recorded in broad, disjunctive categories of walking/standing, feeding, and lying/ruminating.

Additionally, I accumulated 27·5 hours of observations on the diet and foraging behaviour of 120 animals from the three flocks for use in niche analysis. In these observations (method derived from Cody, 1968, 1974), I followed individual animals for 20 minute periods or until I was no longer able to observe the animal. Normal observation distance was 5 to 15 m. For each animal observed, I timed total duration of the observation period on a stopwatch. For each feeding event within this time period, I noted the following information:

1. The species of shrub or category of ground forage eaten. Ground forage includes dried annuals, leaf litter, and seeds, and is either *exposed* on open ground or *protected* under shrub cover.
2. The duration of the feeding event, recorded in seconds on a separate stopwatch.
3. The distance walked to the next food item, estimated by eye.

The focal animals for this study were not chosen randomly, as I could not consistently recognize individuals and marking animals was not feasible. Rather, after terminating an observation I followed the next animal nearest to me at the edge of the flock (so as not to disturb the herd).

In the study of the vegetation, I recorded plant densities and shrub canopy cover in study plots located along radial lines drawn from Sanjari and its *aghol*. These plots were not laid out randomly and do not constitute a sample for statistical analysis. The first series of eight plots (labelled "S" on Map 3(b) begins at the mouth of the milking pen at Sanjari and continues west approximately 3·5 km, at or beyond the maximum range of goat foraging from Sanjari. The second series of four plots ("A") begins at the *aghol* and continues east 1·5 km (meeting the first series). I recorded similar measurements in a series of control plots drawn parallel to the line of the Sanjari study plots, but including no central point of activity. This third series of four plots ("C") begins approximately 2 km south of the settlement, at the same altitude of 1300 m, and continues west 1·5 km.

The study plots were set out at 0·25 or 0·50 km intervals. Each 10 by 10 m plot was laid out in a plains area; ridge slopes and the frequent streambeds cutting through the plains, with their own characteristic vegetation associations, were not studied. At each plot I recorded:

1. Identity of each shrub and the numbers of each shrub species present.
2. The maximum north–south canopy diameter of each shrub.
3. The numbers of annuals in the plots, categorized as exposed on open ground or protected under shrubs.
4. The altitude of the plot relative to Sanjari.

These methods are derived from a combination of a number of ecological and ethological approaches for studying different aspects of the ecology of traditionally-managed sheep and goats in arid ranges. With some modifications, they form the basis of a methodology for extended research projects. The scan sample method of behaviour study, for example, provides information that can be summarized and graphed in a diurnal activity budget. It is ideal for an initial characterization of activity in the grazing cycle, particularly as it permits simultaneous observation of related events, including shepherd's labour. The niche analysis method employed here is based on a focal animal method for observing foraging behaviour which allows for rapid accumulation of quantified data on animals moving and feeding on a range. This method has the advantage that it can be readily combined with simultaneous recording of diet selection. For this reason, diet analysis in this work is based on feeding times rather than on bite counts.

The vegetation study method employed can be used to describe changes in the density and cover of important shrubs over distance from central place. It also shows the varying impact of foraging on herbs that are exposed on open ground or protected under shrubs. A method of vegetation study based on plots distributed at regular points on a line, however, cannot provide more than an initial description of plant distributions and needs to be replaced in future work by a grid sample or a stratified random sample of heavily, moderately, and lightly foraged areas in a series of sites.

C. The Grazing Cycle

The daily summer grazing cycle at Sanjari is distinctly patterned both in space and in time (see Fig. 1 and Map 3(c)). All herds are out on the range continuously except for two periods of the day. The herds return to the spring in mid-morning and again in mid-afternoon for watering, rest, and rumination, and the does are milked at these times. The herds leave the station along radial paths for a short grazing period from noon to mid-afternoon and then for a longer period from evening to mid-morning. Overnight, the animals may rest at heavily trampled campsites or graze if moonlight permits (as in Fig. 1 and Map 3(c)). In the morning foraging period, animals reach the greatest extent of foraging distance from the settlement.

The spring and settlement are a central point of obvious importance to the pastoral system at Sanjari—there is no other free-standing water and no other milking site. Since water and labour resources are localized at a specific point in the rangeland, grazing is restricted in space to the area that can be reached while returning to the core twice daily—a mountain-broken, semicircular area with a maximum radius of approximately 3 km. The implications of this limit on foraging area are significant because the carrying capacity of the

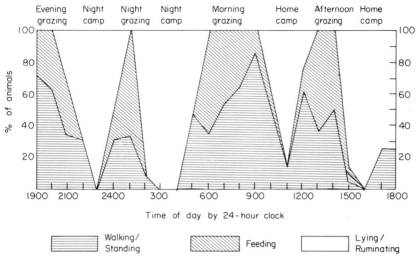

Fig. 1. Diurnal grazing cycle of the goat herd, Sanjari, August 8–9, 1976. Activities are shown as a proportion of the whole and accumulate to 100% for each hour.

range is set by the amount and quality of vegetation near water. Herding and foraging strategies must optimize the use of vegetation within this area. In addition, stress from grazing (and firewood cutting) is placed on the range vegetation in a definite pattern: areas closer to the core are more heavily used while areas at greater distance are relatively unused and undisturbed.

D. Foraging and Management

As the goat herd leaves camp, animals fall into long columns on well-worn paths and march rapidly until they are approximately 1 km from the settlement. The sheep herd also tends to leave camp rapidly, while the progress of the kid herd is more leisurely. Once out on the range, the animals divide into sub-groups and individuals spread out to forage. While grazing, although the animals are dispersed, their movement is parallel to one another and continues to be rapid. During the day, once the animals have been sent out in a particular direction the shepherds for the goat and sheep herds attend to their flocks only in order to prevent the animals from foraging too near to the camp (in relatively depleted pastures that are largely reserved for foraging by kids), and to turn them around halfway through the grazing period. The shepherd for the kid herd, however, is constantly in attendance, keeping the animals near the camp and retrieving strays. All three shepherds guard their flocks closely throughout the night, in case animals should become lost while grazing or be attacked by wolves. The three herds forage over essentially the same range, but data on animal diets and foraging behaviours, presented below, indicate that variations among species and herds may alleviate competition. Observations also indicate that shepherds' labour plays a role in determining where the different herds forage, and therefore how much the herds in fact compete with one another for the same vegetation.

Figure 2 compares major items in the late summer diets of sheep and goats at Sanjari. These data represent a sample of diet selection by all age and sex categories of animals combined by species. Statistical analysis of this material is presented in Table I. These data show that late summer diet selection by sheep and goats overlaps considerably, largely due to heavy feeding on both categories of ground forage, which comprise 54% of goat diet and 59% of sheep diet (the difference between herds in selection of ground forage is not significant). Goats and sheep may eat the same ground forage items (dry annuals, leaf fall, seeds), but this remains undetermined.

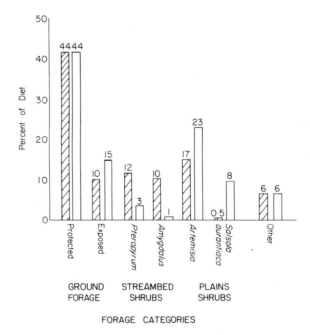

Fig. 2. Sheep and goat diets at Sanjari, summer 1976. 🞕 goat; ⊓ sheep. Based on measurements of foraging times.

Table I. Mean frequency of selected items in the diet records of the goat and sheep species—t-tests between means. At df = 111, Type 1 error of 5%, two-tailed, t = 2·012.

Resource	Goats/Sheep Mean	2 s.d. (±)	Calculated t	Significance
All ground	0·54	0·62	1·024	p > 0·05
forage	0·59	0·42		
Pteropyrum	0·12	0·41	2·141	p < 0·05
	0·03	0·07		
Amygdalus	0·10	0·35	2·547	p < 0·05
	0·01	0·10		
Artemisia	0·17	0·42	1·268	p > 0·05
	0·23	0·44		
Salsola	0·005	0·02	4·889	p < 0·05
aurantiaca	0·08	0·30		

Both sheep and goats also browse extensively on *Artemisia herba-alba*, which provides 17% of goat diet and 23% of sheep diet (the difference between species in selection of *Artemisia* is not significant). Sheep and goats differ, however, in their preference for other shrub species. Foraging goats and sheep tend to move rapidly through the range, stopping briefly (a few seconds) to feed on ground forage or *Artemisia*. When crossing dry streambeds, however, goats may stop and gather to browse on the large *Pteropyrum aucheri* and *Amygdalus lycioides* shrubs located there. These two streambed shrub species together account for 22% of goat diet, but only 4% of sheep diet (the differences in selection of these two shrubs by sheep and goats are significant). Sheep largely ignore streambed vegetation, and specialize on *Salsola aurantiaca*, a succulent sub-shrub occurring on the plains. *Salsola* comprises over 8% of the sheep diet, but is virtually absent in the goat diet (the difference between species in preference for *Salsola* is significant). An additional thirteen shrubs compose the remaining 6% of goat diet, and an additional three shrubs provide the remaining 6% of sheep diet. While goats occasionally forage on dried seeds not yet fallen from *Zygophyllum* (1·5% of goat diet), both sheep and goats essentially ignore this dominant shrub species.

The division of goats into flocks of adults and kids, and the maintenance of a separate sheep herd, brings out a number of additional, if minor, dietary variations (shown graphically in Fig. 3—statistical analysis of some differences in selection are given in Table II). Adult goats eat less exposed ground forage than kids (7% of adult goat diet, 16% of kid diets). However, adult goats eat more protected ground forage than kids (49% of adult goat

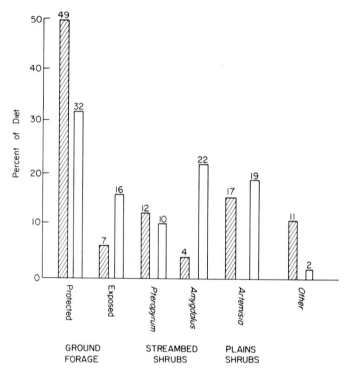

Fig. 3. Adult goat and kid diets at Sanjari, summer 1976. ▨ goat; ⬜ kids. Based on measurement of foraging times.

Table II. Mean frequency of selected items in diet records of the goat and kid herds—t-tests between means. At df = 85, Type 1 error 5%, two tailed, t = 2·012.

Resource	Goat herd/Kid herd		Calculated t	Significance
	Mean	2 s.d. (±)		
Protected	0·49	0·63	2·424	p < 0·05
ground forage	0·32	0·62		
Exposed	0·07	0·23	3·132	p < 0·05
ground forage	0·16	0·36		
Amygdalus	0·04	0·24	4·943	p < 0·05
	0·22	0·42		

diets, 32% of kid diets). As bigger and stronger animals, adult goats can force their way under shrubs and, with effort, feed on a resource that is less accessible to the kids. Additionally, kids specialize on *Amygdalus*, which composes 4% of adult goat diet and 22% of kid diet. This difference in selection of *Amygdalus* occurs largely because the shepherd often herds the kids around and around in a gorge near camp where this species is abundant. These age-specific differences in diet selection, however, contribute little to resource partitioning—adult goats and kids herded in the same area must rely on essentially the same resources.

In general, there are only a few significant differences in late summer diet selection between sheep and goats and among the herds. Goats prefer streambed shrubs and browse on a greater variety of range species, while sheep prefer *Salsola*. The kids and adult goats differ somewhat in selection of ground forage and streambed shrubs, as a result of age-specific behavioural differences and differences in shepherd strategies. A considerable overlap in the resources being exploited is apparently tolerable by the animals at this season.

In addition to diet choices, foraging behaviour must also be considered in niche analysis. Behavioural variations among animals may contribute to resource partitioning, and they also present opportunities for manipulation of herds by the shepherds. Components of foraging behaviour for the species and herds are compared in Figs 4 and 5 on "saw-toothed" graphs (derived from Cody, 1968, 1974). In these graphs, the main slope represents, in Cody's terms, the average speed of progression (m/min). On the teeth, the horizontal line represents mean duration of feeding events (in seconds), and the slope of the tooth is mean percent of time spent feeding.

Comparing data by species, sheep and goat foraging behaviour is similar for all available measures (Fig. 4; Table III). The average speed of progression is

Table III. Foraging behaviour of goats versus sheep (species)—t-tests between means. t values determined at Type 1 error of 5%, two-tailed.

| Behaviour | Goats vs. Sheep | | df/t | Calcu-lated t | Signifi-cance |
	Mean	(±) 2 s.d.			
Average speed of pro-	22·0	21·1	113	1·956	p > 0·05
gression (m/min)	17·5	16·1	2·012		
Mean duration of feed-	14·5	79·5	3383	0·63	p > 0·05
ing events (s)	15·4	40·1	2·012		
Mean % of time	48·9	36·5	111	1·365	p > 0·05
ingesting (%)	43·2	39·5	2·012		

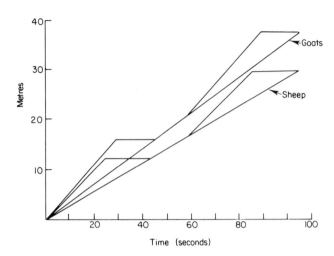

Time (seconds)

Fig. 4. "Saw-toothed" behaviour curves comparing the sheep and goats on average speed of progression (m/min), mean duration of feeding events (seconds), and mean percent of time spent feeding (seconds out of a minute).

Note: The two teeth for each species are duplicates of one another. Each of the three lines composing the teeth are read separately. Main slope—average speed of progression (m/min; read at the one minute point). Teeth—these two components are read on the time axis only. Horizontal line—mean duration of feeding events (the number of seconds covered by the bar). Slope—mean percent of time feeding (expressed as the number of seconds out of a minute covered by the slope, i.e., 30 seconds is 50 percent).

22 m/min for goats (does, geldings, and kids), and 17·5 m/min for sheep (ewes and lambs). Mean duration of feeding events is 14·5 s for goats and 15·4 s for sheep. Mean percent of time spent feeding is 48·9% for goats and 43·2% for sheep. The difference between the average speed of progression of sheep and goats is relatively large, but is not statistically significant. The other foraging behaviours are also not significantly different between species.

Whereas foraging behaviour does not differ between species, some significant differences occur among herds, with kids separated from adult goats, and sheep maintained separately as well (Fig. 5; Tables IV and V). Average speed of progression is 25·1 m/min for the goat herd, 17·5 m/min for the sheep herd, and 14·7 m/min for the kid herd. Mean duration of feeding events is 12·5 s for adult goats, 15·4 s for the sheep, and 18·2 s for the kids. Mean percent of time spent feeding is 46·3% for goats, 43·2% for sheep, and 54·6% for kids. Comparing these measures in analysis of variance, the means for speed of progression and duration of feeding events, but not percentage of time feeding, are significantly different among the three herds (Table IV). All

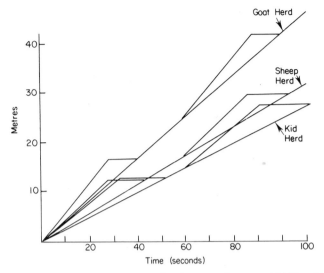

Fig. 5. "Saw-toothed" behaviour curves for the goat, sheep, and kid herds comparing average speed of progression (m/min), mean duration of feeding events (seconds), and mean percent of time spent feeding (seconds out of a minute).

Note: The two teeth for each herd are duplicates of one another. Each of the three lines composing the teeth are read separately. Main slope—average speed of progression (m/min; read at the one minute point). Teeth—these two components are read on the time axis only. Horizontal line—mean duration of feeding events (the number of seconds covered by the bar). Slope—mean percent of time feeding (expressed as the number of seconds out of a minute covered by the slope, i.e., 30 seconds is 50 percent).

Table IV. Foraging behaviour of the goat, sheep and kid herds. Analysis of variances among means for the three components. F determined at Type 1 error of 5% and the indicated degrees of freedom.

Behaviour	Goat/Sheep/Kid Mean	($\pm$) 2 s.d.	df/F	Calcu- lated F	Significance
Average speed of pro-	25·1	20·6	2112	14·45	$p < 0.05$
gression (m/min)	17·5	16·1	3·07		
	14·7	13·9			
Mean duration of feed-	12·5	70·1	2338	7·49	$p < 0.05$
ing events (s)	15·4	40·1	3·07		
	18·2	94·6			
Mean % of time in-	46·3	34·5	2110	2·89	$p > 0.05$
gesting (%)	43·2	39·5	3·07		
	54·6	39·0			

Table V. Components of foraging behaviour of the goat, sheep and kid herds—t-test of components in which there are significant differences as determined by analysis of variance tests (see Table IV). At df = Goats vs. Sheep; Goats vs. Kids, 88; Sheep vs. Kids, 50; Type 1 error of 5%, two-tailed, t = 2·012.

Behaviour	Goats vs. Sheep Mean	t value/ signif.	Goats vs. Kids Mean	t value/ signif.	Sheep vs. Kids Mean	t value/ signif.
Average speed of pro-	25·1	3·21	25·1	4·79	17·5	1·32
gression (m/min)	17·5	p < 0·05	14·7	p < 0·05	14·7	p > 0·05
Mean duration of feed-	12·5	1·13	12·5	3·48	15·4	2·72
ing events (s)	15·4	p > 0·05	18·2	p < 0·05	18·2	p < 0·05

herds spend approximately the same proportion of time feeding, but differences arise from the greater walking speed of goats as compared to sheep and kids, and from longer feeding events among kids as opposed to sheep and adult goats (Table V).

Division of the animals at Sanjari into herds of adult goats, sheep, and kids, rather than into single species herds, results in the creation of groups that behave differently with respect to exploitable resources. Adult goats move faster than kids or ewes plus lambs, and have shorter feeding events. Kids move and eat more slowly than the sheep or adult goats. Essentially, the adult goat, sheep, and kid herds can be described as fast, moderate, and slow foragers, respectively.

Presumably, kids are separated from adult goats to prevent them from suckling after a certain age (ewes and lambs are unseparated, as sheep's milk is basically valued for meat production in lambs), and to divide up the herding labour. Once the flocks have been separated, however, differences among herds in ecological characteristics of animals can be taken advantage of in shepherding strategies. Adult goats march rapidly for the first (and last) 45 minutes of grazing periods away from camp. During this time, one-third of an afternoon grazing period (n = 3), they may cover between 36 and 50% of the afternoon's distance and do as little as 5–8% of the afternoon's grazing. When grazing begins, goats continue to move rapidly. In the overnight grazing period goats usually reach the outermost limit of the grazing range of Sanjari, approximately 3 km away from camp. Adult goats thus exploit areas not used by other herds. In contrast, kids begin foraging before they have travelled very far from camp, and do not move very rapidly as they graze— they are in fact often simply grazed around and around in a small valley nearby (where *Amygdalus* is abundant). Kids and adult goats eat largely the same diet but, as a result of shepherd decisions to herd them at different distances from the spring, they obtain their food in different places in the range. Sheep occupy an intermediate or overlapping position behaviourally and in terms of the way in which they are grazed by shepherds. As stated, however, sheep are partially differentiated from goat competitors in terms of diet selection. Resource partitioning among animals and breadth of resource use by pastoralists, then, is a function of a combination of dietary differences among animals in different herds, and shepherd herding strategies that take advantage of variations in animal foraging behaviour to divide up the range.

Finally, an additional way shepherds at Sanjari influence resource use is to disperse grazing routes, at least during afternoon grazing periods, on approximately a four day cycle (Map 3(c)). As the area around Sanjari is broken by mountains, there are four major routes available out of camp, and the cycle seemed to follow these choices. Plant regeneration, of course, does not occur over a few days in summer, but this herding strategy may equalize grazing pressure on different parts of the range in the most heavily used areas near camp. This strategy is apparently not used in the overnight grazing period when all herds tend to move in the same general direction to the west. Dispersion overnight results more from the different radial distances from camp attained by the three herds.

In summary, then, apparently minor strategies of herd division and the exploitation of animal foraging differences are important ways in which human management efforts combine with animal adaptations to increase the capacity of herds to feed on available vegetation. Such strategies help to increase carrying capacity and may also somewhat diffuse the destructive impact of foraging.

E. Vegetation

Figure 6 compares densities of annuals (categorized as exposed, *Zygophyllum*-protected, and *Artemisia*-protected) as functions of distance from a central place in the Sanjari, *aghol*, and control series of plots (Map 3(b)). In late summer, annuals are infrequent near Sanjari and its *aghol*. The density of shrub-protected annuals increases, however, over a 1·5 to 2·0 km radial distance from these central places (regression lines are shown over these distances). This central place pattern does not occur in the parallel series of control plots, indicating that foraging pressure by settlement-based domesticated herds substantially alters the distribution of annuals around central places, and that the pressure of foraging decreases steadily with linear distance from settlement.

The graphs in Fig. 6 also show that annuals are denser under *Zygophyllum* and *Artemisia* shrubs than on exposed ground. The preference of desert annuals for shrubs may be due to shade or protection from the wind or to the relatively high soil fertility in these locations—the mounds of loose sand that

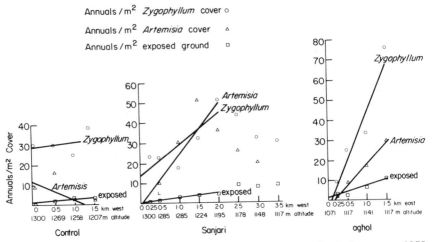

Fig. 6. Density of annuals and distance from a central place, Sanjari, summer 1978.

accumulate under shrubs are high in nitrogen from decomposing vegetal matter and rodent faeces (Garcia-Moya and McKell, 1970). This microdistribution pattern, however, may also be the consequence of herbivory. Annuals germinating on exposed ground are more liable to predation by livestock, while those individuals which germinate under shrubs may be afforded mechanical protection against herbivores.

At 3·0 to 3·5 km from Sanjari, at or beyond the maximum boundary of the grazing range, annuals of most species are relatively frequent in both exposed and protected locations; in other words, no annual species is everywhere totally dependent on shrubs. Closer to camp, however, almost all species are proportionally far more abundant under shrubs than on exposed ground. Those few species that are frequent on exposed ground near camp tend to be relatively very dense under shrubs as well—indicating a successful strategy of satiation. In fact, only two species of annuals are not clearly associated with the cover of shrubs near central places. These are *Astragalus tribuloides* and *Anthemis brachystephena*, members of characteristically toxic genera. In general, these preliminary observations on distribution of annual plants suggest that only those individuals germinating under shrubs are likely to survive to seeding near pastoral camps, as annuals not germinating under shrubs are subject to high predation, unless toxic. The impact of this pattern of herbivory on a population of plants, of course, depends on what proportion of plants are able to set seed before consumption, and whether the seeds are able to survive ingestion.

Figures 7, 8, and 9 graph the relation between central place and the

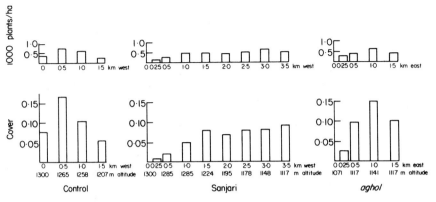

Fig. 7. *Zygophyllum* density and cover, Sanjari, summer 1978.

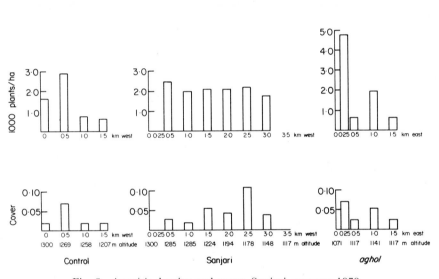

Fig. 8. *Artemisia* density and cover, Sanjari, summer 1978.

distribution of *Zygophyllum*, *Artemisia*, and *Salsola* shrubs in the control, Sanjari, and *aghol* series of plots. Each species shows a different pattern in its relation to central place. *Zygophyllum* (Fig. 7), a dominant shrub virtually never browsed by sheep or goats (although seeds are eaten) but cut extensively for firewood, is very infrequent close to central places. It increases in density and cover, however, over 1·0 to 1·5 km from central place in the Sanjari and *aghol* series. While springtime foraging on seedlings may also be implicated, extensive firewood cutting (for milk processing during spring and summer lactation) may locally increase mortality of *Zygophyllum* to the point where the population cannot replace itself (see Moore and Bhadresa, 1978). The most heavily browsed plains shrub, *Artemisia*, is absent within 250 m of Sanjari, but is unusually abundant within 250 m of the aghol. Beyond this 250 m interval, cover and density of this shrub do not show any patterned variation according to distance from central place (Fig. 8). A study of range ecology near Herat, Afghanistan, also indicated no correlation between *Artemisia* density and water location farther than 0·5 km from a central place (McArthur and Harrington, 1978). These data are also consistent with observations of animal foraging behaviour at Sanjari which indicate that animals forage only very briefly on individual *Artemisia* plants, and occasionally appear to reject plants by smell. Both these lines of evidence suggest the possibility that *Artemisia* produces toxic secondary compounds which, while not preventing herbivory, may reduce it to levels individual plants can sustain, allowing this shrub to persist despite heavy browsing pressure (see Nagy *et al.*, 1964).

Finally, *Salsola*, a plains species preferred by sheep, is patchily abundant both near the central places and at the farthest extent of the grazing range (Fig. 9). Again, this suggests that there is no clear relation between browsing pressure and shrub distribution. *Salsola* is apparently an invader of disturbed areas around pastoral camps, but is also abundant elsewhere and therefore dependent on environmental factors other than disturbance.

In general, degradation of vegetation because of grazing and firewood cutting occurs within 1·0 to 1·5 km of an exclusively pastoral site. The recovery of annual plants grazed by animals, and *Zygophyllum* cut for firewood, occurs as a linear function of distance from central place. Browse

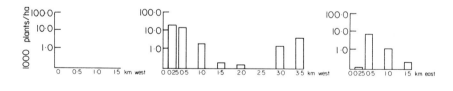

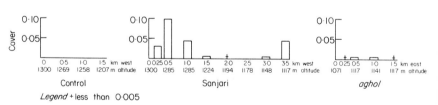

Fig. 9. *Salsola* density and cover, Sanjari, summer 1978.

species, by contrast, appear to be little affected and recover rapidly. Most animals tend to march up to 1·0 km from camp before foraging, although kids are herded in the degraded areas. The fact that resource depletion, particularly of browse, is not greater in a heavily used area of low resource availability, and that relatively depleted core areas remain usable as pastures for kids, is presumably due to plant adaptations minimizing the impact of herbivory, and to animal and shepherd strategies of resource partitioning. It must be reiterated, however, that the available data are not adequate to describe plant distributions and are not amenable to statistical manipulation. Further research on plant community structure, as well as plant chemistry, is clearly needed.

IV. CONCLUSIONS

Range use in this pastoral system involves a complex set of interrelated adaptations by pastoralists, their flocks, and rangeland vegetation. The preliminary data from Sanjari suggest that, in response to regular and heavy browsing pressure from settlement-based domesticated animals, shrubs such as *Artemisia* have adapted through the production of toxic secondary compounds which serve as anti-herbivore defences. While degradation and plant counter-adaptations to herbivory have thus occurred over the long history of Iranian pastoralism, at the same time animal foraging patterns and traditional management techniques have adapted to exploit the remaining vegetation. In particular, shepherds take advantage of the existing dietary and behavioural adaptations among animals to partition the range and optimize exploitation of the available resources. Thus, in the degraded and fully occupied Iranian ranges, management strategies may have evolved in the direction of conservatism. In the study reported here, conservatism is reflected in the role of shepherd strategies for efficient exploitation of the range through dispersal of grazing pressure.

In this chapter, I have drawn on ideas from evolutionary ecology to analyse and explain existing ecological patterns in rangelands, including traditional management practices, foraging by domesticated animals, and vegetation changes in the vicinity of pastoral camps. Traditional pastoralism is an intrinsic feature of the ecology of rangelands in the Middle East and Africa, since vegetation, domesticated animals, and human societies have co-evolved in these ranges from its beginning. Co-evolution in rangelands, however, does not necessarily imply ecosystem stability. It only implies that all organisms in the ecosystem are involved in a continuous process of adaptation and co-adaptation to optimize fitness in the face of intense selective pressures. Thus plant toxicity, animal foraging, and human management strategies are adaptations resulting from the pressures generated by pastoralism, but each of these adaptations has provoked counter-adaptations by affected species. In general, traditional pastoral systems are complexly interrelated with the adaptations and life history strategies of all organisms in rangelands.

The significance of the ecological history of Iranian rangelands for development projects should by now be clear. Under existing cultural and ecological conditions, the introduction of modern technologies to degraded ranges may be extremely difficult or even highly disruptive. Nonetheless, under traditional techniques, these ranges remain productive due to extensive co-adaptations among management strategies, animal foraging behaviours, and plant defences. Development projects which alter previously established relationships among organisms in these ranges are likely to have disastrously ramifying consequences.

Research along the lines reported here can ultimately be used in formulating development strategies that are non-disruptive. These data, however, are preliminary and descriptive, representing research in summer and from one site only, and are as yet inadequate for planning purposes. They do, however, strongly suggest some directions for future work. The chemical relation between foraging animals and range plants, for example, should be studied through experimental feeding trials and laboratory analysis of plant materials. Work on the population dynamics and spatial distribution of range plants under grazing pressure should continue at different types of settlement, and in all seasons of the year, alongside further investigation of animal foraging patterns and pastoralist management strategies. Studies should also include development of a method for quantifying shepherd and owner strategies of resource partitioning, which might include an assessment of their success in terms of animal productivity. Research also needs to be conducted on the perceptions and knowledge of local pastoralists on range quality, herding techniques, and husbanding strategies. Finally, an analysis should be made of the market and subsistence demands on husbanding and production in this area, and the impact on traditional pastoralism of recent social and political changes. Sufficient data on these topics will enable planners to design non-disruptive strategies for development and conservation in this area that are based on readjustment and improvements to the traditional ways of managing herds of sheep and goats.

Endnotes

[1] Most of the data and central ideas of this chapter were first presented in an M.A. thesis (1977) at the University of Pennsylvania. The thesis was subsequently revised for presentation at the Jodhpur symposium on Anthropology and Desertification in 1978 and again for inclusion in the Overseas Development Institute (London) Pastoral Network Papers (1979). Since then, the paper has continued to undergo modifications.

ı

Support for this work in Iran was granted by the Department of Anthropology of the University of Pennsylvania and by UNESCO's Programme on Man and the Biosphere. I wish to thank Dr Brian Spooner and other members of the Turan Programme for encouragement and criticism of my work, and I am especially grateful to Professor Helmut Freitag of the University of Kassel for identifications of plants in the field. Drs Robert S. O. Harding and Daniel H. Janzen of the University of Pennsylvania contributed through criticism and discussion of the manuscript at various stages, and both Dr Spooner and Dr Harding provided invaluable editorial comments. I alone, however, am responsible for all data and interpretations presented here.

[2] The terms r- and K-selection are derived from the Lotka-Volterra logistic equation for population growth in which "r" refers to the intrinsic rate of growth, and "K" designates the carrying capacity. K-selection occurs when resources are stable and competitive ability is favoured, while r-selection occurs when high reproductive rates are favoured under conditions of great resource flux (MacArthur and Wilson, 1967:149ff; Pianka, 1970; Wilson and Bossert, 1971:110–111).

12

Pollen Studies in Dry Environments

P. D. Moore and A. C. Stevenson

University of London King's College
Department of Botany
68 Half Moon Lane, London SE24 9JF, U.K.

The last three chapters in this section, though they are not the only pieces here written by non-social scientists, are the only ones devoted to specifically biological and physical questions. The questions they address, which are both methodological and historical, were nevertheless generated by the same transdisciplinary co-operation, and their results enhance our historical perspective on the interaction of human activities and natural processes in Turan. First, in this chapter, which is a specific example of the use of one of the methods discussed in Dennell, Chapter 3, the authors deal with a basic set of problems in the reconstruction of dryland vegetation history. Pollen studies are also among the methods used in the reconstruction of vegetation history in Rajasthan, cited below by Dhir, Chapter 16.   —Ed.

The balance, or imbalance, between vegetation, climate and land use in arid and semi-arid areas of the world can only be understood when adequate attention has been paid to the time factor. The concept of stability or instability within an ecological system implies a knowledge of the behaviour of that system in time. For this reason, it is essential that studies of man-environment interaction in dry lands should pay due attention to the historical perspective.

Since documentary evidence of past environmental conditions in such areas in extremely scarce, it is necessary to turn to alternative sources, even though they may be less precise than written records. Geomorphological evidence has in some cases been successfully used in the reconstruction not only of past climatic conditions but also at times of human activities insofar as they have affected soils and vegetation. The account of the palaeogeography of the Thar desert by Allchin, Goudie and Hegde (1978), the work of Vita-Finzi (1969) in the Mediterranean, and Chapter 16 by Dhir, below, all provide good examples of this type of work. Another technique which supplies a direct record of vegetation history and which has been used extensively in studies of temperate environments is pollen analysis. Full accounts of the basis and application of this technique are given in such books as Faegri and Iversen (1974), Moore and Webb (1978). The technique is applicable wherever conditions in the past have been suitable for pollen preservation and where a sedimentary process has resulted in a time-related stratification of deposits containing sub-fossil pollen remains. Pollen preservation depends upon low microbial activity and this is most often provided by anaerobic conditions, such as those produced by waterlogging. The vast majority of pollen stratigraphic studies have, therefore, been conducted in sites where lake sediment and peat deposits have accumulated, mainly in temperate, boreal and some tropical areas.

It has also long been recognized that pollen can be preserved in dry deposits, such as the aeolian loess often associated with periglacial climates. In north-west Europe, such loess materials, dating from the closing stages of the last glaciation, are frequent, especially in Belgium, where pollen analytical studies have been carried out by Bastin (1969). However, in such deposits the concentration of pollen grains is often very low, and conventional extraction techniques often need considerable modification before samples suitable for counting can be obtained.

I. POLLEN ANALYSIS IN DRY AREAS

For supplying long and reasonably complete historical records of vegetation development, lake sediments have considerable advantages, but such sites are often not available in desert and semi-desert areas.

Pollen has been successfully recovered and analysed from within aeolian sands in drylands (see for example Price Williams, 1973), but it has also been found preserved in such materials as rock salt (Horowitz and Zak, 1968), subfossil human faeces (e.g. Schoenwetter, 1974a), cave deposits (e.g. Martin, Sabels and Shutler, 1961; Martin, 1973; Schoenwetter, 1974b), buried soil horizons (e.g. Litchfield, 1975), fluvial sediments (e.g. Bonnefille, 1976) and lacustrine deposits (e.g. Niklewski and van Zeist, 1970).

For arid parts of Asia and adjacent areas there are now several published pollen diagrams which trace the history of pollen fallout back over several millennia. For example, Beug (1967) has analysed two lake sites in northern Turkey, lying within the deciduous forest zone to the north of the steppes of the Anatolian plateau. He was able to show a decrease in certain forest trees, particularly *Fagus* (beech) and *Abies* (fir) pollen after about 4000 years ago, which could be a consequence of human land use, but little can be deduced about the pattern of human occupation, or what other changes may have been taking place in the adjoining steppe region.

Van Zeist, Timmers and Bottema (1968), working in southeastern Turkey, consider that human disturbance of the upland vegetation began only 2850 years ago. In south-west Turkey, van Zeist, Woldring and Stapert (1975), who have examined cores dating back to times when glaciers were at their maximum extent in northern Eurasia, show that dry *Artemisia* (sage brush) steppe dominated Asia Minor at that time, although moister phases occurred in which evidence of oak, pine and cedar forest is found. They show that such forest did not expand in south-west Turkey until after 8500 BP, reaching a maximum around 4000 BP and find human interference becoming evident after 3000 BP.

Several reports have resulted from work in the Zagros Mountains of Western Iran, for example van Zeist and Wright (1963), van Zeist (1967), Wasylikowa (1967) and van Zeist and Bottema (1977). These studies centre around two lakes, Lake Zeribar and Lake Mirabad, the former having sediments extending back 40 000 years. From 40 000 to 10 500 years BP pollen assemblages were dominated by *Artemisia,* Chenopodiaceae and Umbelliferae, indicating an open, treeless steppe-desert. Trees (especially *Pistacia, Quercus* and *Acer*) gradually became more prominent towards the end of this period to produce a forest-steppe, not forming a full forest cover until about 5500 BP. The forest expansion here was thus slower than that recorded along the Mediterranean coastlands. Subsequently, they found evidence of some human activity within the forested uplands, which takes the form of peaks in *Plantago major/media* type (broad-leaved plantain) and (especially after 2240 BP) *P. lanceolata* (ribwort plantain).

From Syria, Niklewski and van Zeist (1970) have analysed a lake core which covers most of the last glaciation and the post-glacial period. As in the case of the Zagros studies from Iran, they show an increase in tree pollen (mainly *Pinus* and *Quercus*) in the post-glacial, replacing Chenopodiaceae/Amaranthaceae (Cheno-Ams) and *Artemisia* pollen which dominated the pollen rain of the area during the last glacial episode of higher latitudes.

Even within the tropics there is pollen evidence of aridity during the glacial periods in areas which are now moist. Such results have been obtained by Kershaw (1976) from north-eastern Queensland, Australia. In an area which now receives 2500 mm of rainfall per annum and which in the absence of human interference, would be occupied by tropical rain forest, Kershaw has found sediments dating from the latter part of the last glaciation (27 000–12 000 years ago) in which the pollen of sclerophyllous plants, particularly *Eucalyptus,* dominated the pollen rain.

One of the most detailed studies of sediments within the arid zone of the present day has been conducted by Singh *et al.* (1974) in the Thar Desert of north-west India and Pakistan. They analysed sediments of seasonal salt lakes in Rajasthan and found that the degree of preservation of pollen varied with depth. Often the upper layers of such sites showed bands of low pollen density and poor preservation. Basal deposits consisted mainly of aeolian sands, overlain by lake sediments. The pollen content of these old lake deposits (about 10 000–5000 years ago) is rich in the pollen of grasses, Cyperaceae (sedges), Cheno-Ams and *Artemisia*. This is interpreted as reflecting the development of grass-steppe vegetation following an extremely arid period, during which the aeolian sands accumulated. Between 5000 and 3000 years ago, tree and shrub pollen became much more abundant; such genera as *Acacia*, *Prosopis*, *Capparis* and *Tamarix* were involved. This moist phase was interrupted by an apparently drier period from 3800–3400 BP. Above these horizons, trees and shrubs were not able to survive. The absence of *Artemisia* in these upper layers, could be taken to indicate extreme aridity, for this genus is more typical of steppe than true desert.

The implications of this type of work in the general study of desert history are profoundly important. If Singh *et al.* are correct in their interpretation of the pollen data and other geomorphological evidence, the onset of treeless, desert conditions in the lower Indus region began about 3800 years ago and accelerated around 3000 years ago. The Harappan (Indus Valley) culture of north-west India began to decline after 3750 BP (Singh, 1971); the height of this civilization therefore occurred during the period when Rajasthan was richer in forests. This evidence suggests, therefore, that climatic shifts towards greater aridity in the area were responsible for desertification in Rajasthan and perhaps, also for the decline in the Indus Valley culture.

This interpretation of the data by Singh, which emphasizes the role of climate rather than human factors for the desertification process in north-west India, has been vehemently contested by Vishnu Mittre (1974) whose criticisms are also based upon the use of certain pollen types as climatic indicators. Such types as Cyperaceae and *Artemisia*, which Singh regards as relatively moist climatic indicators, Vishnu Mittre claims in fact represent aridity. The weight of ecological opinion must be on the side of Singh in this debate, particularly since his interpretation is not totally based upon assumptions regarding the climatic indications of individual taxa, but rather upon whole assemblages of pollen consisting of many types and derived from diverse vegetation.

The fact that such diametrically opposed interpretations of a set of data can exist among palaeoecologists does underline one important point, namely the need for a firm basis for the interpretation of sub-fossil pollen assemblages, whether in deserts or elsewhere. The best source of the necessary information is the study of modern pollen fallout in various known plant communities.

II. MODERN POLLEN RAIN IN DRY REGIONS

Individual species of plants vary considerably in the quantity of pollen which they produce, and their pollen grains vary in weight and aerodynamic properties which affect the distance they are likely to be carried from their source. For these reasons, together with various others, such as the time of flowering and the location of the flower (high and exposed, or low and hidden), the pollen fallout from the atmosphere is never a simple reflection of the surrounding vegetation in proportional terms. Further, the degree of survival in soils varies as they are moved and redeposited by water; some species are, therefore, always over-represented and others under-represented.

There have been relatively few studies of modern pollen spectra in the arid zone. The most extensive and intensive is that of Singh *et al.* (1973) in connection with their Rajasthan work. They analysed a total of 114 samples from north-west India, taken in climatic situations ranging from very arid (100 mm) to humid (600–1000 mm). Their most important conclusions for the interpretation of the fossil records of Rajasthan are:

(a) The most arid areas (< 250 mm) have a pollen rain in which trees were poorly represented, *Artemisia* and Cheno-Ams were only sporadically present and *Caligonum* and *Aerva* were the major contributors. These data support their interpretation of increased aridity in Rajasthan after 3800 years ago.

(b) In samples from less arid sites Cheno-Ams and *Artemisia* are more frequent; only in the humid areas (600–100 mm does *Artemisia* reach more than 10% of the pollen rain). Again this result supports the use of *Artemisia* pollen as a climatic index in the fossil studies.

(c) It is evident from the pollen rain data that some pollen grains are carried considerable distances from their source. For example, *Pinus* (pine) pollen is sometimes present at levels of 5% as much as 600 km from its source: some samples still 50 km from pine trees have as much as 50% pine pollen. *Alnus* (alder) and *Betula* (birch) pollen must also have been carried several hundred kilometres. All of these pollen types are found at many levels in the fossil pollen cores from Rajasthan, but these modern pollen rain data indicate that such occurrences may have been due to long-distance transport.

Surface samples from deserts in other parts of the world have also shown a marked representation of tree pollen which has been carried over long distances. For example, Schoenwetter (1974b) found *Pinus* (up to 25%), *Quercus* (up to 10%) and *Alnus* (up to 7%) in surface samples from Oaxaca, Mexico. Rather lower levels of these three trees were found by Schoenwetter and Doerschlag (1971) in *Larrea/Franseria* scrub communities of the Sonoran Desert of Arizona, and also in desert grasslands of the same area by Adam and Mehringer (1975). The strong representation of distant pollen sources in desert areas (see also Hevly *et al.*, 1965; and Hevly, 1968) is probably a consequence of the open conditions and the relatively low local production of pollen by the sparse vegetation. In this respect the pollen spectra of deserts resemble those of the arctic tundra where the long-distance transport of tree pollen is also often a very noticeable feature (Lichti-Federovich and Ritchie, 1968; Birks, 1973; Prentice, 1978).

Obviously, a thorough study of the current pollen rain in any desert area is essential for palaeoecological interpretation of local fossil pollen profiles.

III. CURRENT POLLEN RAIN STUDIES IN NORTH-EAST IRAN

Pollen studies were carried out in 1976 as part of the Turan Programme. Both surface collections of modern pollen rain and fossil deposits from sediments in the area have been analysed. A preliminary species list for the area has been published by Rechinger (1977) and a preliminary statement from more detailed work by Freitag is included in Iran (1977). For the purposes of the pollen survey, the following five vegetation types were selected.

(i) Limestone outcrops. Isolated hills in the area rise to heights of 2280 m and are separated by alluvial plains. Upon these hills scattered trees of *Pistacia khinjuk* grow, and there is an abundance of the shrub *Amygdalus lycioides*. Three sites were examined from this zone.

(ii) *Ephedra* zone. Around the base of the rocky hills a band of vegetation is often found which includes the gymnosperm *Ephedra*. Two species *E. strobilacea* and *E. intermedia* occur in the area, but the latter is more frequent in this zone. Two such sites were examined.

(iii) *Zygophyllum–Artemisia* zone. Extensive areas of the alluvial plains are dominated by the shrub *Zygophyllum eurypterum*, with smaller bushes of *Artemisia herba-alba* growing amongst it. Six sites were studied in this zone.

(iv) Saline areas. In the vicinity of the salt river, around the edges of some undrained basins and especially around the edges of the Kavir, there exist areas with saline soils. These are often dominated by plants of the family Chenopodiaceae (see Zohary, 1973:300). Two such sites were studied.

(v) Disturbed areas. Near human settlements, whether temporary as in the case of sheep pens, or permanent as in the case of villages, a considerable degree of vegetation disturbance has often ensued as a result of grazing pressure and firewood gathering. The vegetation of these areas consists either of annual species or unpalatable herbaceous perennials, particularly *Peganum harmala*. One site was examined.

The precise locations of the sites from which surface pollen samples were taken are shown in Fig. 1. Analysis of the vegetation around these sites is shown in Table I, and the results of pollen analysis in Figs 2 and 3.

The sites examined were chosen subjectively as representative of the vegetation type. Wherever possible the sample area was located in the centre of an extensive stand of uniform vegetation. An area 10 m² was marked out and all shrubs and persistent perennial herbs were counted. Since sampling took place in March, when some perennial geophyte and hemicryptophyte species had not commenced growth, it is possible that some significant species were overlooked. Other species present in the square were listed; these often included many annual species which were growing abundantly at the time of sampling. Other species found in the vicinity, but not within the study quadrat were listed separately.

Adam and Mehringer (1975) working in desert grasslands in Arizona, found a considerable variation in pollen content of surface deposits even when the samples were collected close together and in winter (when local over-representation of flowering species would not be a problem. After discussing various sampling methods that might overcome this problem they conclude that at least five samples are necessary to provide a representative mean value for the major constituents of the pollen spectrum. For this reason we considered that a mixing of ten such samples, followed by subsampling, would provide an adequate representation of the local pollen rain.

From within the square of each site, therefore, we collected ten samples of surface sand with its content of organic debris, each sample being of about 5 cm³ in volume. These were mixed together thoroughly and transported and stored in polythene containers at low temperature (2°C). A single subsample of about 10 cm³ was eventually used for pollen analysis.

Pollen samples were concentrated using conventional techniques as described by Moore and Webb (1978). They were boiled for 10 min in 10% potassium hydroxide solution and then for up to an hour in hydrofluoric acid. They were then washed in 15% hydrochloric acid to remove silicofluorides and subjected to 3 min acetolysis, using a mixture of acetic anhydride and concentrated sulphuric acid in ratio 9:1. This was followed by washing, staining in safranine and mounting in glycerol jelly. Approximately 500 pollen grains were counted within each sample.

The most obvious feature of the pollen data is the high representation of *Artemisia* and the Chenopodiaceae in all samples. This is in agreement with the findings of Wright, McAndrews and van Zeist (1967) in their analysis of surface samples in western Iran. Ten of their samples were taken in the plateau steppe region of Tabriz and eastern Turkey, and their *Artemisia* pollen varied between 10 and 60% (mean, about 25%) and their Cheno-Ams between 20 and 65% (mean, about 45%). These wind-pollinated plants are generally over-represented in the pollen rain.

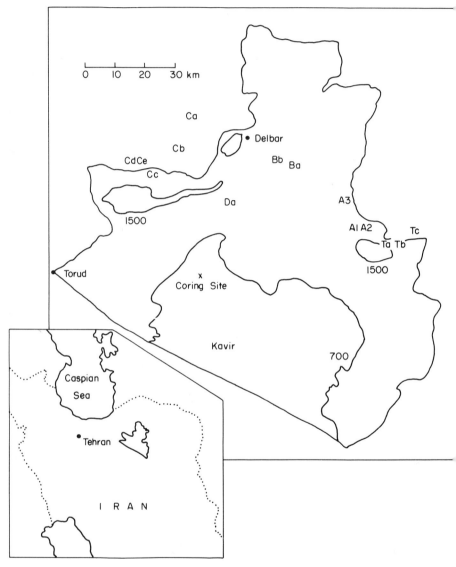

Fig. 1. Map of the Turan area showing the locations of the surface pollen samples listed in Table I and Fig. 2.

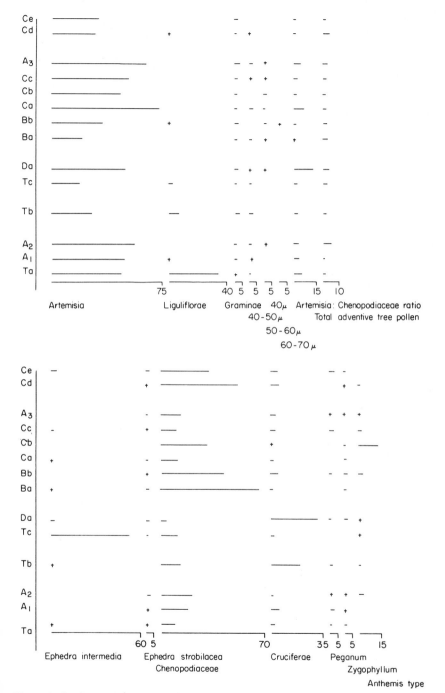

Figs 2–3. Surface pollen spectra from samples taken in the Turan area. Their locations are shown in Fig. 1 and their local vegetation is described in Table I.

Table I. Vegetation analysis of surface sample plots.

	A1	A2	A3	Ba	Bb	Ca	Cb	Cc	Cd	Ce	Da	Ta	Tb	Tc
Anacardiaceae														
Pistacia khinjuk	0	1												
Caryophyllaceae														
Acanthophyllum diezianum	1										1			
Chenopodiaceae														
Halocnemum sp.								+						
Haloxylon persicum				4				+	21	20				
Salsola spp.			6					1						
Unidentified spp.	+	+	+	+	+	+	+	+	+	+	+	+	+	+
Compositae														
Artemisia herba-alba	28	3	5		2	123	53	45	22	1	16	9	0	
Centaurea spp.												2		
Crepis sp.													+	
Echinops sp.							1							
Lactuca sp.	1											9	+	
Cruciferae														
Alyssum sp.	+							+		+		+	+	
Unidentified spp.							+						+	
Cyperaceae														
Carex sp.														+
Ephedraceae														
Ephedra intermedia													3	31
Geraniaceae														
Erodium sp.			1											
Gramineae														
Bromus sp.	+							+		+	+	+	+	+
Poa bulbosa													+	
Stipagrostis sp.				0										

Insect pollinated species, even when abundant, are poorly represented in the pollen samples, with the exception of certain types such as Cruciferae, Liguliflorae and *Scrophularia* type, which achieve high values in some sites. Under-representation is particularly characteristic of some of the shrub species, such as *Amygdalus* and *Zygophyllum* and also of the perennial herb *Peganum harmala*. It is of interest to note that Schoenwetter and Doerschlag (1971) found the same to be true of the shrub *Larrea divaricata* (also Zygophyllaceae) on the scrub communities of the Sonoran desert in Arizona. As in the case of *Zygophyllum* in Iran, *Larrea* is a major dominant of the alluvial plains of the Sonoran desert, but its representation in the pollen rain was found to be normally less than 1% (highest value achieved was 3%). In

Table I—*cont.*

	A1	A2	A3	Ba	Bb	Ca	Cb	Cc	Cd	Ce	Da	Ta	Tb	Tc
Leguminosae														
Astragalus sp.														2
Liliaceae														
Gagea recticulata	0	2												
Polygonaceae														
Calligonum sp.			2											
Pteropyrum aucheri			2											
Primulaceae														
Androsace maxima		+												+
Ranunculaceae														
Anemone biflora	0											+		
Ceratocephalus falcatus	+				0		+		+				+	
Thalictrum isopyroides	39													
Rosaceae														
Amygdalus lycioides	4	2										12	0	
Scrophulariaceae														
Scrophularia striata												3		
Veronica spp.	+												+	
Tamaricaceae														
Tamarix sp.			0											
Valerianaceae														
Valerianella sp.													+	
Zygophyllaceae														
Peganum harmala			5		0	2							0	31
Zygophyllum eurypterum	1		1	1	3	1	2	2			2	0		

All numbers refer to densities of the species 10 m × 10 m quadrat.
+ = Species present in the quadrat, but density could not be readily estimated.
0 = Species not present in the quadrat, but present within 10 m of the quadrat.

Turan, similar values occur for the abundant *Zygophyllum eurypterum*, which has its highest representation in site Bb with 2%.

The pollen of the gymnosperm shrub *Ephedra* is well represented (52%) in one of the samples from the community which the plant dominates, but is also present in most other samples in small quantities. Similarly high values for *Ephedra* have been found in other parts of the world where *Ephedra* communities are known. For example, Schoenwetter (1973) records 26% *Ephedra* from "short-steppe" habitats at an altitude of 3900 m in the Andes of South America. Welten (1957) obtained values as high as 72% in the European Alps, but found a rapid decline in the pollen type as he sampled at greater distances from the plant (cf. site Da). It was long believed, therefore, that *Ephedra* pollen was poorly dispersed, but the finding of airborne *Ephedra* pollen grains over the Great Lakes of North America by Vinje and Vinje (1955) led Maher (1964) to question this assumption. It has since been found in surface samples in small quantities in sites far distant from any source and is generally accepted as a pollen type capable of very long distance transport.

Grass pollen has been separated into various size classes, for cereal pollen grains are large (>40 µm) and can sometimes be identified in this way; cereals are cultivated in the Turan villages. However, many grass genera produce large pollen grains (see Beug, 1961) and this is not, therefore, an entirely reliable means of separation. Grass representation generally is very low, which reflects the paucity of grasses in Turan. Undoubtedly this state of affairs is maintained by intensive grazing. Levels of grass pollen in Turan are lower than those found by Wright *et al.* in western Iran and considerably lower than the various records from Arizona to which reference has previously been made.

The only tree species growing in the area which is represented in the pollen rain (apart from *Platanus orientalis*, *Populus* spp. and *Cupressus* spp. which are planted around some of the villages) is *Pistacia*. Its pollen is found at only 3 of the 14 sites and, at one of these (Ta), a single, fruiting tree was growing nearby. Only a single pollen grain was found, however, but this is not surprising since *Pistacia* is dioecious and the tree was female. Rather surprisingly, 2% *Pistacia* pollen was found at site Cb. Freitag (personal communication) has noted clumps of *Pistacia* in the hills to the south of this site which may provide the source of this pollen.

Since none of the other tree genera represented in the pollen rain are to be found growing in Turan, it must be concluded that these pollen types are being transported from outside the area. The nearest area of deciduous forest is to be found on the northern facing slopes of the Elburz Mountains, about 300 km to the north west of the study area (Zohary, 1973; Sabeti, 1976). The most consistent species, represented in almost all seven sites, is *Alnus*; this genus was referred to earlier as a long-distance transport component in the pollen rain of deserts in India and Mexico. *Carpinus* and *Pterocarya* are the next most frequent genera and these, like *Alnus* are found in the forests of the hill slopes bordering the Caspian Sea. It is interesting to note that neither of these pollen types was detected in the analyses of Wright *et al.* from western Iran, which suggests that the origin of the Turan pollen is indeed the intervening Caspian forests. The records of *Pterocarya* among the long-distance transport component is notable, for this relict genus of trees is known in the fossil pollen floras of the Tertiary deposits of western Europe (see for example, Machin, 1971). Occasional pollen grains of *Pterocarya* were recorded by Turner (1970) in Hoxnian interglacial deposits in eastern England. The likelihood of long-distance transport in these British interglacial sites, however, is not high, for the local pollen productivity of such forested environments could result in considerable dilution of any distant source. It is interesting to note, however, that *Pterocarya* is a pollen type capable of such movement.

Fagus pollen probably derives from *F. orientalis*, the only Iranian species of the genus. This is restricted to the Elburz range (Sabeti, 1976). *Cedrus* is a genus not native to Iran, but it is occasionally planted.

In Fig. 2, total adventive tree pollen is shown as a percentage of the total pollen for all sites. Values fall between 0·8 and 5% total pollen. This high proportion of tree pollen in an unforested environment has often been observed in tundra sites (e.g. Lichti-Federovich and Ritchie, 1980) and on oceanic islands, such as the Shetland Isles, where Tyldesley (1973) was able to associate periods of tree pollen arrival with air mass movements from mainland Scandinavia. The sites with highest arboreal pollen input are A2 (an exposed ridge in the eastern part of the reserve) and Cd (a low lying but western site). One cannot ascertain whether the high tree pollen rain is a consequence of geographic locality, aspect and topography, or low local pollen production. All three factors probably play a part.

One of the aims of surface pollen studies is to permit the recognition of a plant community or vegetation type on the basis of its associated pollen rain. This, if it is possible, facilitates the process of interpretation in the examination of fossil pollen assemblages. From the limited number of surface samples which have been analysed from Turan, it is evident that there is a considerable variation in the pollen spectrum of sites within the same vegetation type (e.g. Ta and Al, or Ba and Ca). Also, many of the most important species, in that they account for much of the biomass of the site, are not well represented in the pollen rain (e.g. *Amygdalus*, *Peganum*, *Zygophyllum*). Consequently, the separation of the communities on the basis of their pollen spectra is very difficult. One could look to "indicator species" which show a certain degree of fidelity to certain communities, such as *Thalictrum*, *Acanthophyllum* and *Ceratocephalus* type, but the frequency of these taxa is low, hence large sample counts and numerous replicates would be necessary to provide an adequate statistical basis for such an interpretative technique.

The abundance of the *Chenopodiaceae* and *Artemisia* pollen types has led to an examination of the possibility of using their ratio as a guide to the vegetation source. This type of approach has been found to be of value in various other palaeo-ecological work. In north-west Europe, the ratio of tree to non-tree pollen has proved useful in studying prehistoric forest clearance, and Turner (1964) has developed an arable to pastoral index based on the ratio of the sum of various indicator species. Young and Schofield (1973), in analysing habitat changes on the Kerguelen Islands, used the ratio of *Azorella* to *Acaena*. High values, they considered, represented a cool, upland climate, and low values a warm, lowland one. Martin (1973) used the ratio of Myrtaceae (*Eucalyptus* and *Malaleuca*) to Chenopodiaceae pollen in her studies of Quaternary changes in the vegetation of the Nullabar Plain of South and Western Australia. High values were characteristic of mallee scrub and low values showed a dominance of chenopods over the scrub vegetation. She showed an increase in the ratio over the past 18 000 years.

In the Turan studies, the *Artemisia* to Chenopodiaceae ratio has been calculated for the 14 surface samples analysed (see Fig. 2). The ratio is very variable within each vegetation type, especially within the *Zygophyllum/Artemisia* zone and the *Ephedra* zone. The saline sites, having strong, local contingents of chenopod species, have very low ratios. Undoubtedly the picture is made difficult to interpret because of the presence of ruderal members of the Chenopodiaceae in many of the non-saline locations.

IV. ANALYSIS OF FOSSIL POLLEN FROM A CORE IN TURAN

Turan is situated at the north-east extremity of the kavir. The geomorphology of the kavir has been described by Bobek (1959) and Krinsley (1970). The sediments are sands and silts which are derived by fluvial inwash from the surrounding slopes and hills during the spring melting of snows. Evaporation through the summer months leads to deposition of gypsum layers in the sediments which may become thick and concreted.

A core of 3 m depth was obtained from a site approximately 1 km into the kavir beyond the last vegetation (*Haloxylon persicum* and other Chenopodiaceae). The core was obtained by a combination of excavation and the use of Russian-type corer (Jowsey, 1966). It was transported and stored in rigid plastic tubes and kept in cold (2°C) conditions in King's College, London, for approximately six months before analysis began. Samples were taken approximately every 4 cm and pollen was concentrated by the same techniques as those described for surface samples. Approximately 500 pollen grains were counted at each level.

The concentration and the state of preservation of the pollen varied with depth, and was generally countable except between 210 and 270 cm. The pollen diagram is given in Figs 4–6. No wood or other organic, macroscopic material was found in the excavated section and the organic content of the sandy sediment was low. Consequently, it has not been possible to obtain radio-carbon dating for the profile.

The overall uniformity of the pollen data with depth is striking, but certain

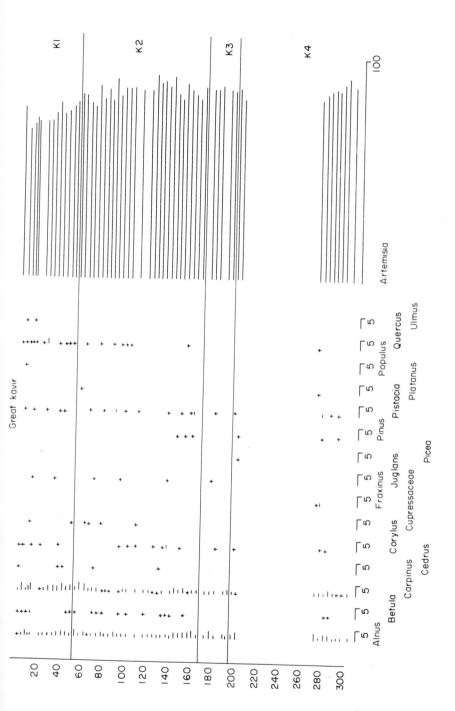

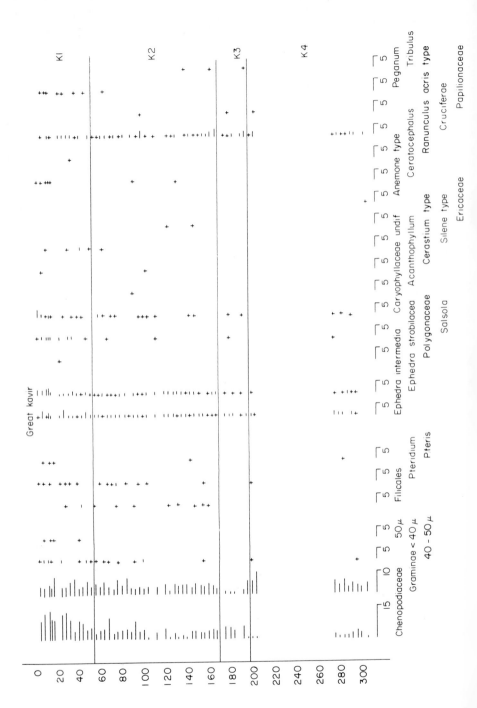

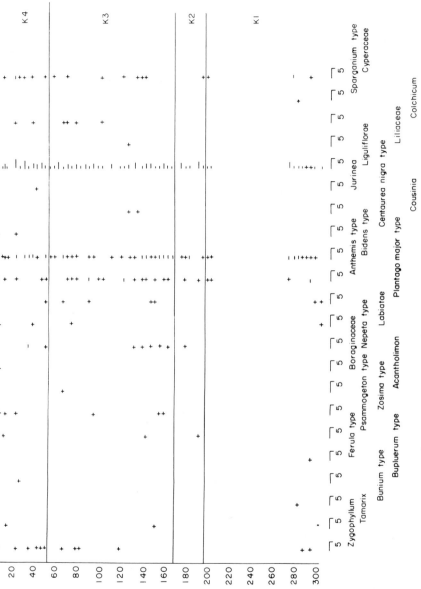

Figs 4–6. Pollen diagram from the Great Kavir, Iran.

changes within the profile are evident which permit a tentative zonation. The main features of the zones are:

K$_1$ local pollen assemblage zone, 302–194 cm.
Very high *Artemisia* pollen, but low Chenopodiaceae especially in the upper part.

K$_2$ local pollen assemblage zone, 194–164 cm
Marked by an abrupt rise in Chenopodiaceae and a concomitant fall in grasses. *Artemisia* remains steady.

K$_3$ local pollen assemblage zone, 164–50 cm.
Grasses rise almost to their former level and Chenopodiaceae fall gradually, but they rise gradually later in the zone.
Fern spores appear and *Pteridium* becomes a regular member of the assemblage. *Quercus* also becomes regular in its occurrence in the upper part of the zone. *Artemisia* remains fairly steady through most of the zone, possibly declining slightly near the top.

K$_4$ local pollen assemblage zone, 50–0 cm.
Artemisia shows a distinct decline through this zone apart from the top sample (in which an anther of *Artemisia* was found). Chenopodiaceae rise to their highest levels. Grass grains of over 50 μm are present, which is suggestive of cereals. *Ephedra* spp. and *Salsola* type become generally more frequent, as do certain other types such as *Silene*, *Ceratocephalus* type and *Peganum*.

Interpretation of this profile is difficult even with the aid of the modern pollen rain data. The lower diversity of pollen taxa in the lower layers could be the effect of differential survival, but this is thought unlikely, for some delicate pollen types such as *Pistacia* have survived even near the base. Also, there are changes in the proportions of the more robust and easily recognizable taxa such as *Artemisia* and Chenopodiaceae, which are not likely to have suffered in this way.

Both zone K$_2$ and K$_4$ are marked by increases in Chenopodiaceae (compare the Syrian pollen diagram of Leroi-Gourhan, 1974). This could be a response to increased salinity or to increased vegetation disturbance by man and his domesticated animals. In the case of K$_2$, the latter alternative would also explain the decrease in grasses, since increased grazing intensity would reduce the flowering of grasses if not their total populations. In zone K$_4$ the rise in Chenopodiaceae is accompanied by a variety of other pollen types indicative of human activity, such as *Ceratocephalus* type and *Peganum harmala*. This latter species has been described as widespread in overgrazed areas of Iran by Rechinger and Wendelbo (1976) and is very abundant near villages and sheep pens in Turan at present. *Ephedra* is known to be poisonous and its increase in the upper part of the diagram could be a response to more intense grazing in recent times. *Pteridium* (bracken) is a fern which invades many temperate grassland and woodland habitat in response to disturbance and grazing, though it is sensitive to very intense trampling. Wendelbo (1976) describes how *Pteridium aquilinum* is currently invading clearings in the Elburz forests, and its increasing frequency in the upper part of zone K_3 and in K_4 could reflect greater disturbance within these distant forests, since it is unlikely to have grown in the drier lands closer to the kavir site.

If these pollen types are indeed indicative of increased human disturbance and grazing pressures, then one must come to the conclusion that there has been an increase in the rate of vegetation change in the period represented by the upper 50 cm of core. The length of time involved cannot be known in the absence of absolute dates.

Pollen analysis of upper layers of sediments in semi-arid areas can thus provide evidence of vegetation changes which should give rise to a cautionary response in those responsible for range management. It cannot, of itself, predict at what stage such changes could result in environmental instability, but it underlines the need to conduct controlled experiments, such as exclosure studies, to observe the consequences and the rate of vegetation adjustment to the stress imposed. More analyses of this type from other areas, combined with further surface pollen studies may lead to greater precision in their interpretation.

13

Desert Shrubs: The Implications of Population and Pattern Studies for Conservation and Management

R. Bhadresa and P. D. Moore

University of London King's College
Department of Botany
68 Half Moon Lane, London SE24 9JF, U.K.

The state of the range, or even of the vegetation generally—in dry lands—when it appears in decline is too easily explained in terms of human activity, especially when that activity is not "current". The study reported in this chapter shows how other causes may in some cases be equally or even more important, with significant management implications for both grazing and cutting.   —Ed.

Professional attitudes to shrubs vary in different parts of the arid and semi-arid world. In North America, McKell (1975) has referred to shrubs as a "neglected resource"; their forage value, he claims, has been underrated and their drought resistance makes them a useful forage material for grazing animals in dry regions. In Africa and Asia, their value has long been recognized and has often been over-exploited (Cloudsley-Thompson, 1974), with consequent reduction in both tree and shrub cover and the replacement of palatable species by unpalatable ones (see Gupta and Saxena, 1971). In addition, shrubs often form a cheap and accessible source of energy for human populations in dry lands. Such exploitation has been largely casual and uncontrolled, though the possibility of managing shrub populations efficiently as an energy resource is currently receiving serious attention (see, for example, Kalla, 1977).

These pressures upon the shrub populations of dry areas have often led to their decline. Ecological surveys of areas in which shrubs predominate generally look into the current state of populations and the demands being made upon them. Sometimes it is possible to determine what factors are

269

limiting shrub population growth without recourse to socio-economic studies of human populations in an area. It may be that these limiting factors are physical or biological (apart from human demands and domestic animal grazing) in which case the human influence must be regarded as a secondary limitation only. Studies on two woody perennial species—a *Zygophyllum* and an *Artemisia*—from Turan illustrate the potential value of such data in the formulation of policy.

I. *ZYGOPHYLLUM EURYPTERUM* BOISS. & BUHSE

Zygophyllum eurypterum is the major dominant shrub of the alluvial plains of Turan. Several hundred square kilometres are occupied by a homogeneous cover in which this species, together with interspersed plants of *Artemisia herba-alba*, is the most frequent and conspicuous. These gently sloping expanses dissected by wadis, are grazed by the mixed goat/sheep flocks of both the permanent village pastoralists and, in winter, the transhumants.

A square, one hectare plot was selected as an exclosure for long-term studies within this vegetation, and the *Zygophyllum* bushes within the plot were counted and measured (see Moore and Bhadresa, 1978). The height of each bush was measured, together with its widest diameter and its diameter at right angles to this. A "volume index" was then devised, being equivalent to the volume of the shrub assuming its form to be cylindrical. This does not provide a true volume measure, but a volume-related index which was found to show a linear relationship with above ground biomass. It therefore provided a means of estimating shrub biomass non-destructively. The result obtained was about 2600 kg per ha.

The number of shrubs within the hectare plot was 531, and their height and volume frequency distributions are shown in Fig. 1. Most natural and healthy populations of plants and animals contain an excess of juvenile over mature classes; a lack of juveniles would indicate a possible failure in the regeneration process. Both these histograms in Fig. 1 show a paucity of small individuals. Since these low height and low volume classes contain the juvenile individuals, this paucity can be taken to imply a lack of recruitment into the shrub population.

A similar situation has been described at Koonamore, South Australia, by Crisp and Lange (1976) for the shrub *Acacia burkitii*. On certain sites they found a virtual absence of plants under about 50 years old. This condition was related to the build-up of intensive sheep grazing in the area which could have resulted in the destruction of seedlings and saplings. Even in areas where sheep grazing has ceased, the continuation of rabbit grazing has maintained the pressure upon *Acacia* seedlings. Rabbit grazing alone has resulted in

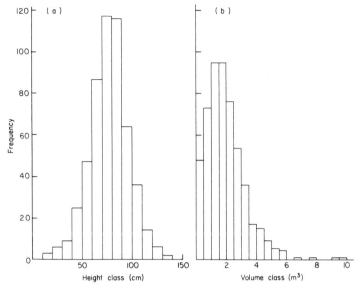

Fig. 1. Frequency distribution histograms of (a) height classes and (b) volume classes (where volume is taken to be $\pi r^2 h$) for *Zygophyllum eurypterum*.

reduced recruitment, but has not caused the total failure of regeneration which is seen where it is combined with sheep grazing.

In the light of this comparable situation in Australia, it is tempting to consider the low recruitment in *Zygophyllum* also as the outcome of high grazing pressures. There are, however, alternative explanations:

(i) Climatic changes (for example, increased drought) could have reduced the flowering, fruit ripening, germination or seedling survival. Of these, flowering and the production of viable seed are known to have occurred by observation. It cannot be discounted that germination and survival may have been adversely affected, but wood increments of the mature plants (which are likely to be influenced by spring water supply) are relatively large for the two years preceding the population analysis.

(ii) Rodent predators of the seeds or seedlings may be preventing regeneration.

(iii) Intra-specific competition in the population may be limiting recruitment. This could take the form of competition for a limiting resource (such as water) or it could result from autotoxicity, or the inhibition of germination among the plant's own offspring by a chemical secretion. This has been reported from some plant populations, such as in *Typha latifolia* (McNaughton, 1968).

If a competitive factor is important, it might be evident from a study of pattern in the population, since intense interaction would produce regular dispersion of individuals (see Woodell *et al.*, 1969 and King and Woodell, 1973). Pielou (1962) has shown that even within clumped populations, the detection of regularity of spacing of individuals within clumps is possible and can provide a means of locating intra-specific competitive interplay.

Two tests are employed for analysing pattern, the nearest neighbour test of Clark and Evans (1954) and the ratio of variance to mean in sample density counts (Greig-Smith, 1964). The first of these gave a result which showed a significant departure from randomness at $P = 0.001$. The second (Fig. 2) also gave a significant departure (at $P = 0.05$) from randomness and gave indication of regularity ($s^2/\bar{x} < 1$).

These results show that *Zygophyllum* bushes are regularly dispersed, which, in turn, suggests that the primary limitation upon the population is intra-specific competition. This being so, it is entirely possible that the lack of juveniles can be accounted for in these terms and without recourse to explanations involving grazing pressure. It is also quite possible that the *Zygophyllum* population is healthy, in the sense that as bushes senesce (the oldest ring-counted individual was 88 years old) and die, a gap in the regular pattern will result, allowing the establishment of one or more juveniles. Thus, despite appearances, the population may be perfectly stable.

It is possible, however, that there are secondary limitations on the population, such as grazing, which will prevent the establishment of seedlings even when the biological constraints are lifted. A success rate, however, of only one seedling per adult per century might be sufficient to maintain the population at its present density. The existence of such a secondary limitation could only be revealed by experimentation.

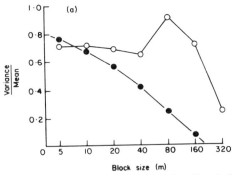

Fig. 2. Variance: mean ratio at various block sizes for *Zygophyllum eurypterum* (open circles). Solid circles represent the lower 95% confidence intervals. The 5 m × 5 m sample size falls outside this interval, demonstrating a significantly regular dispersion pattern.

II. *ARTEMISIA HERBA-ALBA* ASSO

A second study was carried out on the lower slopes of a limestone ridge about 10 km north west of the first study area. Here *Zygophyllum* was present at a lower density and *Artemisia herba-alba* was the most abundant species present. Again, counts and measurements of height and diameter were taken and, again a relationship was found between volume and biomass, but variability was high and a power curve provided the best fit (see Fig. 3). On this basis,

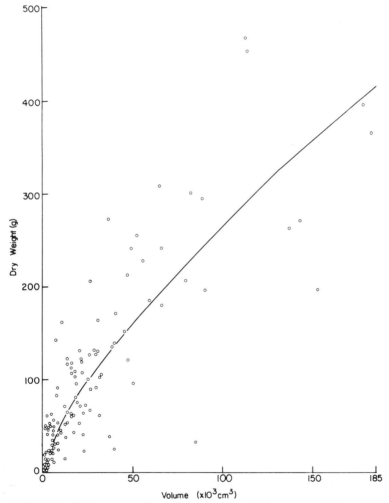

Fig. 3. Relationship between volume index ($\pi r^2 h$) and above ground dry weight for *Artemisia herba-alba*. The line represents a power curve where $y = ax^b$ (a = 0·066 and b = 0·721).

the density of *Artemisia* plants was estimated at 14 500 per ha and the above-
ground biomass 1170 kg per ha. This compares favourably with a value of
1300 kg per ha for *Artemisia/Poa* communities in Syria (Rodin and Basilevich,
1967).

The frequency distribution of size classes (Fig. 4) in the case of *Artemisia*
shows a preponderance of small individuals, which indicates the absence of
severe restraints upon regeneration. Nearest neighbour and variance to mean
ratio tests showed no significant departure from a random dispersion pattern
(Fig. 5).

One must conclude, therefore, that these *Artemisia* plants do not exhibit the
symptoms of intra-specific competition which were evident in the *Zygophyllum*

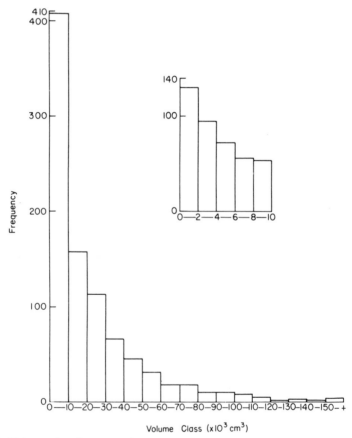

Fig. 4. Volume class frequency for a population of *Artemisia herba-alba* plants within
a 25 × 25 m sample plot. The inset shows a breakdown of the sub-population within
the first volume class.

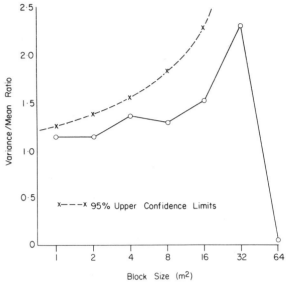

Fig. 5. Variance: mean ratio at various block sizes for *Artemisia herba-alba*. Dashed line represents 95% confidence interval, which is not exceeded at any sample size. The population can therefore be regarded as randomly dispersed.

population. Yet Friedman and Orshan (1975), when investigating *Artemisia herba-alba* populations in the Negev Desert, Israel, found that few seedlings germinated within 10 cm of an adult plant and that there was intensive mortality of seedlings up to 50 cm from the adult during the water-stressed conditions of the summer. The density of adults in the Negev studies was between 4 and 5 plants per m², whereas in the Turan study it was between 1 and 2 plants per m². Evidently, the Turan population has not expanded to a level at which environmental resources are being fully exploited. This suggests that some constraint other than intra-specific competition is holding the population in check. This constraint could be a physical one, or it could be the effect of predation upon the population by domestic or wild herbivores.

III. CONCLUSIONS

Although studies of the population structure and the spatial pattern of plant communities may not provide complete answers to questions relating to the stability of vegetation and the nature of limiting factors, they do supply indications which may be relevant to the solution of these problems. The age structure of populations (or the size structure, where the two are related) provides information on seedling recruitment. A deficiency in a particular age class may indicate a failure to regenerate, but the causes and the effects of such a deficiency are several. A knowledge of the longevity of the species and its density should indicate the recruitment rate which will be necessary for the maintenance of the population. If this is not met, then the population will enter decline, even if only temporarily. Where the cause is a self-regulatory mechanism such as germination inhibition, or high seedling mortality close to adults, the overall recruitment rate is not likely to fall below the essential level, unless some further stress is placed upon it. The existence of self-regulatory mechanisms in a plant population are often marked by a regularity in dispersion; so, an analysis of pattern, coupled with survivorship curves, should elucidate this situation.

A survivorship curve which exhibits a progessive decline in frequency from young (or small) individuals to old, shows that recruitment is taking place, but it is still possible that recruitment is inadequate for the maintenance of a population. If so, in time the survivorship curve would move towards the ordinate of the graph. Only a long term set of observations on the population can provide an answer to this question. Pattern analysis in such a situation is likely to produce a random dispersion (especially if there is some area of inhibition around the adult plant) or clumping (especially if there is no inhibition). Pattern, however, is also influenced by factors other than those associated with population dynamics and growth strategies (see Kershaw, 1974). Underlying pattern in the environment, caused by such inequalities as substrate depth and texture, or microtopographic variations resulting in spatial patterns of water availability, may have a profound influence upon the dispersion of a species in an arid environment. But randomness is a characteristic of immature communities in which a species has not achieved its potential population level (Greig-Smith, 1964). The factors which prevent, or have prevented, this from occurring may be habitat perturbations, some restraining force such as predation, or competition from another species.

A final solution to these problems can be achieved only by experimental studies involving the reduction of the variables present. A useful starting point is the experimental exclusion of grazing animals from certain areas and such exclosures were established in Turan in 1976. Experimentation involving the selective removal of plants has proved a valuable tool for investigations into the patterning of *Larrea* and *Ambrosia* in the Mojave Desert, California (Fonteyn and Mahall, 1978). In one area in Turan, all *Zygophyllum* plants have been removed. If intra-specific competition is the only restriction on germination and establishment, then this removal of adult plants should release this constraint.

14

The Significance of Salinity[1]

Siegmar-W. Breckle

University of Bielefeld
Fakultaet fuer Biologie
Postfach 8640 D-4800 Bielefeld, F.R.G.

No treatment of dry lands is complete without some attention to the problem of salinity, which seriously reduces the productive potential of a large proportion of them. The final chapter in this section explains the salt cycle in general, and discusses the relationship between it and human activity in Turan.   —Ed.

I. THE PROBLEM

A. The Source

Salinity is the presence of excessive amounts of salts, especially sodium chloride (NaCl), in the soil. The main source of sodium and chloride in continental areas is in igneous rocks, where they are bound in small amounts. By weathering and leaching the water-soluble NaCl is formed, dissolved and transported with the flow of water.

Salt deposits are also formed by evaporation from sea water basins, and these deposits provide a second source. A third source, which is often overlooked, is so-called airborne salt. Probably most of the salinity found naturally is airborne, brought in by rainwater and accumulated during geological periods of tens or hundreds of thousands of years (Breckle, 1981). It is attempted in the following to explain the role of salt both generally and in the specific case of Turan. The discussion of related problems in soils which are alkaline, for example where the NaCl toxicity is enhanced by Na_2CO_3 is

277

omitted. Other types of salinity are produced by excessive amounts of SO_4^{2-} ions, especially in gypsum soils but also in soils rich in Na_2SO_4. It is therefore, important to distinguish between chloride and sulphate salinity.

B. The Cycle

Salt is transported and accumulated in the process of the hydrological cycle. Water cycles continuously by evaporation, precipitation and flow partly from the sea and partly from land. Annually some $100\,000\ km^3$ of water cycle through the atmosphere, consuming a high proportion of the total solar energy available.

The ratio between evapotranspiration ET (evaporation E from the ground plus transpiration T by plants) and precipitation P distinguishes dry from humid areas. Dry areas are characterized by an annual deficit. About half of the world's land surface suffers from annual water deficit to some extent, some authors even say two-thirds (see Table I). About one third suffers seriously. The dryness is often accompanied by salinity. Where precipitation is concentrated in one season—especially in Mediterranean and Monsoon areas—the rain may wash away and leach out the salt, and so prevent accumulation, even in areas where annual water balance indicates severe aridity.

In dry areas, however, the water cycle is often not complete, but interrupted; thus the area is endorrheic. The erosion base is not the ocean, but the lowest point in an inland drainage basin, and instead of being returned to the sea, salts accumulate in the basin (see Fig. 1).

The salt cycle on land differs from the water cycle in some aspects. Pure rain water generally contains small amounts of salts, an average of 5–15 ppm. In coastal areas and in deserts sometimes this proportion is considerably

Table I. Arid and humid areas of the world (calculated according to Henning and Henning, 1976).

Continent	Total area $10^6\ km^2$	Arid portion $10^6\ km^2$	Percentage (%) of arid area
Asia	44·2	31	70
Africa	29·8	25	85
North-America	24·2	14	58
South-America	17·8	7·8	43
Europe	10·0	4·2	42
Australia	8·9	7·3	83
Total	134·9	90	66

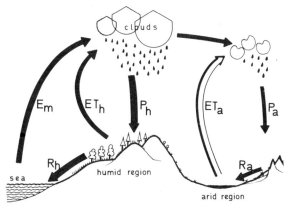

Fig. 1. Hydrological cycle; on left half in humid regions, on right side in arid regions with short-cut water-cycle. E_m: marine evaporation; ET_a: evapotranspiration in arid regions; ET_h: evapotranspiration in humid regions; P_a: precipitation in arid regions; P_h: precipitation in humid regions; R_a: run-off (above and below ground) in arid regions; R_h: run-off (above and below ground) in humid regions.

exceeded through mixture with additional saltspray or fine dust (see Fig. 2) (Ruttner-Kolisko, 1964, 1966; Ruttner and Ruttner-Kolisko, 1973; Waisel, 1972).

Within a particular area it may be more useful to consider the salt balance rather than the salt cycle. The balance can be expressed by the following equation:

$$S_p + S_i + S_r + S_d + S_f = S_{dw} + S_{pp} + S_{cr}$$

where S_p is the salt in natural precipitation, S_i is the salt in the irrigation water diverted into the area (often the most important input), S_r is residual salt in

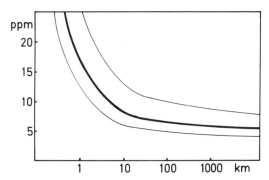

Fig. 2. Decrease of salt content in rain with increasing distance from the sea. Abscissa-axis is in logarithmic scale. Data from Waisel (1972).

soil (which may be considerable even in unirrigated drylands), S_d is salts percolated from substrate minerals and rocks or deposited by ion exchange and dissolved into the soil solution, S_f is salt in applied fertilizers (usually negligible); S_{dw} is salt in drainage or ground water flow, S_{pp} is salt chemically precipitated in the soil or absorbed by ion exchange from substrate, and S_{cr} is salt taken up by a crop (or removed by grazing). For crops, however, not only the whole amount of salt in the soil is important, but also the distribution within the soil profile and at the main root zones (Carter, 1975).

C. The Ecology

Salts affect plant life, and the ecosystem generally. We can distinguish the following different levels:

(1) Biochemical effects, e.g. enzyme activities; investigated by biochemistry

(2) Effect on membranes in cells, e.g. membrane-bound enzymes, electrical potentials; investigated by transport physiology

(3) Effect on cell organelles, e.g. photosynthesis in chloroplasts, respiration in mitochondria; investigated by physiology

(4) Interaction with cells and tissues, e.g. formative effects, succulence; investigated by ecophysiology

(5) Interaction with individual plants, e.g. ecological adaptations; investigated by autecology

(6) Interaction in and with ecosystems, e.g. salt cycling, budget; investigated by ecosystem science, synecology

(7) Interaction in biomes, in biosphere, e.g. cycling, accumulation; investigated by climatology, geology and global ecology

In discussion of desertification levels 6 and 7 are the most immediate, but the other levels must also be kept in mind. In particular, salt affects membranes of cells by osmotic and by toxic effects of ions. Metabolic processes are modified, resulting in lower growth (Caldwell, 1974), lower dry matter production and slower development, and reduced competitive force. These mainly physiological characteristics, varying between minima and maxima, are the base for ecological adaptation between competitors in a plant community. The ecological optima (not the physiological) describe a typical niche formation. In the long term evolutionary trends and genetics are involved, as more salt tolerant ecotypes are selected. However, various strategies are used by plants for success in growing on saline substrates. Mechanisms increasing salt tolerance in plants and variously used by various plant types are as follows:

1. *Tolerance mechanisms*
    osmotic adaptation of cytoplasm
    ion-specific adaptation of cytoplasm
    compartmentation by increased membrane activity
    proline accumulation
    increased absorption of ion antagonists $(K^+, Ca^{2+}; SO_4^{2-})$
2. *Avoidance mechanisms*
    (a) regulation mechanisms
    selective salt-, ion-absorption
    salt leaching
    decreased xylem flow
    increased succulence
    backflow of salt by phloem (?)
    salt recretion by salt glands (external)
    salt recretion by bladders (internal)
    defoliation
    (b) evasion strategies
    short time life cycles (e.g. geophytes, therophytes, hemicrypto-
    phytes) active only during low salinity seasons
    defoliation

Tolerance mechanisms need a real adaptation of major parts of the plant or plant tissues or cells. Avoidance mechanisms are strategies of the plant to evade direct effects of the stressing factor. In some strategies a clear distinction is not possible, as in compartmentation, which means that salt is unevenly distributed over parts of a cell, a tissue or a whole plant; or in decreased xylem flow, which means accumulation in roots and stems and reduction of ion concentrations in leaves. Some types of halophytes may be distinguished by the criterion of growth response (Kreeb, 1965; Walter and Kreeb, 1970; and Fig. 3). Various other parameters can be used in defining types of halophytes (Waisel, 1972). One important factor is the strategy of the regulation of the internal salt content (Fig. 4).

Plants take up salt from the soil in very different amounts. Thus different types of halophytes can be distinguished by this criterion, too. Plants with a high salt uptake are the euhalophytes and the crinohalophytes (Fig. 4). The cycling of salt that occurs mainly in the latter type of halophytes leads to a redistribution in the soil profile and the formation of a mosaic pattern (Sharma and Tongway, 1973). An opposite type of mosaic is shown in Fig. 5, an example of a saline location with high salt cycling in an *Atriplex confertifolia* community in western U.S.A. This semidesertic dwarf shrub community is widespread around Great Salt Lake and in Nevada. High amounts of salt are cycling within the community (Fig. 6). A similar process of salt cycling within

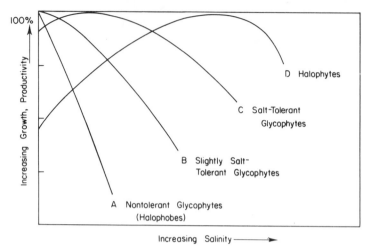

Fig. 3. Schematic graph showing growth of various types of plants under salt-stress (Kreeb, 1965).

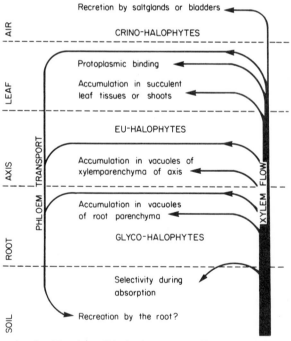

Fig. 4. Schematic classification of halophytes according to strategy of regulation of internal salt concentration (Henkel and Shakhov, 1945; Soloviev, 1967; Breckle, 1976).

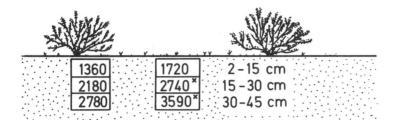

Fig. 5. Mosaic of salt concentration in soil shown by significantly different Cl⁻-values between and below halophytic shrubs (*Atriplex confertifolia* in Northern Utah, after Breckle, 1976). x: statistically significant higher value, 0.5% level, 10 parallels.

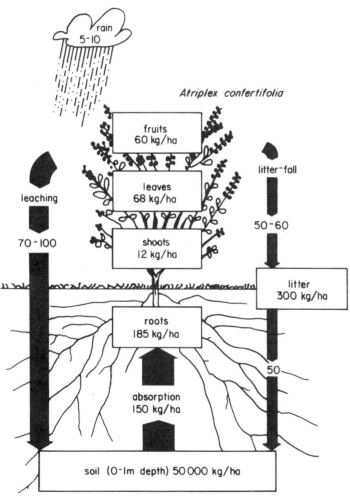

Fig. 6. Standing phytomass and cycling of sodium from an arid saline location in Northern Utah, cold shrubby semidesert with *Atriplex confertifolia* (Larcher, 1980; Breckle, 1976).

vegetation types in Turan is found only in *Reaumurea fruticosa* communities, which cover only small areas and in the *Tamarix* stands.

Little is known about the cycling and redistribution of salt in other areas, although there is a certain amount of information from Australia (Jones, 1970), from Israel (Waisel, 1972), and from Central Asia (Rodin and Basilevich, 1967). The various semidesertic and desertic vegetation types more or less affected by salinity are characterized by Walter (1974) for most parts of non-tropical Asia and by Zohary (1973) for the Middle East. The flora of the Iranian highlands generally is treated by Rechinger (1963 ff.).

II. THE TURAN CASE

Turan is part of a large endorheic system with several independent erosion basins. The 200 m isohyet passes east-west roughly through its centre (see Map 4, p. 142). A preliminary general description of the flora is given by Freitag (1977). Further details may be found in Rechinger (1977) and Rechinger and Wendelbo (1976).

The zonation around the northern edge of the kavir is fairly uniform, although the vegetation belts vary greatly in width. The general catena is shown schematically in Fig. 7. The salt content of the soil exhibits a strong gradient horizontally and vertically. Within many vegetation types the Chenopodiaceae play a major role. This plant family includes many genera and species which have apparently adapted to extreme environmental conditions. This adaptability can be derived partly from the peculiar mineral metabolism of its members.

Because of the apparent importance of the Chenopodiaceae for the salt budget within some desert ecosystems, a classification was carried out (based on field research and collections in the summer of 1976 and 1977) which has some systematic–taxonomic relevance (see Table II). The halophytic type and the character of ion accumulation are to some extent correlated. Euhalophytes accumulate very high amounts of salt, mainly Na and Cl (Wiebe and Walter, 1972; Moore *et al.*, 1972; Mirazai and Breckle, 1978). They exhibit the typical halo-succulence of leaves and/or stems. In contrast glykophytes (that is, non-halophytes, halophobes) have much more affinity for K, which is an important nutrient element. It should be pointed on the fact, that many species exhibit their specific ion pattern in cell sap, even on rather varying soil conditions.

The occurrence of typical bundle-sheaths is an interesting anatomical feature, which gives some evidence on the photosynthetic pathways used by the plants. There are two main categories: the C3 plants, which synthesize a C3 compound (such as glycerolphosphate or dioxiacetonephosphate) as one of

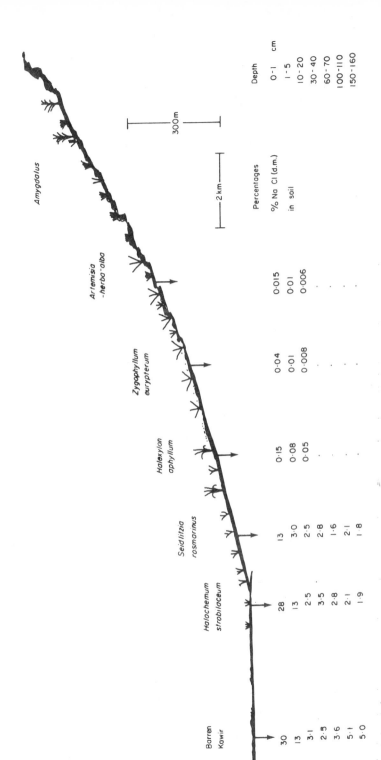

Fig. 7. Schematic vegetation profile of the northeastern fringe of the Great Kavir with dominant species and salt content of soil (%NaCl) to 1·6 m.

Table II. Ecophysiological characteristics in genera of Middle Eastern Chenopodiaceae in systematic arrangement.

Genus		Halo-type	Ion-type	Bundle-sheath
fam.: *Chenopodiaceae* subfam.: Cyclolobeae tribe:				
Beteae	*Beta*	G	K, Na	—
Chenopodieae	*Chenopodium*	G, C	K	—
Atripliceae	*Atriplex*	C, G, E	Na, K	+, −
	Spinacia	N, G	K (Na)	−
	Eurotia (*Ceratoides*)	G	K	−
	Ceratocarpus	N	K	−
Camphorosmeae	*Camphorosma*	G	K, SO$_4$	−
	Panderia	N, G	K, SO$_4$	?
	Kirilowia	G	K, SO$_4$	+
	Londesia	G	K, SO$_4$	+
	Kochia	G	K, Na	+
	Bassia	G	K, Na	+
Corispermeae	*Corispermum*	G, P	K, Ao	−
	Agriophyllum	G, P	K, Ao	−
	Anthochlamys	G, E	K, Ao	−
Salicornieae	*Kalidium*	E	Cl	−
	Halopeplis	E	Cl	−
	Halostachys	E	Cl	−
	Halocnemum	E	Cl, Na	−
	Salicornia	E	Cl	−

the first intermediate assimilative products in their chloroplasts; and the C4 plants, in which the first intermediate chemical compound is a C4 structure, such as malate, oxalacetate or aspartate. The efficiency of C4 plants on average seems to be greater: they can produce the same amount of organic matter while losing less water than similar C3 plants. It is significant that many C4 plants are halophytes.

An interesting example demonstrating correlations between specific ion patterns and systematic position of species is seen in the genus *Salsola*. This genus exhibits a broad spectrum of halophytic adaptation and strategies, though the systematic sections correlate with ion ratios to a limited extent only (see Table III). The section Kali differs greatly from the other sections, and two species have especially strong affinity for K. On the other hand *S. pungens* is now transferred to the genus *Horaninovia* (Botschantzev, 1978)— which is corroborated by consideration of the Na/K ratio (Na:K). The sections *Heterotricha* and *Physurus* are typical euhalophytes with succulent leaves, and high proportions of organic acids which like Halogeton (Dye,

Table II—*cont.*

Genus		Halo-type	Ion-type	Bundle-sheath
Spirolobeae				
Suaedeae	*Suaeda*	E	Cl	+, −
	Bienertia	E	Cl	+
Salsoleae	*Salsola* (+ *Climacoptera*)	E, G, C, P	Na, K, Cl	+, (−)
	Noaea	G,	Na, K	?
	Aellenia	G, E	Na, Cl	(+)
	Traganum	E	Na, K	?
	Cornulaca	E, P	Na, K	?
	Horaninovia	G, P	(Na), Ao	?
	Seidlitzia	E	Na, Ao	+
	Girgensohnia	E, P	Na, K	
	Anabasis	E	Na, Ao	+
	Esfandiaria	E	Na, Ao	+
	Arthrophytum	E	K, Ao	+
	Haloxylon	E, P	Na, K	+
	Hypocylix	E	Na, (K)	+
	Nanophytum	E	Na, K	?
	Petrosimonia	E	Na, Cl	?
	Halocharis	G, E	Na, SO$_4$	? (+)
	Halimocnemis	E	Cl, Na	+
	Halarchon	E	Na, SO$_4$	+
	Halanthium	E	Na	+
	Halogeton	E	Ao	+
	Gamanthus	E	Na, SO$_4$	+

C	Crinohalophytes with bladders	Ao	organic acids
E	Euhalophytes (± succulent)	Cl	chloride
G	Glycohalophytes	K	potassium
N	None-halophytes	Na	sodium
P	Psammophytes	SO$_4$	sulphate
		( )	not pronounced

1956) poison pastures by the free oxalic acid in cell sap of the plants. Furthermore, plants from most sections have alkaloids in their cell sap. More chemical details cannot be given here.

Little is known so far on phytomass and productivity in Turan. As is known from studies of other halophytic communities moderate salinity can enhance growth and increase competitiveness of halophytes, which is an important factor in their natural maintenance. In Table IV some preliminary estimates of phytomass and salt percentage are given which both stress the importance of Chenopodiaceae in the area, and allow an approximation of the amount of salt bound by the various species.

The ecological behaviour of the various chenopod species in Turan may be expressed by an ecogram (see Fig. 8), in which salinity and particle size of soil

Table III. Sodium/Potassium ratio (Na:K) in Iranian and Afghan *Salsola* species (n = number of analysed samples; Na^+/K^+ = ionic ratio; S = standard error).

Section and species	n	Na^+/K^+	S
Aleuranthus			
carinata, sclerantha (leptoclada, turkestanica)	14	6·2	3·6
Anchophyllum			
arbuscula (arbusculoides)	4	4·7	1·3
maracandica	3	9·3	8·9
montana	1	5·5	—
richteri	4	1·0	0·1
Belanthera			
aucheri	2	13·4	3·6
gemmascens	6	6·3	2·1
gossypina	3	1·9	0·8
Caroxylon			
dendroides	1	7·7	—
incanescens	2	3·1	0·2
nitraria	8	8·5	4·7
orientalis	8	6·9	4·0
tomentosa bungeana	1	6·0	—
tom. lachnantha	2	3·1	0·7
tom. tomentosa	8	14·8	3·9
Heterotricha			
brachiata	1	16·9	—
Kali			
aperta	1	3·0	—
iberica (ruthenica, pestifer)	3	0·14	0·16
paulseni	1	0·013	—
pungens (= Horaninowia p.)	2	3·9	1·3
Physurus			
ferganica, turcomanica	15	18·2	4·6
longistylosa	2	10·3	3·3

are taken as variables and some attention is given to water supply. Of course other factors should ideally be taken into account. Water balance, nitrogen supply, alkalinity for example, all modify the relationships shown in the ecogram. The special place of a species within the ecogram is derived by the interference of interspecific competition which, together with intraspecific competition, leads to a typical canopy structure in each of the basic vegetation types (cf. Bhadresa and Moore, Chapter 13).

Besides the osmotic effects of salts, the toxicity of ions must also be taken into consideration. Accompanying ions, such as sulphate (SO_4^{2-}), mag-

Table IV. Main plant communities in Turan with estimated cover, biomass percentages, and salt content (partly after Freitag).

No.	Community (ecotope)	Area (km²)	% in TBR	Approx. % plant cover	Community biomass (t/ha)	Chenopod biomass	% NaCl (d.m.)	t NaCl in TBR
1	Barren Kavir	8400	45·5	0	0	—	—	—
2	Halophytic	1500	8.1	30				—
	a *Halocnemum strobilaceum*	250	1·4	10	1·5	1·4	29	8100
	b *Halostachys caspica*	50	0·27	50	4·5	3·5	17	2800
	c *Seidlitzia rosmarinus*	800	4·3	30	2·5	2·0	16	25 600
	d *Petrosimonia, (Cressa)*	30	0·17	20	1·0	0·4	13	160
	e others							
3	*Zygophyllum-Salsola*	5700	30·9	15	3·0	0·2	c. 8	9100
	a *Salsola tomentosa*	70	0·35	10	2·0	0·5	10	320
	b *Aellenia glauca*	120	0·65	10	2·5	0·7	9	750
	c Others							
4	*Haloxylon aphyllum*	750	4·0	40	6·0 ?	4·5	1·4	4700
5	*Artemisia-Amygdalus*	1900	10·4	15	1·5			
	a *Hypocylix kerneri*	700	3·9	20	2·0	0·8	15	8600
	b *Gamanthus gamocarpus*	120	0·65	10	1·0	0·5	21	1400
	c *Eurotia ceratoides* s.l.	350	1·9	15	2·0	1·2	0·8	500
	d Others							
6	*Calligonum*	70	0·4	20	3·5			
	a *Haloxylon persicum*					0·1	1·2	10
	b *Salsola richteri*					0·5	4·0	150
7	*Astragalus-Amygdalus*	110	0·6					

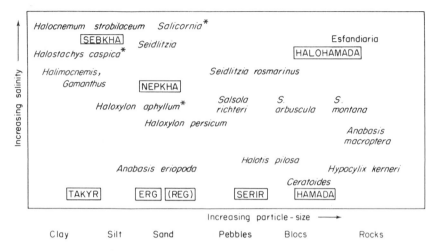

Fig. 8. Ecogram of Chenopodiaceae species and correlation to various desert types (mainly for Northeastern Iran and West- and North-Afghanistan, cf. Breckle, 1982). * With additional water supply.

nesium (Mg^{2+}), and boron (B as BO_3^-), each affect the ecological position of species. Boric acid especially, which is an essential trace nutrient for all plants, appears—in excess—to accelerate salinity effects. Borate salts partly accompany excessive salinity.

Finally with regard to salt accumulation, it is interesting to consider briefly a small kavir known as shaqq-e-Biar on the western boundary of Turan. It has an area of about 14·8 km². Within the 1·6 m soil layer about 500 000 t NaCl are accumulated (based on average 1·2% NaCl). The salt balance formula (above) suggests that most of this amount is derived from precipitation, considering the local conditions, and it can therefore be calculated that the accumulation time, depending on the necessary assumptions, is within the range of some tens of thousands of years (more details of this calculation are given in Breckle, 1981).

III. CONCLUSION

Excessive salt accumulations prevent or limit agricultural and pastoral production over millions of hectares. The problem is especially difficult in dry endorrheic basin systems, which tend to be naturally saline. It is most serious economically where dry lands are under irrigation, and about 25% of the irrigated areas of the world already suffer from salinity to a greater or lesser extent (cf. Merrey, Chapter 5).

Human activity has brought immense amounts of NaCl to agricultural areas. It has been calculated that the Tigris alone has contributed two billions of tons of NaCl to the soils of southern Mesopotamia as the direct result of irrigation (see Kreeb, 1964). A model calculation can illustrate how salt accumulates through irrigation:

salt content of irrigation water (e.g. Tigris)   300 ppm = 300 mg/l
amount or irrigation water applied annually per
    unit area $1000 \, l/m^2$ = 1000 mm
salt input annually per unit area $300 \, g/m^2$
    after 50 years $15 \, kg/m^2$
salinity percentage after 50 years within the 1 m
    soil depth   1%

One percent salinity is beyond the limit for productive agriculture.

Most processes develop over long periods. In some soils a tremendous quantity of salts can be stored between the root zone and the ground water for many years. However, the storage capacity may ultimately be filled, and then large quantities of salt can be expected in effluents for many years causing both local desertification and new salinity problems elsewhere.

Human activities interfere generally with the natural cycles of water and salt. The use of water for irrigation certainly constitutes the greatest interference. In the long term irrigation in arid areas is successful only if enough water is available, not only for irrigation but also for drainage and soil leaching. Removal of organic matter by collecting (see Horne, Chapter 10) for fodder or by grazing is in this respect only of minor importance—though it may change thermal properties, albedo, or evapotranspiration values substantially, with unpredictable effects.

Within central Iran human activities are hindered by the vast expansion of the kavir basin system. With the exception of a few small oases cultivation is not possible, and even an extremely high input of water would not change this situation significantly. Grazing is also greatly disadvantaged, both because of low productivity and because of salt. But it is worth pointing out that grazing on non-saline or slightly saline areas does not lead to increased salinity.

In Turan generally salinity causes only minor problems, because of the low salinity of the water used for irrigation, which is mainly from qanats and to a lesser extent from run-off at the higher levels of the plain, and because of good drainage of the small irrigated fields on terraces. Salinity could, however, become a serious problem if more use were made of run-off at lower levels on the plain where there is more salt and poorer drainage.

Each year more is understood about managing salt affected soils in relation to the adaptations of various wild and domesticated plant species. But the problem is far from being solved, especially because of the tight correlation with water balance and climatic conditions.

Endnotes

[1] The field research on which this chapter is based was partly supported by the exchange programme between the universities of Bonn and Kabul, the DAAD (German Academic Exchange Service), and various other authorities in Germany and Afghanistan during and after stays in Afghanistan. In Iran the research was supported by a Schimper-Fellowship for extra-European Ecological Studies and by the Department of the Environment (Tehran). All this support is acknowledged with gratitude. Also, the assistance of Dr Mirazi, Mr Rasik Bhadresa, Dr K. A. Schmitt, Ms Irmingard Meier, and especially Ms Uta Breckle was invaluable, and I am grateful to Dr Brian Spooner for his help and advice, and to Professors Heinrich Walter and Yoav Waisel for discussions, advice and criticism.

Section B
The Central Arid Zone Research Institute (India)

H. S. Mann

The earliest civilization in India reached its peak in the arid northwest of the subcontinent at the beginning of the second millennium BC comparable in many ways to similar developments a millennium or so earlier in Mesopotamia and Egypt. Its decline a few centuries later has not been satisfactorily explained, but a number of natural and cultural factors are likely to have been involved and the arguments for them (some of which have been pursued with some vehemence by various scholars over the last few decades) are rehearsed briefly in the second chapter of this section. One or another form of desertification is likely to have been an important contributing factor.

Arid and semiarid conditions have continued to present a challenge to Indian society throughout its history, not only in the arid northwest which now constitutes—within India—the western half of the state of Rajasthan, southern Punjab and Haryana and northern Gujarat, some 320 000 km² in all, but throughout the broad zone of territory classified as semiarid which stretches from the state of Punjab in the north to Tamilnadu in the south, altogether nearly 920 000 km² (see Map 1 and Table I). As a result of the irregularity and unreliability of precipitation which is one dimension of these conditions, famine, as is well known, has been a constant threat throughout the area not only since the recent population explosion but from early historical times—which implies that the pressure of human use on the renewable natural resources of India's drylands must have been both continual and ecologically and historically significant.

It is, however, also significant that despite the vastly increased population, famine is probably less of a threat to the Indian farmer now than ever before. The causes of this improvement lie not only in technology but in administration or organization—especially the organization of storage and distribution of food. The implicit ecological question—to what degree has the productivity of the land suffered as the result of the ever increasing pressure of cultivation and grazing—is difficult to answer definitively for the area as a whole, but it is certain that the land is now producing more than ever before

293

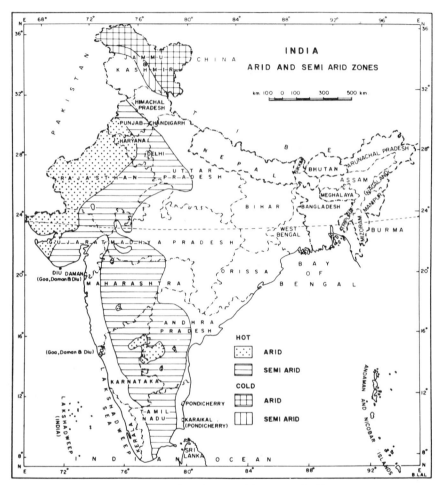

Map 1. Arid and semi-arid zones in India.

to the point where, contrary to all predictions in the 1960s, at the end of the 1970s India had once again achieved self-sufficiency in food grains. But the ultimate desertification question—the outlook for the ratio of population to productivity in India's dry lands in the long term—can be answered only through careful synthesis of work on the relevant natural and social factors focused on the target of developing more efficient and more acceptable systems of organization. Until recently, in India generally, as elsewhere, attention has concentrated mainly on technological solutions, but the papers which follow demonstrate that a sociological approach is also followed and is gaining ground in the scientific dialogue.

Table I. State-wise area of arid and semi-arid zones in India.

Region	State	Area in sq. km		% of area under each state	
		Arid	Semi-arid	Arid	Semi-arid
North	*Cold* Jammu and Kashmir	70 300	13 780	—	—
	Hot Uttar Pradesh	—	64 230	—	7
	Madhya Pradesh	—	59 470	—	6
North-west	Rajasthan	196 150	121 020	61	13
	Gujarat	62 180	90 520	19·6	9
	Punjab including Haryana	27 350	58 650	9	6
South	Maharashtra	1 290	189 580	0·4	19
	Mysore	8 570	139 360	3	15
	Andhra Pradesh	21 550	138 670	7	15
	Madras	—	95 250	—	10
	Total	387 300	970 530	11·80%	29·57%

Total area of India—3 282 016 sq. km

Until recently most of the success in obtaining higher productivity in dry lands came by methods of trial and error and by learning from successful experience. Although the environmental challenges are very similar in the various arid areas of the world, different societies have reacted to them and evolved in markedly different cultural and institutional ways. The most spectacular technological successes have been in large scale irrigation, but these successes also provide examples of one of the most serious problems in dryland development. The economic and demographic growth which accompanies them apparently often occurs too fast for the structure of the society, which consequently undergoes a certain amount of *ad hoc* adaptation but in the process tends to lose flexibility and remains unsuited for the long term efficient exploitation of the new technological system. Where irrigation is introduced into dry lands the resulting production systems become highly specialized, so that if the system goes wrong it cannot easily be returned to its former condition. Technological development generally, and in dry lands especially, tends to lead to loss of flexibility in both resources and in social forms. Planners need to keep the implications of this loss of flexibility fully in view, and social and natural scientists should pay more attention to it in their development oriented research in India especially.

In the Indian, or "Thar", desert and its margins the consciousness of these problems began to develop in the fifties. The situation has fortunately not reached a point of no return with regard either to desertification or to flexibility of land use or social forms. Since early historical times this area had been divided among a number of relatively small polities, known during the British period as Princely States. Up to the end of the Moghul period, warfare was endemic, and no one ruler was able to organize sufficient numbers of people or sufficient investment to bring about a situation in which excessive pressure would be brought to bear on the land over a significant period. Until India attained independence no unified approach to the systematic exploitation or development of the area could evolve. It is even claimed by some historians that some rulers discouraged development in order to avoid attracting alien forces. With unification of over 20 Princely States to form the State of Rajasthan in 1949, the importance of the Indian desert and its vital role in the economy and development of the country began to be realized. As the global significance of drylands attracted more and more attention among food planners in international development agencies in the 60s and 70s, India's attention to her own dry lands also grew. In the Thar desert and its environs, because of a combination of a number of natural and historical factors, India effectively has a "field laboratory" for dryland research that is unique. The most significant of these factors are depth of soil, variety of well adapted grasses, shrubs and trees, density of population, richness of historical records, and relative flexibility in both land use capability and social forms.

Conscious of the significance of its drylands the Government of India established a Desert Afforestation Research Station at Jodhpur in Central Rajasthan in 1952. In 1957 this station was expanded and became the Desert Afforestation Research and Soil Conservation Station, and in 1959 it was upgraded to the status of a fully-fledged research institute at the national level and renamed the Central Arid Zone Research Institute (CAZRI). The UNESCO Advisory Committee for Arid Zone Research played an important role in this process, and a long term relationship of technical co-operation was established with the Australian Commonwealth Scientific and Industrial Research Organisation (C.S.I.R.O.) through the Colombo Plan. The process was also encouraged within India by the National Institute of Sciences (later renamed the Indian National Science Academy) which organized a symposium on "The Rajasthan Desert" in 1958. When in 1966 the Government of India reorganized agricultural research, CAZRI became one of thirty-three research institutes under the Indian Council of Agricultural Research.

The Central Arid Zone Research Institute today is the result of nearly thirty years institutional and intellectual evolution in response to developments in the ecology of arid lands at the local, national and international levels.

As CAZRI rose in status it grew also in size of scientific staff (who now number over 120—all Indian, and largely Indian trained) and expanded its disciplinary coverage. After Desert Afforestation, Soil Conservation and supporting disciplines in the physical and biological sciences, in 1960 a Human Factor Studies Division was established. Since this Division was probably the first autonomous cell of social scientists in an institution of this type anywhere, it deserves special notice. The insight that problems of the arid and semi-arid region in India are generally problems of *human* ecology was given organizational recognition. This recognition was based on the implicit argument that since the results in the physical and biological sciences at CAZRI would provide methods to influence land utilization, domestic economy, habits and values, a social science division was necessary in order to facilitate adoption of these results. Not only must research results be of a kind that are acceptable and applicable to the changing society, but the society has to be conditioned to accept them. At the time this orientation set CAZRI conceptually at the forefront of applied ecological research.

The Human Factor Studies Division at CAZRI began in the early sixties with the investigation of existing social situations in the arid region of Rajasthan and sought ways to mediate between those situations and the recommendations of the other (physical and biological science) divisions, according to the following explicit terms of reference:

(1) To conduct socio-economic surveys of settled, semi-nomadic and nomadic populations in relation to demographic structure, caste differentials, household characteristics, occupational structure, production pattern and livestock population which could aid in formulating development plans for raising the standard of living of the population;

(2) To study the nature and extent of interaction between man, land, livestock and environment, and the extent of symbiotic relationship between settled and nomadic groups;

(3) To estimate the economic efficiency of various components of arid land management technology;

(4) To incorporate the new technology matrix into revised resource planning in the Indian arid region, and ;

(5) To impart statistical, service and data processing advice to various divisions of the Institute.

The Division was expanded in 1973 to include economics and statistics, and there are now plans to split it into two divisions of which one would focus on developing, maintaining and improving the socio-economic data base for the Indian arid zone as a whole and the other would conduct intensive studies of selected communities with the aim of formulating explanations of change in the relationship between human activities and natural processes.

Meanwhile, the biological and physical sciences expanded into five divisions: Basic Resources, Plant Studies, Animal Studies, Wind Power and

Solar Energy Utilization, and Soil-Water-Plant Relationships, and in 1974 a division of Extension and Training was added, with specific responsibility not so much for extension in the usual sense but for communicating CAZRI's research results to its immediate farming public, monitoring their application, investigating problems and providing feedback by means of various activities including an Operational Research Project. This structure is illustrated in the attached chart (Fig. 1), which shows that besides being organized into seven divisions, each of which encompasses several disciplines, CAZRI also now includes centres for seven All India Coordinated Research Projects.

Unlike the Turan Programme in Iran, which was a largely informal association of scientists all focusing on one general research problem from different disciplinary angles, CAZRI is responsible for basic research in all disciplines bearing on dryland problems. As might be expected from the chart climatology, geomorphology, geology, hydrology, pedology, botany, agrostology, range science, forestry, plant ecology, plant physiology, plant genetics, agronomy, animal ecology, animal physiology, animal nutrition, solar energy utilization, sociology, social anthropology and economics are all represented. Each year individual scientists propose projects through their Division Heads to a co-ordinating council chaired by the Director. The problems of application of the findings of these disciplines to problems of soil conservation, pastoral, horticultural and agriculture production, and pest management are investigated on research farms and range management stations at Jodhput and elsewhere in Rajasthan at Pali, Jaisalmer, Jhunjhunu, Barmer, Samdari, and Borunda (see Map 2), and in the Operational Research Project with volunteer farmers in a cluster of five villages 16 km northeast of Jodhpur.

In this latter programme CAZRI scientists investigate operational constraints in the transfer of their technological innovations, and work in collaboration with State Development Departments. Sand dune stabilization, afforestation and shelterbelt plantation, pasture and grassland development, watershed management for stable crop production, improved management of rainfed and irrigated crops, optimum water utilization methods (drip and sprinkler irrigation), rodent and pest management, solar energy and bio-gas utilization, and community organization constitute the major emphases of the Project, which has the added advantage of increasing farmers' awareness of scientific assessments of their resources and of the significance of ecological trends on time scales other than those of their own perception.

Since its initial establishment in 1952, and more so since its later expansion and strengthening, CAZRI has been responsible for a number of technological advances and improvements, and has been associated with the development of others, a number of which are now being implemented by state development agencies and central government departments not only in

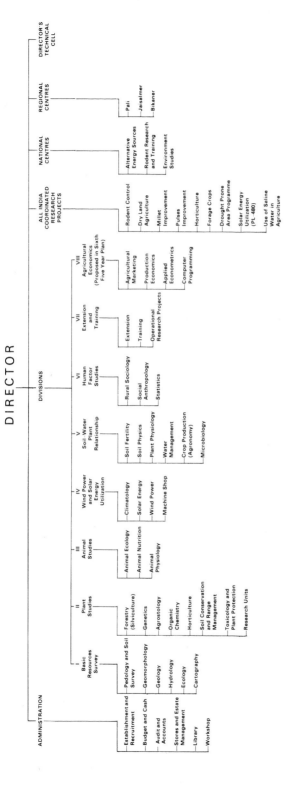

Fig. 1. Organization Chart. Central Arid Zone Research Institute, Jodhpur.

Map 2. Range management areas of CAZRI in Western Rajasthan.

the arid but also in the semi-arid regions of the country. The following is a summary of the highlights of CAZRI's work:

1. Integrated Basic Resource Surveys. These surveys are multidisciplinary ecological descriptions and assessments of administrative districts, together with statistical summaries of their human populations. They have been commissioned so far for Chitradurg (Karnataka), Santhalpur (Gujrat) and Mahendragarh (Haryana) by the respective state governments, and recently the Rajasthan government has requested a similar survey of the Upper Luni Basin. As part of its basic research on the arid region of northwest India CAZRI plans to carry out surveys of the whole region and has so far completed about a third, radiating out from Jodhpur. The importance of these surveys and their attraction for state governments and planning agencies lies in their co-ordination of data from the physical, biological and social

sciences. This part of CAZRI's programme is run from year to year by an interdisciplinary council formed for this specific purpose.

2. Sand dune fixation and afforestation. For this purpose CAZRI has worked mainly with *Acacia tortilis* and has had considerable success. The states of Gujrat, Haryana and Rajasthan have applied CAZRI's methods in selected areas, and the Rajasthan Government has created a Directorate of Afforestation and Pasture Development for this purpose.

3. Horticulture. A modified propagation technique has been developed by which budded *ber* (*Zizyphus mauritiana*), a fruit tree well adapted to local conditions, can be raised within a period of 5 months. This technique is being adopted by local farmers with considerable economic success.

4. Range Management and Livestock Production. Productive ranges with optimum wool/meat/milk return in local conditions have been developed and demonstrated in 52 paddocks in different parts of the State. Forty of these have been transferred to State Government agencies and work is continuing in the remainder. The results are being applied in the national Drought Prone Area Programme, which is being executed by the states with World Bank collaboration.

5. Rodent Pest Management. As a result of success in this area CAZRI has been designated the National Centre for Rodent Research and Training and a national programme has been launched.

6. Solar Energy Utilization. A number of prototypes of small appliances suitable for rural areas have been developed for water heating, cooking, drying, and distillation of brackish water.

CAZRI's involvement in training and in the communication and publication of work in the ecology of arid lands has grown considerably over the last decade. Besides the international linkage referred to in the preface, several national and local symposia and training courses have been held, and several volumes of scientific papers and syntheses have been published. The journal, Annals of Arid Zone, published by the Arid Zone Research Association of India (based at CAZRI) is now in its 20th year. In all linkages—both domestic and international—CAZRI is the only institution of its type that puts together the various disciplinary perspectives on drylands from the point of view of a developing country.

Like all other research institutions of this type CAZRI continuously faces the problem of how to integrate the findings of its various discipline-based teams into significant packages or messages that will be successfully adopted or assimilated by the public for which they are meant. The problem is all the more difficult when the product conflicts in some way with the values or the preconceptions of that public. The scientists at CAZRI are particularly aware of this problem and make piecemeal headway against it as and where they can. At another level, however, CAZRI has accepted the challenge to lead public opinion in the direction dictated by its own disinterested collection and systematic analysis of data on two significant themes—both of which have to do with the world debate on desertification.

Firstly, in the face of the continuing appearance of statements in the media to the effect that the Indian desert is advancing eastwards at an identifiable and inexorable speed of so many kilometres per year, CAZRI has consistently expounded the view that the desert is not advancing, but rather desertification is a spatially discontinuous—spotty—process that appears here and there as the result of specific identifiable human activities in particular conditions. In this way CAZRI has sought to demystify and de-sensationalize desertification and turn people's attention to concrete cases *that can be combatted.*

Secondly, CAZRI has thrown its weight behind a synthesis of statistical data (see Sharma, 1972) to the effect that although (1) the population of the arid region of Rajasthan has grown faster over the past 60 years than the population of the State of Rajasthan as a whole, (2) the population of all Rajasthan has, over the same period, grown at a higher rate than the population of India as a whole, (3) the animal population growth rates parallel the rates for the human population, (4) the cultivated area has increased similarly, and fallow periods have decreased and (5) figures for agricultural production show a decline in productivity per hectare for all crops, nevertheless (6) the general standard of living of the population of the arid region of northwest India has steadily risen. The immediate and obvious reaction to this synthesis is that such a combination of processes cannot continue indefinitely: sooner or later a point will be reached when the cultivated area cannot increase further and productivity will drop below demand with disastrous consequences. We do not seek to contravert the logic of this argument. Rather, we point to the fact that this situation in Rajasthan is unique in the comparative study of the ecology of arid lands, and that it demands the most careful investigation, in order to make the most of it for development planning in dry lands generally. Whatever the ecological future of Rajasthan may be, the area is so far the most densely populated and the most successfully exploited arid region in the world. And it is also the only such region in the Third World that contains its own established specialized ecological research institute.

The four chapters of this section have been selected with the aim of representing the social dimension of CAZRI's work. The selection begins with a review and interpretation of the social mechanisms that mediate the relationship between human activities and natural processes before the industrial era. Chapter 16 extends the history of that relationship back into archaeological periods in order to form the broadest possible basis for assessing future prospects. Chapter 17 investigates the administrative relationship between the land user and government. The final chapter deals with ideas and perceptions, and attempts to unravel the process whereby deprivations of drought and famine have been experienced and rationalized as a common sense basis for future action. With this combination of social, historical, administrative and cultural analysis we present a summary of where we stand in our research in relation not to technical solutions but to the human response to desertification in western Rajasthan.[1]

Endnote

[1] CAZRI has enjoyed important support and encouragement from several quarters throughout its history; and this is an appropriate place to make acknowledgement. UNESCO played the essential role of facilitator and catalyst for vital international linkages. The Government of Australia helped consistently by financing equipment and international visits. We owe a special debt to Mr C. Christian who provided the rationale for upgrading the Institute of 1959 and reorganizing it into interdisciplinary divisions, including one devoted to the Human Factor. Within India, and more recently, CAZRI has especially benefited from the administrative encouragement and scientific guidance of Dr M. S. Swaminathan, who was Director-General of ICAR throughout most of the seventies. We are very happy to offer our sincere gratitude to each of these benefactors. Finally, I myself must acknowledge the support I have enjoyed, during my tenure as Director, from CAZRI's scientific staff. Though relatively few are named as authors in this section, much of the work that is represented here derives from co-operative interdisciplinary projects, for which credit must be given to the whole team.

15

Desertification and the Organization of Society

S. P. Malhotra and H. S. Mann

Central Arid Zone Research Institute
Jodhpur (Raj.), India 342003

Perhaps the most difficult problem in ecological research is to integrate social and natural science approaches into a unitary argument. Invariably, interdisciplinary co-operation generally—but especially in this field—produces theoretically un-balanced interpretations, with only one discipline providing the explanatory con-clusions, however integrated the presentation of data may be. The first chapter of this section is the product of co-operation between a social and an agri-cultural scientist. Though brief, it presents a disciplinarily balanced view of the relationship between human activities and natural processes in Western Rajasthan. It is representative of the interdisciplinary tradition that CAZRI has established, and provides an appropriate introduction to this section on the Indian contribution.   —Ed.

Desertification is not a new phenomenon. Human activity has generally had a deleterious effect on the productivity of natural resources, but the rate of deterioration has in most cases been slow—so slow that it has not been impractical to think in terms of equilibrium. Many studies in the ecology of "traditional" dryland societies have emphasized the balance between natural and social processes. For several decades now the balance has been disintegrating. The general causes of demographic, social and economic change are often cited, but the details are rarely studied. Western Rajasthan is an interesting case for such a study, because of the variety of specialized social groups whose interdependence seemed to be self-regulating, until it was disturbed by the effects of changes in the larger society. In the new situation of relatively rapid social change, desertification has become a serious threat, which can no longer be counteracted by internal mechanisms.

I. INTERNAL BALANCE

Choice of resource is the most significant factor underlying social differentiation. In western Rajasthan it results in division of the population into two major groups: nomadic and sedentary. Pastoral nomads inhabit the areas which are too arid for cultivation and supply livestock and pastoral products to the sedentary population in exchange for cereals and other crops. Other nomadic groups provide specialized services to the sedentary population in return for grain (Malhotra, n.d.). Both nomadic and sedentary populations are further broadly grouped on the basis of their traditional occupations. There is a definite division of labour, a pattern of exchange relationships and a distributive use of the eco-system among nomadic groups, between the nomadic and sedentary populations, and among sedentary groups. Based on their functional specialization, the nomads may be categorized into four groups: the pastoral nomads, who are Raika, Sindhi, Parihar, and Billoch; the trading nomads, Banjara, Ghattiwala Jogi and Gowaria, who deal in salt, combs-beads-mirrors, and cattle, respectively; artisan nomads, Gadoliya Lohar, Sansi, and Sattia, who serve as blacksmiths and general repairers; and miscellaneous groups, such as Nat and Kalbeliya Jogi, who are concerned with entertainment. Each named group has the status of a sub-caste in traditional India society.

The sedentary population had a heterogeneous caste composition. Each named group performed certain functions, which were interdependent and complementary. On the basis of these traditional occupations, sedentary caste-groups may be grouped into five broad categories: agricultural groups, such as Jat, Bishnui, Rajput; craftsmen, including Suthar, Lohar, Bambi, Sonar, Nai, and Darji, who are associated with tools, hair dressing, gold and silver work, tailoring and cleaning; religious practitioners, including Brahmin, Sad, Swami, Sewak, Dakota and Dadi; money-lenders and traders, such as Mahajan, and Khatri; and sedentary groups concerned mainly with animal husbandry, including Raika and Muslim groups such as Gopera, Kharla and Billoch. The interdependence of the different caste-groups in the economic organization of the village is most conspicuous at marriages and other festivals, when each group has a specific contribution to make. The services of the craftsmen groups especially are determined by what is known as the *aat* or *jajmani* system, under which every family caters to the needs of a group of families for a particular type of service, payment for which is fixed by tradition and made on an annual basis either in cash or in kind at harvest. The amount to be paid depends upon the amount of work, and families with large agricultural holdings or large number of members generally pay more. Sometimes a part of the services are governed by aat and a part by other transactions.

This functional specialization and intricate exchange system not only knitted the society together, but also facilitated a more homogeneous exploitation of the ecosystem. Only the pastoral and agricultural castes put direct pressure on the land. The socio-economic structure provided for a balanced use of the eco-system for the benefit of the society as a whole. Within this framework, both the nomadic and sedentary population has further evolved mechanisms for the maintenance of resource use equilibrium.

The pastoral nomads—semi-nomads would be more correct, since they use permanent habitations for part of the year—contributed to the economy of the region by providing milch cattle, mutton, ghee (clarified butter) and wool to the sedentary as well as to other nomadic populations. They utilized the areas which were more arid and less suitable for cultivation and had developed unique methods of water harvesting, for the most effective utilization of the range and for ensuring its revival and growth in the following year.

The administrative district of a typical pastoral village encompasses an area of about 40 km². During the rainy season the settlement (*abadi*) is deserted. With the start of the rains the population divides into different caste groups and disperses to small dug-out pools (*toba*) in different parts of the village territory along with their livestock. When a toba is exhausted, rather than return to the village the people move to another toba. Only when the water in all the tobas is exhausted does the entire population along with their stock return to the settlement and use the water in the village tank and the lush growth of grasses around the settlement. In many cases it would be possible to dig the toba deeper so that it could store more water, but no one was prepared to do this. This reluctance appeared to be due to the convention of allowing all to come and use not only water but also the grazing resources around the toba. Deepening the toba would put additional pressure on the range and possibly lead to long term reduction of palatable species and desertification.

In the low rainfall years the pastoralists moved their livestock to adjacent semi-arid and even humid regions where the stubble from winter crops could be grazed. En route, they sell milk, ghee and animals to the sedentary population. If they pen their livestock in the fields during the night the farmers provide food and money in return for the manure. The livestock return at the start of the monsoon and again disperse to the tobas within the village territory.

Non-pastoral nomads similarly moved according to definite cycles in small bands with their own system of social control and relationships of interdependence with the sedentary population.

Most farmers also raised livestock as an insurance against years when they are unable to harvest their crops. The animals were able to graze on the unharvested grain. However, like the nomads, the livestock raisers among the sedentary population also resorted sometimes to migration to better rainfall areas. For similar reasons farmers were generally of the opinion that the more parcels of land they held the greater their chance of harvesting at least some grain in bad years. Since land was shown only after it received rain, fragmentation of plots had the added advantage of allowing a greater percentage of the land to be fallowed in dry years. In any case farmers customarily fallowed their land for 3–4 years every 4–5 years, enabling it to regain fertility. These fallow lands also provided grazing resources for the livestock which, in turn, provided fertilizer to the fields.

Rotation and mixed cropping provided a further strategy. After lying fallow, guar (*Cyamopsis tetragonoloba*) was usually grown in the first year and the land was ploughed three times. This was followed by a mixed crop of bajra (*Pennisetum amerucanum*), moth (*Phaseolus aconitifolius*); moong (*Vigna radiata*), and til (*Sesamum indicum*) in the ratio of 20:1:1:1.

Finally, soil moisture was exploited to the utmost by the cultivation of "*khadin*", which are natural depressions found in the rocky and stony terrain partially filled with fine sediments resulting in the development of localized silty clay-loam and clay-loam soil pockets in otherwise sandy and rocky territory.

The khadins are usually flooded during the monsoon and, where necessary, a band is built to dam the run off. As the water recedes they become ploughable in the winter (*rabi*) season and are cultivated. In a few instances of lighter and better drained soil pockets, and also in cases where smaller catchments with less run off are involved, khadins are cultivated in the monsoon (*kharif*) season also.

High fertility was an important strategy for both the human and the animal populations. Early marriage and child bearing were important cultural values. Divorce was rare and widowhood quickly ended in remarriage. In animal holdings also quantity was valued over quality.

In order to have an effective and balanced use of the ecosystem, and to utilize different types of grasses, shrubs and tree foliage, each household would herd a mixture of animal species including cattle, camels, sheep and goats. Only in very rare cases did households raise only one species. In the course of interviews, many farmers claimed that keeping different types of livestock provided them some security during the prevalence of certain diseases which affect some species and not others.

It needs to be mentioned that farmers in the arid areas have always raised goats—which are often branded by ecologists as the most dangerous destroyers of ecosystems. The importance of goats, according to the farmers, lies not merely in their productivity but in their role in shepherding. Goats often act as foster mothers to lambs as well as assisting the shepherd in herding the sheep. Also, in the case of attack by wild animals only the goats bleat and alert the shepherd. The rate of mortality among sheep is higher than among goats in bad years. According to a common saying, during scarcity the camel will leave only *Callotropis* while the goat will leave only pebbles (*"cont chodde aakra aur bakri chodde kankra"*)—which implies that the goat can survive on the scantiest vegetation. The goat is also described as the poor man's cow because of its milk yields.

Since most of the population lives in scattered homesteads, and the man-land ratio was relatively low, there was little difficulty in procuring firewood and the demand for firewood had no serious impact on the vegetation cover. There were, however, customs that helped to conserve vegetation on certain types of land, and certain species of tree. For example, it was religiously prohibited to cut any vegetation in the immediate vicinity of temples or other religious places. Similarly, women daily watered the Pipal tree (*Ficus religiosa*) as a pious act. This practice indirectly encouraged conservation and provided good shade and soil protection.

From these examples it can be seen that the rural society coped well with the climatic vagaries and other harsh realities of the arid ecosystem.

II. EXOGENOUS IMBALANCE

The foremost development of modern technology which initiated change in the traditional network of relationships was the increase in the means of communication and opening up of hitherto inaccessible areas to rail and road transport. This development reduced the importance of the distributive activities of trading nomads, many of whom have sought sedentarization as the villagers have ceased to depend on their services. The economic interdependence of pastoral and non-pastoral populations also declined. Before independence the pastoralists migrated as far as the Bahawalpur region (now in Pakistan) between April and June. The closing of the border has led to additional pressure on the desert range. Where they were earlier welcomed the pastoralists are now looked down upon and the sedentary population has begun to dislike their visits (Malhotra, 1977).

As a result of this disruption in the relationship of mutual dependence of different social groups, those whose economy suffered have turned to cultivation, increasing the pressure on agricultural land. The net effect, therefore, was to upset the former balance both between man and man and between man and land, leading to increased desertification.

The introduction of medical science and improved health programmes in the region has stamped out many diseases, increased the span of life and greatly reduced infant mortality. There has been a widening gap between birth and death rates, leading to accelerating increase in population. The increase in population in only 10 years (1961–1971) was 63% of the figure for the beginning of the century. The growth rate during the same decade was 28%, compared with less than 9% during 1901–1911. As a consequence, there is a high dependency ratio. The potentialities of the expansive future growth rate of the population are further exhibited by the present broad based age pyramid, the marital status of the population, the presence of a high percentage of women in the earlier half of their reproductive period, and last but most important, the prevalence of high fertility norms and sanctions in almost all sectors of the population. Family planning and planned parenthood have yet to make any impact on the society. The man-land ratio is declining fast.

This account indicates that the traditional fabric of exchange relationships in western Rajasthan had inherent mechanisms and concepts which facilitated the rational use of resources. The spread of new technologies has increased human and livestock population pressure on resources and led to social disintegration. There is an urgent need for programmes that would reverse this trend and create conditions conducive to the continued evolution of traditional institutions. One of the grestest untapped potentials in present times is human resourcefulness. The people still are the repository of detailed knowledge and profound understanding of their environmental limitations and adjustment mechanisms. Intensive micro-cosmic anthropological studies pertaining to traditional relations of production are likely to provide the basis for the evolution of the traditional institutions and for envisaging plans for their change and their response to change.

16

The Human Factor in Ecological History

R. P. Dhir

Central Arid Zone Research Institute
Jodhpur (Raj.), India 342003

Societies are inseparable from their histories, and this constraint is generally acknowledged. The relationship between ecosystems and their histories receives less attention. The synthesis of ecological and social history is particularly difficult to achieve. Chapter 15 describes Rajasthan society as adapted to the conditions of its natural environment. This description suggests that the ecology might also have been modified by the effects of human activity over time, in which case the history of this interaction should prove instructive for planning future development. This second chapter in the section reconstructs the history of ecological change in Rajasthan in relation to what is known of the history of human activity from the earliest times to the present. —Ed.

Western Rajasthan stands out among physically similar regions of the world by virtue of the present high density of human and livestock population and of the long history of human settlement. According to the last census (1971), over an area of nearly 0·2 million km² the human population density was 48 persons per km² (Malhotra, 1977). This figure may be assumed to have increased substantially during the last decade. The density of livestock is even higher. The livestock census for the same year shows 16·44 million for the region, which is equivalent to 80 head per km².

There is ample evidence of human settlements starting from the beginning of the historical period. Over time these settlements have grown and multiplied. The area, therefore, provides an excellent arena not only for the study of human adaptation to an inhospitable environment but also for analysis of the impact of prolonged human exploitation on drylands. The subject matter derives particular significance from the present day concern to assess desertification in its historical context as a basis for the design of culturally acceptable strategies to combat it.

I. NATURAL RESOURCES

Western Rajasthan is the eastern extremity of a much larger region of low precipitation that stretches from the Sahara through the Arabian Peninsula and the Iranian Plateau to the Indian subcontinent. Within this larger zone,

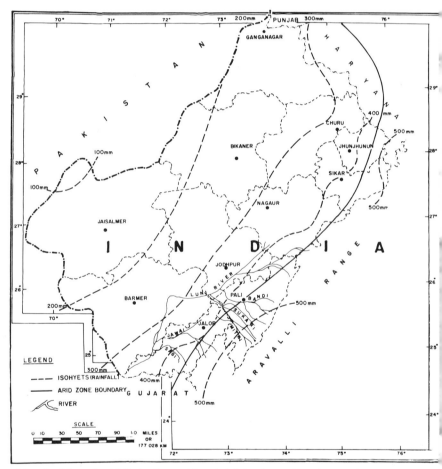

Map 1. Map of arid Rajasthan with rainfall iso hyets. Based upon Survey of India map with the permission of the Surveyor General of India. Government of India copyright, 1980.

however, the Indian part is distinguished by its pronounced monsoonal rainfall regime. Almost the entire rainfall, which ranges from about 450 mm on its eastern boundary to less than 100 mm in the extreme west, the district of Jaisalmer, is received during the monsoon, from early July to September (see Map 1). Only in the north west, in Ganganagar, is winter rainfall of some significance.

This hot arid area is situated in subtropical latitudes and covers 285 680 km², of which 196 150 km² or 69% is in Rajasthan; the rest lies in Gujarat (22%), Haryana (4%) and Punjab (5%) (Krishnan, 1968). In Rajasthan the Aravalli mountains almost form its eastern boundary, though this association is coincidental and should not be allowed to give the impression that these mountains are an obstruction to the inflow of moisture-laden monsoon air from the east. Similarly, although in the past rivers originating from the Aravallis together with those from the Himalayas have contributed to the development of the vast alluvial plains which are a dominant land form of the region, their present contribution is small. The organized drainage from the Aravallis traverses only 30 000 km² of the south east of the area, and in the extreme north west the area under the influence of drainage from the north is even smaller.

A. On the Origin of the Desert

The occurrence of an arid tract in the Indian subcontinent, despite the strong monsoonal regime, is a climatic peculiarity—especially when the north-west is compared to the north-east where the same latitudes receive the highest rainfall in the world. This comparison has generated a number of ideas about the causes and origins of the aridity. These ideas have been further stimulated by the archaeological discovery of a series of pre-historic "Indus Valley" or Harappan cultures. Krishnan (1952) considered that the present pattern of rainfall must have been set during the Tertiary when the Himalayas rose high enough to become an obstruction and cause establishment of the monsoonal regime, but that the general lowering of temperatures during the Ice Ages and the presence of a fore-deep, fed by the Himalayan rivers, must have kept Rajasthan moist till perhaps subrecent times.

Others have posited a post-glacial origin of aridity (for example, Wadia, 1960). Pigott (1950) concluded from various pieces of evidence, such as paved drains in the Harappan settlements, timber used in the dwellings and the presence of animal remains, that the climatic conditions during that period (*c.* 300–1500 BC) must have been appreciably more moist than today. Bryson and Barreis (1967) also advocated a recent onset of aridity which they would attribute to over-exploitation of vegetation resulting in accelerated wind erosion, increased atmospheric dust and consequent reduced precipitation.

Not all writers agree with such a late origin of the region's aridity. Roy and Pandey (1971) stated that desert conditions set in with the rise of the Himalayas and they do not consider that the decline of prehistoric cultures signifies any change of climate. They argue that the various archaeological sites are located along river courses, and that shifts in these courses, which are not uncommon historically, could alone bring about their collapse. Similarly, Meher–Homji (1973) believed that the type and thickness of calcareous pans in the soils are indicative that the climate in the area has been basically dry for the past several thousand years, and possibly even more. It has also been suggested that the profile organization and genetic features of soils, and even of those land forms that have remained stable during Holocene wind activity, do not show evidence of a humid climate (Dhir, 1977a). Although there are certain indications of moister conditions, nevertheless the general soil formation in the area is characteristic of arid and semi-arid conditions only.

The Quaternary in arid Rajasthan is characterized by the accumulation of alluvium and aeolian sands. Ghose et al. (1966) and Singh et al. (1971) have shown old aggraded alluvial plains as the dominant geomorphic unit in the Central Luni Basin. Subsequent surveys in the Jodhpur and Bikaner districts have also shown extensive occurrence of these plains even in areas that are presently extremely arid. Indeed, the thickness of alluvium, often 5 to 20 m or more, indicates massive alluvial activity during the Pleistocene (Dhir and Kolarkar, 1977). Further, sedimentation seems to have taken place in a climate appreciably wetter than the present but still in overall moisture-deficit conditions. These conditions are inferred from the fact that masses of alkaline earth carbonates have remained accumulated, though with considerable spatial redistribution (Dhir, 1977a). Desertic conditions appear to have set in during the late Pleistocene and led to a breaking down of the extensive light textured alluvial plains. Only patches of medium and fine textured plains, for example in Pali and Nagaur, escaped this deflation and have remained stable. The few detailed studies on this later period are reviewed below.

Allchin et al. (1972, 1978) and Goudie et al. (1973), through a multidisciplinary study based on analysis of land forms, fossil soils and cultural remains, have come to the conclusion that aridity in the region is not a Holocene phenomenon but rather an ancient one. Their work further shows that the region had undergone a series of major environmental changes in the late Pleistocene. From the presence of rolled lower Palaeolithic artefacts in colluvium associated with major aeolian sand sheets, they show that the pre-Middle Palaeolithic period was a major dry phase. The succeeding Middle Palaeolithic, extending from about 45 000 to 25 000 BP was characterized by wetter conditions as evidenced by substantial increase in human occupation and stabilization, decalcification and weathering of dunes. These conditions seem to have persisted till the early Upper Palaeolithic, after which another major dry phase set in, to be followed by a moist phase around 10 000 BP. For the succeeding period we have some pioneering studies by Singh (Singh, 1971, 1977; and Singh *et al.*, 1974), who uses pollen analysis from different stratigraphic units of lake sediments, coupled with radiometric carbon dating, to show that the period from 10 000 to 4000 BP enjoyed substantially higher rainfall than at present. This interpretation is corroborated by the absence of intercalations of sand in the clay sediments (indicating stabilization of dunes and absence of loose sand generally) and the presence of pollens of a vegetation rich in grasses and sedges and poor in halophytic species. The maxima of mesophytic elements are seen in the period for 5000 to 3800 BP. Thereafter, with minor recoveries, conditions of progressive aridity seem to have set in leading to the present climate. This sequence is similar to that put forward by Grove (1973) for northern Africa, the Arabian Peninsula and the Iranian highlands.

There is ample evidence based on stratigraphic analysis of land forms that fluctuation in environmental conditions has indeed occurred (Misra and Rajaguru, 1975; Kar *et al.*, 1977; Singh, 1977; Ghose *et al.*, 1980). The present author has also come across sites showing a number of fluvial and aeolian cycles in the late Pleistocene sediment deposits. However, after traversing the length and breadth of the western arid zone of Rajasthan I have not come across any soil formation characteristic of a humid environment, even on landscapes that have remained stable during late Pleistocene climatic fluctuations. Nor do the dunes of the old cycle exhibit any sign of differential weathering. The maximum manifestation is some redistribution of alkaline earth carbonates and some reddening of the surface extending down just a few centimetres, and the latter is not very common and is limited to fossil dunes on the margin of the present day arid area. It appears, therefore, that the humidity of the Late Pleistocene and Holocene could not have been as high as has been made out in the studies mentioned above: climatic conditions conducive to stabilization of aeolian features and weak pedogenesis did occur but it is not necessary to invoke a humid environment to explain them.

B. Present Natural Conditions

1. Land forms and soils

As stated earlier, the region comprises a vast stretch of Quaternary alluvial plain and wind sorted sands. Its mean elevation in the east at the foot of the Aravalli mountains is generally 350–450 m above sea level, and from there the plains slope to the west and south west down to about 100 m in the west and 20 m in the south west towards the Rann of Kachchh. However, this general slope is broken at a number of places by hills, both low and high, and outcropping of igneous and sedimentary rock of the Early Palaeozoic to the Late Tertiary.

The most spectacular land forms are the dunes, which are present over 58% of the region. Within this 58% nearly 30% has 60–100% dune cover, and another 58% has 20–60% cover (Raheja and Sen, 1964). The preponderance of dunes is in the northern and western half of the region, but even here there are dune free corridors, such as the area from Pokran through Bap to Bikaner. A variety of dune types has been recognized: coalesced parabolic, longitudinal, transverse, barchan and obstacle. Amongst these, parabolic and coalesced parabolic dunes are dominant (Ghose *et al.*, 1977). All the types are very variable in height—the common range being 10–80 m. The highest are in Jodhpur and Barmer districts. They are highly sandy and contain only 1·8–4·5% clay and 0·4–1·3% silt. The inter-dunes and associated plains are also light textured, though with a higher percentage of silt and clay, and have varyingly developed concretionary strata at depths commonly of 40 to 120 cm. There is an appreciable area also of old surfaces with well developed calcrete often with some well rounded rock fragments underneath or exposed at the surface (Dhir, 1977c).

The central and southern part of the region is made up of medium and fine textured soils developed from *in situ* or alluvial parent material, with almost no signs of deflation or of blown-in aeolian sand. These soils are well aggregated and have a good moisture-retention capacity, but like those of the dunes and sandy plains they are somewhat low in nitrogen, though well provided with phosphorus, potassium and various other nutrient elements. Although the nitrogen level is low, this condition does not limit the establishment and maintenance of good vegetation cover. However, arable farming requires the application of fertilizer for optimum yields.

2. Climate

Mean annual rainfall in the region ranges from 450 mm in the east to less than 100 mm in the extreme west. Rainfall and potential evapotranspiration of some stations are given in Table I. The isohyets run mostly from the south west to the north east. Of the total rainfall 83–95% is received during the monsoon between early July and late September. This period also has high atmospheric humidity with mean relative values of 60–70%. These two features, namely a concentrated rainfall together with high humidity, which also has the effect of reducing temperatures, make this period favourable for vegetative growth—which largely explains the capability of the Indian arid zone to sustain its uniquely high pressure of human and livestock population.

The pre-monsoon summer from April to June is characterized by high temperatures and very low atmospheric humidity. Mean monthly maxima

Table I. Mean monthly rainfall and potential evapo-transpiration at selected stations.

Stations	Jan.	Feb.	March	April	May	June	July	Aug.	Sept.	Oct.	Nov.	Dec.	Total
					Normal Rainfall in mm (1901–1970)								
Ganganagar	7·5	7·3	11·4	5·4	9·2	26·0	66·9	69·2	30·0	1·6	1·3	5·6	241·4
Bikaner	5·5	7·7	6·6	5·0	12·0	26·1	79·6	88·9	47·1	6·8	1·7	4·0	291·0
Barmer	2·1	3·0	4·1	1·5	8·0	19·7	88·1	104·0	34·1	3·0	0·9	1·5	270·0
Jaisalmer	2·6	5·2	3·4	2·5	7·1	12·8	57·6	66·9	20·3	1·2	1·5	1·6	182·2
Jodhpur	4·4	4·9	4·6	2·5	9·3	29·1	111·3	128·6	57·4	7·3	2·2	1·6	361·2
Pali	3·0	3·6	2·9	1·3	5·3	39·6	128·3	165·9	62·3	5·8	1·4	1·4	420·8
					Normal Monthly potential Evapotranspiration (mm)								
Ganganagar	43·0	65·6	111·3	164·6	217·3	247·9	236·3	193·7	161·4	115·6	60·7	40·4	1662·2
Bikaner	53·3	75·3	131·4	172·4	236·8	258·3	220·2	196·6	177·7	124·7	68·3	48·8	1772·4
Barmer	81·0	100·5	159·7	206·7	259·4	239·4	178·1	152·3	163·6	141·1	92·2	77·1	1857·7
Jaisalmer	78·6	91·6	153·2	203·5	281·8	317·4	247·6	210·7	192·0	145·0	83·9	64·1	2863·2
Jodhpur	83·7	101·3	156·3	192·3	264·7	260·3	184·0	145·2	156·1	134·3	87·7	76·5	1843·0
Pali	58·0	83·0	145·0	182·0	240·0	240·0	170·0	128·0	132·0	125·0	82·0	65·0	1650·0

during this period are in the range of 41–42°C with relative humidity of only 15–25%. This period is also characterized by a strong wind regime. The mean wind speeds of the windiest month, June, are 27·2, 25·6, 18·5, and 14·2 km per hour at Jaisalmer, Phalodi, Jodhpur and Barmer respectively. The winters are very mild with mean monthly maxima and minima mostly in the range of 22–25°C and 5–10°C respectively.

Though the monsoon season does set in with incredible precision, the actual amount of rainfall received varies greatly from year to year, and the coefficient of variation is high, around 40% in the north east at Churu and Sikar, and over 70% where the average is least at Jaisalmer. There are also situations of delayed monsoon or of extended droughts during the monsoon period (see Table II).

Table II. Aridity index and percentage frequency of droughts at some stations in arid Rajasthan for the period 1941–1960 (Krishnan and Thanvi, 1971).

Station	Mean aridity index %	Slight	Moderate	Severe	Disastrous	Total
			Droughts			
Bikaner	66	15	25	15	30	85
Barmer	64	20	35	30	5	90
Jodhpur	52	30	10	15	5	60

Categorization of droughts is based on aridity index in the crop growing season of the monsoon period Slight, moderate, severe and disastrous droughts correspond to aridity index values of 50–60, 60–70, 70–80 and over 80 respectively.

3. Causes of aridity

Atmospheric humidity during the monsoon is high—in fact comparable to that found in neighbouring semi-arid and sub-humid regions—but rainfall is poor. The causes of this apparent anomaly are not well understood, but there are two principle arguments. The first attributes it to the low level divergence and associated subsidence of air, which does not allow clouds to grow and instead dissipates them. It is only when monsoon depressions move close to or over the region that widespread heavy rainfall occurs. The frequency of such depressions moving over the region has been shown to be 24 during the period from 1891 to 1960 (see Datta and George, 1964).

The second argument comes from Bryson and Barreis (1967) who believe that atmospheric dust is the factor responsible for low precipitation. According to them western Rajasthan has an unusually dusty atmosphere which influences the infra-red cooling of the mid-troposphere and subsequently adversely affects the subsidence and precipitation pattern in this region.

4. Surface and ground water resources

As a result of the low rainfall, surface water resources are also scarce. In the extreme south east there is an organized drainage emanating from the western slope of the Aravallis that covers about 30 000 km². Here also the actual flow is limited to a few days only, in response to rain. In the north west there is another small stretch under the influence of the Ghaggar river. In the remaining major part of the region there is no organized drainage, though points of concentration have been intercepted for the use of the region since the beginnings of human settlement. Depending upon the rainfall, this technology can meet the needs of people and livestock for periods from a few days to months, after which, obviously, recourse must be had to groundwater. A few decades back the water from outside the arid zone, from the Indus basin, was diverted to provide irrigation to a part of Ganganagar district in the north. Presently, other major projects are underway, which on completion will bring a net area of 1·14 million ha under irrigation.

The tracts under the direct influence of integrated drainage or those that are recharged through subsurface flow have reasonably good potential for human exploitation, much of which is already being realized. Such tracts add up to nearly 30% of the arid zone (Mehra and Sen, 1977). In the rest of the arid zone ground water is too little or too deep, and is often too poor in quality. Even so, through combined use of surface and ground water some populations are able to sustain themselves.

II. HUMAN SETTLEMENT

A. Historical Development

1. Palaeolithic

Recent investigations have uncovered the existence of Palaeolithic activity in Rajasthan. Misra (1962) while surveying along the course of the Luni river came across Middle Palaeolithic artefacts mixed with river gravels. Mohapatra *et al.* (1963) have identified remains of this culture at a site between Jodhpur and Bikaner and Allchin *et al.* (1978) found a major lower Palaeolithic factory site near Hyderabad in lower Sind (Pakistan). Very recent explorations (Misra *et al.*, 1980) have unearthed another site of this period at Jayal (45 km east of Nagaur), which is regarded as the richest in the present day deserts of the world. Similarly, a number of investigations have demonstrated the presence of Mesolithic activity. Misra (1977) has found Mesolithic sites on or close to hills and rock outcrops, as well as on sand dunes in Pali and Barmer districts. Subsequent work has shown cultural remains of the same period at a number of locations in Jodhpur and Nagaur districts as well as at sites adjoining natural depressions further west in Bikaner and Jaisalmer. Allchin *et al.* (1978) compared microliths here with those from the surrounding more humid areas and, noting that these sites are thinly distributed, argued that the size and variety of tools suggest they belonged to pastoral nomads, or even to Chalcolithic merchants who were able to penetrate the desert.

The sequence need not imply unbroken continuity of human occupation. The arid phases that probably preceded and followed the Middle Palaeolithic must have led to the withdrawal of human occupation with the exception of a few locations within daily reach of a watering point. The onset of congenial conditions on the other hand would have drawn in populations from the adjoining regions.

2. Protohistoric

The extreme western part of arid Rajasthan, along the banks of the now dried up Ghaggar river, was the stage of the emergence of the earliest rural and urban centres in the area. Stein (1942) was the first to discover a string of these settlements, but it was the systematic explorations by Ghosh (1952) that unearthed 25 Harappan and a larger number of later settlements in this area. Since then works too numerous to cite have showed the whole range of cultures—pre-Harappan and Harappan, Ochre-coloured pottery and Painted Grey Ware—that flourished here one after the other from 4500 to 2500 BP. It is now well established that the pre-Harappan settlers (4500–4100 BP) used copper for making axes. The discoveries of a terracotta cart with a single sided hub and of a ploughed field are noteworthy. The Harappan culture that flourished between 4100 and 3800 BP was a major step in our record of social evolution. The people are known to have had domesticated cow, buffalo, pig, elephant, ass and camel, and they cultivated barley. These developments were associated with the blossoming of urban life in well laid out cities, with rich material culture, social differentiation and religious observance (Misra, 1977). However, neither these nor the succeeding Iron Age cultures appear to have expanded out from their ecological niche away from the river courses to the larger tract of arid Rajasthan further south (ibid.).

3. Historical

Since the vast arid expanse south of Ghaggar is blank on the Chalcolithic and Iron Age maps of India, it seems that human occupation of the region remained thin and depended on hunting and limited pastoralism. Historians are generally of the view that significant immigration into the area began in the 4th century BC. In the words of one specialist:

"Alexander never reached Rajasthan. Nevertheless, his Indian incursion profoundly affected the source of its history. Weakened by the Greek onslaught and yet desirous of retaining their independence, some of the republican tribes which fought against Alexander namely the valiant Malloi, the Sibis and the Aggalassoi or Arjunayanas thought it best to migrate to Rajasthan. With them probably came other tribes also. And thus arose some of our great republican States" (Sharma, 1966:49).

This process of migration from the fertile Indo-Gangetic plains into this environmentally hostile, but otherwise secure region, in response to waves of invasions from the west, seems to have continued during the early mediaeval and mediaeval periods. The settlements increased and expanded, and by the sixth to seventh centuries of the Christian era much of arid Rajasthan had been not only colonized but politically organized. According to local tradition, Mandor (8 km north of Jodhpur) was already in existence in the fourth century AD. It was held first by Nagas, then in the 6th century the Pratiharas took it (Jain, 1972). The Chapa kingdom with its capital at Bhinmal was already in existence in 628 AD. Jalor was a flourishing town in the 8th century. Similarly, the Nagaur region was ruled by a clan of Rajputs as early as the 7th century, but it is likely that a Naga settlement was there earlier. Even in the driest parts of Jaisalmer, 16 km to the north west of the town, is a site called Lodorva which was founded by a Rajput clan and in the tenth century taken by another clan. Thus, the entire arid region including its most arid parts has been well occupied since early mediaeval times.

In the year 1212 AD when scions of the Rathor Rajputs journeyed into the area, Nagaur had as many as 1440 villages and a large area around Jodhpur had a well settled rural society (Tod, 1832:2–9). According to local records in 1459 the very dry habitat of Bikaner had a settled pastoral community living in 2670 villages (ibid.), which must have been spread over an area of about 45 000 km². The present author's cursory observations in the Jodhpur-Nagaur tract show that the majority of villages are from 500 to 1200 years old. More authentic record is available from Mahnot Nainsi (Bhati, 1969) that the majority of today's villages existed also in his time (1658–1662 AD). By the mediaeval period, therefore, a thick network of human settlements had become established, which in the succeeding period may simply have grown in size.

4. Late Mediaeval

Although socio-economic conditions in Rajasthan seem to have been better than in many arid regions, reliable data on many topics, such as population, are difficult to come by. Concentrated scholarly effort would, however, no doubt uncover considerably more information than is available at present. The best demographic data presently known from this period are from

Mahnot Nainsi, a chief revenue official of Raja Jaswant Singh who ruled
Marwar State (which was roughly equivalent to the present day districts of
Jodhpur, Nagaur, Jalor and Pali) from 1638 to 1681. Nainsi during the period
1658–1662 traversed Marwar State and made a detailed record of, among
other things, the location of villages, the number of households, ponds, wells,
and the type and size of crop. These data allow an estimate of the population.
Table III shows such an estimate for some villages in the Phalodi area.

Table III. Population growth in selected villages of Phalodi area.

Village name	1660[a]	1891	Census year 1921	1941	1971
Lohawat	2432	5366	4663	5537	9855
Naneo	72	898	451	565	1459
Sanwrij	1200	957	623	1094	2449
Hopali	440	432	503	610	1062
Au	648	2721	1555	1797	2789
Bheeyasar	640	1533	818	823	1196
Mungasar	488	1039	627	791	1518
Banasar	240	814	521	562	1035
Kelausar	344	976	528	790	1685
Jaislo	520	852	631	975	2173
Ranisar	184	629	458	411	813
Khichan	376	2449	1977	2658	2795
Dadarwalo	200	661	252	211	502
Mithio	120	356	174	119	231
Chhila	320	530	438	428	816
Mayakor	400	99	68	104	251
Total	8624	23 312	14 287	17 475	30 629
Phalodi Town	5226	3942	12 801	17 689	17 379

[a] Calculated from the number of households assuming 8 members per household.

The results show a variation in size of village similar to the present, from 72
to 2432. More importantly the population in most villages was appreciably
less than the present. The population of all the villages together recorded for
1658–1662 is less than half the mean for the census years 1891, 1921 and 1941.
Thus, over a period of 250 years before the recent acceleration in growth rate,
the population doubled. This increase seems to have taken place only in the
first 150 years or so since from 1820 to 1921 the population of Marwar State as
a whole seems to have remained around two million.

A process of phenomenal growth seems to have set in from 1921 onwards, as
is shown by the figures of subsequent censuses for Jodhpur district and for the
area of erstwhile Marwar State, in Fig. 1. The population almost doubled in

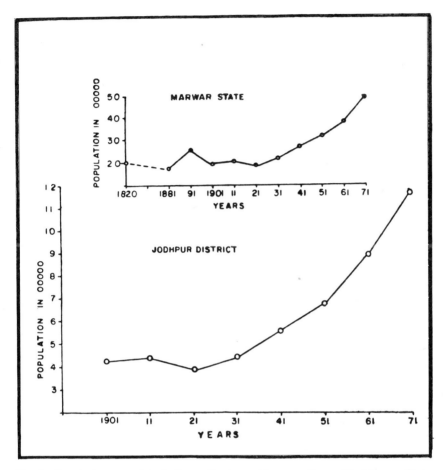

Fig. 1. Population growth in Jodhpur district and in area of erstwhile Marwar State.

the forty-year period from 1921 to 1961 and then increased a further 30% in just a decade. This growth seems to be in conformity with the trend in the larger region (see Table IV).

Arid Rajasthan has an even higher growth rate than Rajasthan as a whole, and Rajasthan has a higher rate than the whole of India. Malhotra (1977) has enumerated high fertility norms originally developed in high mortality conditions, considerations of lineage, social values and the sharp decline in mortality as the factors responsible for this extraordinary growth rate.

Table IV. Decennial percent variation in population in arid zone of Rajasthan and India 1901–1971 (Malhotra, 1977).

Year	Arid zone of Rajasthan		Rajasthan		India	
	Population (millions)	Growth rate (%)	Population (millions)	Growth rate (%)	Population (millions)	Growth rate (%)
1901	3·57	—	10·29	—	236·28	—
1911	3·88	+8·70	10·98	+6·70	252·12	+5·42
1921	3·58	−7·80	10·29	−6·29	251·35	−0·303
1931	4·28	+19·81	11·75	+14·14	279·02	+11·00
1941	5·28	+23·16	13·86	+18·01	318·70	+14·22
1951	6·2	+16·73	15·97	+15·20	361·13	+13·31
1961	8·0	+29·84	20·16	+26·20	439·24	+21·64
1971	10·23	+27·95	25·77	+27·83	547·37	+24·66
Overall increase from 1901 to 1971	6·66	+186·78	+15·47	+150·30	+311·08	+131·65

Based on decennial census of India, Manager of Publications, Govt. of India.

B. Land Use

The present system of land use and management is based on agroclimatically well adapted crop species, mixed cropping or scheduling of crops to cope with conditions of delay or failure of monsoon rains and an efficient mixed farming system to satisfy human needs. This combination bespeaks many generations' experience. Useful top-feed species are maintained as an insurance against harvest failure, and a system of farming based on run off collection in small eroded basins known as *khadin*, in the desertic conditions of Jaisalmer are other striking features indicative of the long history of adaptation in resource management.

Since land use and agriculture are everyday affairs and taken for granted, there is scarcely any specific account in the extant historical records. Consequently, it is necessary to a large extent to rely on indirect evidence. It is logical to assume that the process of agricultural growth started with the establishment of permanent settlements around the 4th century BC. Excavations at the ancient regional capital of Mandor (8 km north of Jodhpur) have unearthed a huge earthen jar, 3 m in circumference, of the Gupta period (AD 350–500) that was used for storing food grains. During the Mughal period, this region like the rest of the country had an elaborate system of land revenue in which separate rates were fixed for winter and rainy season crops as also for cotton, vegetables and opium (Bhadani, 1980). Therefore, although settlement moved away from the rivers relatively late (*c.* 400 BC) agriculture has probably been practised continuously since about that time.

1. Late Mediaeval

The unique account of Mahnot Nainsi for the period 1658–1662 gives detailed information for each village throughout Marwar State, including main crops, numbers of wells and ranks, quality and depth of water, lift mechanism, incidence of revenue taxation on individual crops and livestock and revenue assessment. For many villages it also gives the number of households and area of cultivated land. The area until used by him is "*dharti halwa*", a name still in use, meaning the area that can be ploughed by a pair of bullocks in a single day, corresponding to 1·6 acres. On this basis, Dhir (1977b) has shown for the Phalodi area, which has its centre 120 km northwest of Jodhpur, that cultivated land per household during the period ranged from 0·7 ha to 2·9 ha with a mean value of 1·5 ha (see Table V). The size of holding is generally too small to produce surplus for barter—which suggests that animal husbandry must have been an important complement to the average farming operation. However, in agro-climatically favourable localities, such as "Pargana" Merta (70 km to the south east of Nagaur) agriculture seems to have been more extensive. In this area of about 4200 km² over 6500 wells were in use for irrigation, of which many had lift mechanisms. A variety of crops including opium, cotton, aniseed, caraway, vegetables and cucurbits were raised. Then, as now, there seems to have been considerable variation in the incidence of agriculture, consonant with the variation in rainfall, soil and groundwater availability.

2. Recent changes

Although in relatively favourable areas such as Merta a highly evolved agriculture seems to have been in operation for the past few centuries, in the rest of the Marwar State or arid Rajasthan land use seems to have undergone appreciable development more recently. Table V shows that compared to a mean holding of 1·5 ha per household in 1658–1662 the size of holding for the same villages in 1971 had increased to 7·4 ha. However, this sharp increase in area under cultivation must have been very recent since in 1925–1927 the cropped area constituted on average only 20% of the total area in the crown villages of Marwar State. This is corroborated in Fig. 2 which shows the cultivated area for the year 1931 and that added by 1958. A very large increase in the area under cultivation has taken place in the intervening period. From the year 1951–1952 regular land use records are also available for individual districts of Rajasthan. Table VI summarizes this information for one district, namely Jodhpur.

By 1951–1952 over 80% of the area had been brought under cultivation.

Table V. Incidence of arable farming in late mediaeval time and now in selected villages in Phalodi Tehsil of Jodhpur district.

Name of Village	Year					
	1660			1971		
	No. of house-holds	Cultivated land in ha	Cultivated land in ha/ household	No. of house-holds	Cultivated land in ha	Cultivated land in ha/ household
1	2	3	4	5	6	7
Lohawat	304	389	1·3	1598	7567	4·7
Naneo	72	130	1·8	222	2140	9·6
Sanwrij	150	291	1·9	411	3622	8·8
Au	90	259	2·9	454	5490	12·1
Bheeyasar	80	194	2·4	394	4110	10·4
Mungasar	81	130	1·6	244	1616	6·6
Banasar	30	32	1·1	182	1425	7·8
Kelansar	85	65	0·8	262	3173	12·1
Jaislo	65	58	0·9	322	3015	9·4
Ranisar	45	32	0·7	157	754	4·8
Khichan	47	39	0·8	526	907	1·7
Dadarwalo	45	32	0·7	91	832	9·1
Mithis	15	39	2·6	36	1072	29·8
Rayada	27	39	1·2	23	308	13·4
Mayakor	50	65	1·3	60	782	13·1
For all villages	1186	1794	1·5	4973	36 923	7·4

Table VI. Recent changes in landuse in Jodhpur District (in ha).

	1951–52	1961–62	1971–72
Reported Area	2332	2253	2265
Area under cultivation	1893	1854	1812
Area actually cropped	808	1027	1202
Area under fallow	1085	827	610
Culturable waste	109	51	99

Figures vary over time because of changes in administrative boundaries and expansion of land under settlement.

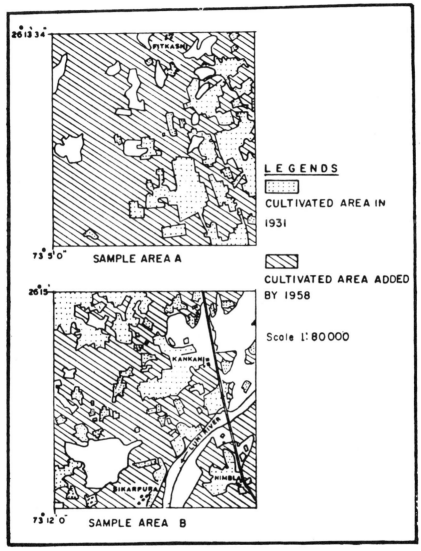

Fig. 2. Recent increase in incidence of cultivation in two randomly selected sample areas based on study of year 1931 G.O.I. 1 in = 1 mile topo-sheets and year 1958 aerial photographs.

But there has been no further increase, presumably because there is hardly any scope for further expansion since most of the remaining land is either under settlements, public utilities or is rocky and stony waste.

Between the late 20s and the early 50s, therefore, there was a new and significant trend of expansion of cultivation almost to the limit of the total cultivable area. However, the predominant farming system still maintained nearly half of the area fallow in any particular year. Since then, the change has been for a reduction in fallow in order to increase the annually cropped area.

Barring the extremely arid districts of Jaisalmer and Bikaner, the picture in the remaining arid districts is very close to that of Jodhpur (see Table VII). Annual cultivation is by far the dominant practice and little land is left fallow. Even so most households also maintain a sizeable herd of livestock which is sustained on crop residues as well as natural vegetation on both fallow and cropped lands. Though the present land management is quite efficient, it is obviously somewhat over-exploitative of the natural productivity (see Fig. 2).

Table VII. Incidence of cultivation in arid districts 1977–78.

Name of the district	Mean annual rainfall in mm	Reporting area in ,000 ha	Cultivated[a] area in %	Net area sown in %	Area of fallow lands in %
Barmer	277·5	2817	71·7	52·6	19·1
Bikaner	263·7	2741	39·5	28·9	10·6
Ganganagar[b]	253·7	2063	79·1	72·8	6·3
Jaisalmer	164·0	3831	6·4	5·4	1·0
Jodhpur	318·6	2256	83·6	51·0	32·6
Nagaur	388·6	1764	86·8	64·5	22·3
Churu	325·5	1686	89·9	75·0	14·9
Jalor	404·4	1056	81·5	58·9	22·6
Jhunjhunu	456·4	775	79·8	63·7	16·1
Pali	415·3	1230	69·0	45·0	24·1

Source: State Land Record Office, Jaipur
[a] Comprises net sown and current and long fallow
[b] Partly canal irrigated.

III. PRESENT LAND MANAGEMENT AND ENVIRONMENTAL DEGRADATION

The study of ecological change requires comparative analysis of sites with different degrees of known history of exploitation. Unfortunately, there are no sites in this region that have not undergone some change as the result of human activity. In the absence of such benchmark sites the best approach appears to lie in the comparative study of examples associated with human activity according to distance from centres of activity, as an indication of the intensity of impact. Under both the dominant types of land use—arable farming and open grazing—there is ample evidence to show the type and to some extent the magnitude of ecological change. One very characteristic feature of the village setting in the climatically less favourable western half of the region is that lands surrounding the village centre appear more desolate in that they have much less vegetation and a preponderance of barchans as compared to lands away from the village. Fence line hummocks, general increased component of loose sand and depleted degraded condition of grazing lands are some of the other major manifestations associated with intensive usage of land resources of the region.

The increase in the loose sand component in the landscape is a conspicuous feature—though difficult to quantify. Masses of sands saltating across cultivated fields, piling up behind shrubs and fences, blocking roads and railways and choking entrances and courtyards are a common sight during pre-monsoon summer months. Though the body of a dune may be stable its crest and leeward flank and to a small extent also the windward flank are covered with fresh deposit. In the farm lands on the sandy plains the farmers construct brushwood fences, sometimes green, along field boundaries in order to stop stray cattle from wandering in and the soil from blowing out. As the sand piles up at the boundaries, the fences are raised and give the fields a peculiar saucer-shaped look. Another interesting feature of these lands is that the surface soil layer is distinctly more sandy as compared to the soil below. Dhir *et al.* (1977) on the basis of an analysis of 46 sites have shown that whereas the mean content of clay and silt was respectively 7·9 and 6·3% in the subsoil, their content was only 4·1 and 2·7% in the surface. It seems that frequent preparatory ploughing, a common practice in order to remove unwanted vegetation and conserve moisture, has interacted with a strong wind regime to cause the surface instability that is visible all around.

These wind-sorted sands and hummocks have a relatively lower content of nutrient elements, both in available as well as reserve form (see Table VIII). They are particularly low in humus and somewhat low in phosphorus and potash (Dhir, 1977b). Furthermore, though tillering and growth of individual crop plants is good on these sand features, the stand of crop is much poorer because of the surface instability. Slight wind action at the early stages of the crop causes burial or exposure.

Over-exploitation of natural vegetation is another characteristic feature. Some manifestations are: (a) change of climax community to seral stages that are dominated by low yielding annuals as unpalatables; (b) deformation and diminution of useful shrubs and trees; (c) meagre production of propagules

Table VIII. Major nutrient elements in a light textured soil and associated loose sand features in Jodhpur area.

	Potassium (mg/100 gm)		Phosphorus (mg/100 gm)		Org. C. %
	Total	Available	Total	Available	
Chirai series	1032	11·4	18·6	0·78	0·17
	1128	12·8	24·3	0·63	0·13
	1232	10·7	30·7	0·93	0·15
	1047	9·4	27·3	0·77	0·17
	1008	8·2	22·5	0·63	0·18
	1114	8·1	20·1	0·45	0·13
	1276	9·2	95·3	0·64	0·17
	1133	11·7	21·1	0·52	0·19
Mean	1121	10·2	23·73	0·67	0·16
Chirai series	938	8·4	16·2	0·52	0·13
hummocks	1007	9·2	18·7	0·53	0.14
	808	7·6	21·2	0·60	0·11
	913	7·8	29·3	0·62	0·15
	718	8·5	21·7	0·72	0·14
	926	8·7	18·6	0·49	0·13
Mean	885	8·03	21·0	0·58	0·13
Stabilized dunes	827	7·9	19·3	0·53	0·12
	708	6·8	19·7	0·54	0·13
	787	8·1	18·7	0·51	0·11
	832	7·3	18·3	0·62	0·09
	838	7·4	17·4	0·59	0·09
Mean	798	7·5	18·7	0·56	0·11
Fresh sand	811	7·0	17·7	0·48	0·04
	785	6·8	18·2	0·52	0·03
Mean	798	6·9	18·0	0·5	0·035

(Saxena, 1977). However, much of the data on loss of productivity parameters comes from comparison of plant cover in the open grazing lands with that in the reseeded and well managed experimental rangelands. Such comparison shows that whereas the latter have a basal cover of 11·5 to 24%, the open grazing lands have only between 0·1 and 0·5% in the worst situations and up to 5% in others (Gupta and Saxena, 1972). Some estimates of biomass production (CAZRI, 1976; Mann et al., 1977) show that overgrazed lands, such as by far predominate in the region, produce only 10–25% of that obtained under optimum management.

There is evidence of over-exploitation also of other resources such as surface and ground water. In one case uncontrolled increase in the use of underground water has led to virtual depletion. The villages concerned—Tarnau, Pharrod and Deri (50 km east of Nagaur)—had scores of dug wells a few generations ago. Presently, all but four lie abandoned because of lowering of the water table and deterioration of water quality. However, this is an extreme case: the dominant picture is gradual depletion only.

Obviously skill and ingenuity over many generations seem to have resulted in a land management regime that is highly adapted to the vagaries of nature and fairly efficient in meeting the range of human needs—but not without a certain cost, in the form of reduced primary productivity. Fortunately, the process is gradual and the stage of irreversibility has not yet been reached. But if the relationship between the local populations and the natural resources of arid Rajasthan, and between arid Rajasthan and the larger society, are to continue for long without serious stress, careful planning appears to be urgently required, to minimize further loss in the productivity of the renewable natural resources of India's dry lands.

17

The Role of Administration in Desertification[1]
Land Tenure as a Factor in the Historical Ecology of Western Rajasthan

N. S. Jodha

International Crop Research Institute for the Semi-Arid Tropics
Patancheru P.O., Andhra Pradesh 502324, India

The application of any desertification solutions depends on administrative measures. It is regrettable, therefore, though perhaps not surprising, that investigation of the causes of desertification is rarely focused on the mechanics of administration. This chapter considers administration in its function of formulating and implementing land tenure policy. Attention is drawn most significantly to the way bureaucratic dynamics encourage formal classifications that militate against the integration of scientific—let alone local—understanding into the administrative process, with results that exacerbate desertification problems. This study corroborates the underlying argument of the book: that the ultimate social causes of desertification generally lie outside the immediate vulnerable area, in the political and economic centres of the larger society. —Ed.

Land tenure, or the legal framework for rights to land, is conditioned by a number of diverse factors, none of which is predictably determining or predominant. The most significant of these factors are: climatic—temperature, water balance, wind, radiation, and seasonal variation; edaphic—the depth and quality of soil; technological—the available repertoire of techniques; economic—the quantity and certainty of returns, and the availability of capital for investment and markets for produce; demographic—the density and structure of population; social—the dynamics of group formation and individual co-operativeness; political—the degree of stratification or egalitarianism; and cultural—such as values that favour

333

particular activities or foods. The mix of these factors varies over time and space as one or another becomes more pressing, producing the actual history of man–land relationships in a given region: the distribution of rights, formal and informal leasing arrangements, and taxation, which may or may not be a direct reflection of current legal provisions.

Since the control of land has historically often been a key factor in socio-political dynamics, governments have generally treated land tenure as an important sphere of policy. Depending upon the importance attached to the productivity of the land and other technical considerations, policy is concerned not only with the distribution of the land but with how the land is used. In the long term, the power to control land use can be more important than the power to control its distribution—especially in inherently less productive ecosystems, such as arid lands, where some uses which may appear economically desirable in the short term may lead in the long term to ecological decline which would result in a reduction in the significance of the power to control or distribute.

The physical characteristics of a region, therefore, not only condition the productivity of land but also determine the suitability of institutional arrangement to harness that productivity. In what follows I begin by reviewing the physical conditions of the arid districts of Rajasthan; I then assess the implications of these conditions for long term effective land tenure policy. Finally, I show that the actual record has been very different, for reasons that are significant in the context of the general debate on desertification.

I. THE PHYSICAL CONDITIONS

Some relevant details of the natural factor endowments of 11 arid districts in Rajasthan are presented in Table I. The districts are broadly arranged in descending order of aridity and deficiency of the land-resource base. The impact of the deficiencies of the land-resource base is broadly reflected in the relative position of different districts in terms of density of population, size of farm holdings, and importance of livestock in the district. The classification of use capability was made at the level of *tahsil* (sub-district) on the basis of information on soil, vegetation and water resources gathered by a team of soil scientists, agronomists and conservation experts (State Land Utilization Committee Report, 1960). The committee subdivided the whole arid region into three zones:

Zone 1. All the tahsils or parts thereof characterized by lands belonging to FAO land classes VI and VII which are suitable only for pasture and range development.

Zone II. All tahsils or parts thereof which possess mainly lands of FAO classes VI and VII, but a few pockets of land of class IV which could be put under cultivation in a very restricted manner involving two crops interspersed by long (3 to 4 years) fallowing and various conservation practices.

Zone III. All tahsils and parts thereof, having different proportions of lands belonging to FAO class III (land suited to cultivation) and class IV.

These three zones account for 56%, 23% and 21% respectively of the total area of the arid region (Jodha and Vyas, 1969). If one ignores the small pockets of good land within Zones I and II, nearly 79% of the land in the arid region is unsuited to the relatively high intensity land use involved in crop farming—unless transformed by irrigation.

Hence any attempt to increase the intensity of use by putting the land under the plough, particularly on a regular basis, or by overgrazing, tends to expose the land to greater erosion hazards and leads to a fall in productivity even in terms of forage. Further, given the limitation of these lands, especially in the context of climatic variability in the region, crop-farming cannot offer high and stable yields on a sustained basis (cf. Kaul and Misra, 1961; Seth and Mehta, 1963; Jodha and Vyas, 1969; Jodha and Purohit, 1971; and Jodha, 1972).

In Zone I, therefore, limits on the intensity of use, the low productivity, and the high instability of crop-farming tend to impart a comparative advantage to pasture-based livestock farming in large parts of the arid region. In Zone II restricted cultivation with conservation practices characterized by rotations between cropping and long fallows can be encouraged. Zone III can support annual cropping.

II. IMPERATIVES FOR LAND TENURE POLICY

For tenurial policy and land management, the situation described above implies the following imperatives:

(a) Land use planning and tenurial legislation must follow comprehensive classification of the land according to use capability.

(b) The intensity of land use in each class must be determined and regulated at the level of operating units, such as farm, village, or pasture.

The provisions by which these imperatives may be incorporated into policy are indicated broadly in Table II. The essence of this Table is that an individual's rights in land and his decision about the mode and intensity of land use as well as his obligations as a landowner or user have to be assessed in keeping with the requirements of different land classes.

Unfortunately, this classification is completely neglected in practice (cf. Jodha, 1970). The only classification to be found in the official land records is

Table I. Details indicating land-resource base and related aspects in the arid districts of Rajasthan.

District	Mean aridity index[a]	% area irrigated[b]	% area by use capability classes of land[c]			% area given for private cultivation[d]	Av. size of optimum holding (ha)[d]	Density of population (No./km²)[e]	% of rural population	Density of cattle per		Density of sheep and goat per	
			VI & VII	VI, VII & IV	IV & III					100 ha	100 persons of rural population	100 ha	100 persons of rural population
Jaisalmer	89	0·16	100·00	—	—	13·26 (42·86)	24·60	4	84·40	2	46	13	353
Bikaner	81	0·03	100·00	—	—	38·68 (57·72)	22·08	21	58·63	9	75	12	98
Barmer	82	0·92	85·58	9·06	5·36	74·26 (67·20)	19·72	27	92·74	7	27	59	230
Jodhpur	78	1·96	69·04	15·74	15·22	77·77 (80·59)	12·83	50	68·05	21	60	60	173
Churu	78	0·02	—	76·10	23·90	88·85 (87·64)	14·73	52	70·42	25	68	54	148
Nagaur	73	1·80	—	54·83	45·17	85·54 (96·62)	8·52	71	87·72	40	65	91	139

Jalore	72	7·10	—	61·09	38·91	80·59 (93·26)	8·93	63	95·58	30	49	84	140
Jhunjhunu	72	2·91	—	52·72	47·28	80·60 (97·57)	4·52	157	82·56	55	42	102	79
Pali	71	17·22	—	22·86	77·14	65·51 (86·71)	5·14	78	88·82	53	75	101	142
Sikar	69	5·53	—	49·10	50·90	80·25 (97·10)	4·67	135	82·97	60	54	131	117
Sirohi	—	23·34	—	—	100·p0	42·83 (86·51)	3·27	82	82·09	58	83	94	136
Total	—	3·22	56·33	22·52	21·15	59·01 (80·12)	9·29	44	81·30	22	88	54	223

[a] Prepared by using Thornthwaite's method of climatic classification. For details, see Krishnan *et al.* (1969).

[b] Net irrigated area as percentage of net cultivated area. The details relate to private operational holdings. Calculated from district data of *Report on Agricultural Census 1970–71 in Rajasthan*, Government Press, Bikaner, Govt. of Rajasthan, 1975.

[c] Land classes following the FAO use-capability classification of lands for soil conservation (for details, see text). The calculations are based on the *tahsil*-level data, following the findings of *State Land Utilization Committee Report* (1960). The report, however, did not include the Sirohi District, but we have treated it as indicated above.

[d] Calculations based on the details of private land holdings (excluding State farms) culled out from the *Report on Agricultural Census 1970–71, op. cit.* Figures in parentheses indicate the percentages of net cultivated area to the total land area assigned to ploughing.

[e] Details relating to human population are based on Census of India, 1971 and those related to animal population are based on *Livestock Census*, 1972. Cattle include a very small proportion of buffaloes also.

Table II. The features of the three zones in the arid regions of Rajasthan and the suggested provisions of tenurial and land-use policies.

Features	Zone I	Zone II	Zone III
1. Land classes	VI, VII	VI, VII, IV	III, IV
2. Share in arid region (%)	56·33	22·52	21·15
3. Farm activity favoured most by the land classes	Pasture-based livestock-farming	As Under Zone I, and restricted cultivation	Normal cultivation and restricted cultivation (on land class IV)
4. Objectives of tenurial policies/arrangements, as influenced by land classification	Conservation and development of land resource base for efficient livestock-farming	As under Zone I, conservation and development of land for restricted cultivation	High farm productivity with high intensity of land use on land class III and restricted cultivation on land class IV
5. Pattern of land use (Land use intensity)	Highly extensive type of land use through natural forage production on pasture/range lands	As under Zone I, and slightly intensive use through crop-fallow rotation system	Intensive cropping on land class III, crop-fallow rotation on land class IV
6. Major area of public policy	Zoning of areas for specific land uses, law relating to development, utilization, conservation\of pastures/range¹ lands	As under Zone I, and policy for periodical land retirement from cultivation	Usual land distribution, tenancy regulation policies; land retirement policies for class IV lands

7. Ownership	Collective or individual with provision for collective action, when required	As under Zone I, with special provisions for crop lands	Individual
8. Size of holding	Extremely large (i.e. pasture holdings)	As under Zone I, and large holdings for cultivation to facilitate long fallowing	Normal (as permitted by ceiling laws in similar dry areas)
9. Need for public regulation of land use	Very high, at district, village and pasture levels	Very high at pasture level, and high at farm level	High at farm level for land class IV
10. Issues for public regulation	Land–animal ratio, grazing rights, rotational grazing, seasonal migration, conservation measures, other issues of pasture policies	As under Zone I, and issues related to crop-fallow rotation and conservation measures on crop lands	Conservation measures and crop-fallow rotation on land class IV. For land class III, similar to other dry areas
11. Basis of taxation, penalty/rewards related to land	Number of animals	Number of animals and land size, crop produce	Land-holding size, crop produce
12. Land users' obligations	Adherence to land-use regulation and conservation laws	As under Zone I	As usual, as in the case of other dry areas

on the basis of present use—an inventory of the actual use of the land, with no reference to capability.[2] Some historical illustration of this situation follows.

Land distribution. The tenurial situation on the eve of the formation of the Rajasthan State in 1949 was characterized by a variety of formal and informal arrangements governing the control and use of lands in different princely states. There was a variety of intermediaries—for example, *jagirdar, biswedar, muafidar, rajvi,* and rack-renting and tenurial uncertainties were common.[3] The early land legislation of the State was designed to regulate rents, protect tenants and finally abolish the intermediaries. The actual tillers of the lands were made landowners and brought into direct contact with the State. Further, vast areas of submarginal lands were distributed as private holdings for cultivation under the various legislations, in particular the Rajasthan Land Revenue (Allotment of Land for Agricultural Purposes) Rules 1957. However, while making the tenants owners of the land as well as distributing new lands to private land-holders, the State did not give any consideration to the use capability of the land. Nor did it impose any obligation on the beneficiaries regarding use or conservation. In other words, with regard to ownership, agrarian legislation in Rajasthan (except for the land ceiling laws) treated the desert lands or submarginal lands in arid areas on a par with the lands in well endowed areas of the southern and southeastern parts of the State.

Land-use regulation. No land legislation dealt specifically with the determination and regulation of the use intensity of land. The Rajasthan Tenancy (Government) Rules 1955 and the Rajasthan Land Revenue Act 1956 contain some provisions for the regulation of land use, but their approach is entirely different from what has been described above. They merely stipulate that it is the duty of the revenue officials to ensure that a given piece of land is put to the same use to which according to the revenue records it was put in the past. Thus, through the above laws, the State attempts to regulate the use of the land according to the revenue records and not according to its physically desirable level of use intensity. In effect this procedure tends to perpetuate the existing maladjustments in the land use pattern.

Indeed, the Rajasthan Agricultural Lands Utilisation Act 1954 goes a step further. The traditional practice of rotating cropping and fallowing is a compromise between the intensive and extensive type of land use in the arid area. But the above Act (Section 4) completely ignoring the traditional wisdom of the desert farmer, empowers the district collector to prohibit the fallowing of the croplands. If the landowner fails to cultivate the land and obstructs the alternative arrangements for cultivation made by the collector he is liable to penalty up to Rs. 500 per case. Though no cases of implementation of this provision have been noted, the law itself illustrates the government's disregard of the nature of the problem of arid lands.

Pasture policy. The economy of the arid region, in general, and its north-western parts, in particular, is dominated by livestock farming. Practically every village had some common pasture lands. But the State has no legislation reflecting its concern for these grazing lands and there is no coherent pasture policy. The State Government, through certain provisions of the Rajasthan Panchayat (General) Rules 1955, has shifted the responsibility for pasture development and management to the village *panchayat* (council). The panchayats have neither technical competence nor political and administrative power to introduce any measure to regulate stocking rates, grazing rights, or rotational grazing. An evaluation study (Anon., 1963) revealed that not a single panchayat had taken up any step to manage or develop village pastures and forests. Thus, as in the past, the utilization of grazing lands is governed by informal institutional arrangement under which uncontrolled and un-restricted grazing, without any obligations to the grazing land is the rule.

Conservation measures. In arid lands especially conservation should be an essential part of any land-related policies and programmes. The only legislation regarding land conservation is the Rajasthan Soil and Water Conservation Act 1964. The second schedule of this act enumerates a large number of conservation measures, which should form part of the schemes to be undertaken by the Rajasthan Soil and Water Conservation Board and other agencies. However, Chapter 2 of the act makes it clear that the approach of the legislation is very specific and narrow. In fact, the act provides for the framework for initiating and operating the small scale departmental conservation schemes in any notified area. It contains no provision, whereby the farmer or a group of farmers can be compelled or persuaded to adopt conservation measures on a "catchment" of watershed basis. Nor does it have any provision whereby conservation measures are made an essential part of the farming system in the arid areas.

III. REASONS FOR NEGLECT

The absence of inducement to conservation, development and efficient utilization of arid lands in the agrarian legislation of Rajasthan can be attributed to the following factors:

Policies related to land tensure emerged as part of the land reform programmes in the State during the 1950s. The primary objective of land reforms was to replace the feudal agrarian structure by a more equitable and just agrarian system in keeping with the ideals of a democratic state. The feudal agrarian order, which had developed over a long period in the princely states of Rajasthan, had a number of exploitative features. The production, distribution and exchange relations between the landowner and the land-user were exploitative. The land use pattern it encouraged was not related to the land use capabilities and hence was already over-exploitative. The land reforms were only concerned with the exploitative features of the feudal order, as they related to the tillers of the land, and not to the land itself. Thus the need for preventing very intensive use of land was completely overlooked, and the conservation needs of the resource base never figured in the land reform laws. Indeed, the real consequence of agrarian reforms through the distri-bution of additional submarginal lands has been to accentuate the process of over-exploitation.

The neglect of the conservation needs of the arid lands in the land reforms was partly due to the relatively poor political pay off in the context of the immediate socio-political objectives of land reform. Further, in a young state with a long feudal background, the new policy makers did not fully grasp the technical factors, and it was easier to execute common land policies for the whole state rather than have specific policies for different ecological areas. Consequently, the land policies designed for the State as a whole have been applied to the arid region (including Zones I and II) as well.

IV. CONSEQUENCES

Land distribution. The immediate consequence of the absence of tenurial policies which would have helped to treat the arid lands according to their use capabilities is the indiscriminate distribution of submarginal lands for cultivation. As shown in Table I the extent of the area under private cultivation alone far exceeds the extent of the land in classes III and IV (Zone III) in all the districts, except Sirohi and Pali. Further, as the bracketed figures in the same table show, in most of the districts more than 80% of the area of private holdings is actually cultivated. This implies that the possibility of restricting the already distributed land for less intensive use (such as fallowing for grass) is quite limited. Even in the driest districts—Jaisalmer, Bikaner and Barmer (where lands suited to cultivation are too limited)—the extent of the area of private holdings put under cultivation ranges from 43% to as much as 67%.

Even after the distribution of submarginal lands to private owners, their use intensity can be kept at a low level by adhering to a rotation comprising a crop followed by a long fallowing. But this rotation requires a fairly large size of holding. The smaller the holding, the shorter the duration of the fallowing and, therefore, the less the chances of maintaining a low degree of land use intensity. This fact is clearly indicated in Table III where the extent of the old

Table III. Details indicating the land-use intensity by size of operational holdings in the arid districts of Rajasthan, 1970–71.[a]

| District | Percentage area under important land-use categories[b] on operational holding group (hectares) | | | | | | | | | | | |
| | Net cultivated area[c] | | | | Fallow lands other than current fallow | | | | Cultivable wasteland[d] | | | |
	<0·5– 5·0	5·0– 10·0	10·0– 20·0	20·0 and more	<0·5– 5·0	5·0– 10·0	10·0– 20·0	20·0 and more	<0·5– 5·0	5·0– 10·0	10·0– 20·0	20·0 and more
Jaisalmer	82·28	64·35	46·85	26·23	3·13	6·22	9·94	16·45	13·12	29·02	42·55	52·76
Bikaner	88·40	77·16	69·58	51·82	4·02	6·04	7·62	9·38	7·10	16·35	22·38	37·38
Barmer	89·78	81·56	74·35	63·03	8·18	14·64	18·92	22·75	1·90	3·54	6·56	13·86
Jodhpur	97·14	92·57	87·19	74·23	2·50	6·66	12·21	24·02	0·16	0·17	0·20	1·29
Churu	97·70	95·21	90·73	84·51	1·42	3·35	6·02	9·54	0·64	1·08	2·43	4·58
Nagaur	99·01	97·60	97·56	94·32	0·66	1·85	2·05	4·77	0·16	0·16	0·18	0·52
Jalore	96·97	94·96	93·23	90·30	2·76	4·61	6·30	8·18	0·13	0·21	0·23	0·85
Jhunjhunu	99·14	95·58	98·68	97·01	0·24	0·42	0·69	1·05	0·37	0·38	0·38	1·07
Pali	96·26	92·63	86·81	78·02	3·08	6·10	11·69	18·23	0·19	0·45	0·68	2·19
Sikar	98·27	97·92	97·29	94·62	0·41	0·88	1·37	2·60	0·66	0·57	0·55	1·63
Sirohi	92·65	86·72	82·33	75·30	1·65	10·73	14·49	19·27	0·97	1·15	1·64	3·14

[a] Data culled out from *Report on Agricultural Census 1970–71, op. cit.*

[b] The total of the following three categories in each holding class does not add up 100, as the area put to non-agricultural use, etc., is excluded.

[c] The net cultivated area included the net sown area and the current fallow. For explanation see text.

[d] I includes other uncultivated land (excluding fallow), the proportion of which, however, does exceed even one per cent of the area in any of the groups of holdings in any district, except the largest group of holdings in Jaisalmer.

fallows (that is fallows other than current or one year fallows), representing the extent of the resting of the land or periodically using it for forage rather than for crop production declines with the decline in the size of the holding.

Table III suggests that in the absence of statutory measures to regulate it, the use intensity of arid lands can be indirectly influenced by distribution. In other words, farmers can be induced to keep the use intensity at a lower level through the provision of bigger land holdings. This inference implies that no newly allocated land holding should be less than a specific size. Similarly, the existing holdings should not be permitted to be fragmented below a certain level. This amounts to fixing a floor limit for arid lands similar to the ceiling limit already imposed generally. However, the distribution pattern of land holdings obtaining in different arid districts (see Table IV) suggests that land distribution policies are not concerned with such provisions. Contrary to the general impression, nearly 45% of the land holdings in the arid region are less than 5 hectares. In several districts the extent of such holdings ranges from 48% to 83% of the total. Even in the driest districts—Jaisalmer and Barmer— the proportion of holdings lower than 5 hectares is more than 20%. On such holdings or even on the holdings below 10 hectares (which account for nearly

Table IV. Distribution of operational holdings by size in the arid districts of Rajasthan 1970–71.[a]

District	Percentage distribution of holdings				Extent % of rented in the total holdings[b]	
	<0·5– 5·0	5·0– 10·0	10·0– 20·0	20·0 and above	Holdings	Area
Jaisalmer	20·65	20·42	25·00	33·92	10·39	2·53
Bikaner	8·34	20·88	33·38	37·39	2·73	2·23
Jodhpur	37·03	21·05	21·84	20·09	6·11	2·80
Barmer	20·90	19·01	25·24	34·85	5·61	2·43
Churu	16·58	26·48	32·17	24·77	4·82	3·35
Nagaur	44·32	25·38	20·96	9·33	5·66	4·57
Jalore	48·94	23·32	18·05	9·69	4·71	3·17
Jhunjhunu	67·78	21·17	9·33	1·72	1·99	1·50
Pali	72·62	13·00	9·31	5·08	13·21	10·60
Sikar	68·61	20·14	9·40	1·86	7·23	5·17
Sirohi	82·88	11·22	4·48	1·43	22·14	15·16
Total	44·86	19·02	23·59	12·51	7·10	3·79

[a] Based on details from *Report on Agricultural Census 1970–71 in Rajasthan*, Government Press, Bikaner, Government of Rajasthan, 1975.
[b] Includes both partly and completely rented in holdings.

60% of the total holdings in the region), adherence to a rotation comprising cropping followed by a long fallow is quite difficult. Owing to low and unstable crop yields, even a farmer with 10 hectares cannot afford to keep a large portion of the area under fallow for 2 to 4 years.

Under existing tenurial policies, therefore, both the opening of the submarginal lands for ploughing through the distribution of such land as private holdings and the permitting of the holding size to become smaller than required by the use capability of the land tend to encourage the higher use intensity of the land. Rapid, large scale tractorization has further reinforced this trend.[4]

The degree of land use intensity encouraged by the pattern of land distribution is not altered by temporary land transfers through private tenancy. As indicated in Table III (last two columns) except in the better irrigated districts, Pali and Sirhoi, only 2–10% of all holdings rent any land from others. The extent of the area falling under the rented category is still smaller and ranges from 2% to 5%.

Resource depletion. The consequences of tenurial policies that evolved during the feudal system and those initiated during the post-Independence land reforms are reflected both in terms of the depletion of the land resource and of the falling productivity of land.

Regarding resource depletion, no details covering the whole of the arid region are readily available. Yet the evidence available from different locations clearly indicates depletion. At several locations, the following consequences of mismanagement of arid lands have been noted: the deterioration of fertile lands due to the removal of top soil or the submersion of fertile land under shifting sand dunes (Ghose *et al.*, 1968; Anon., 1965); the conversion of fertile lands into patches of saline wasteland in the low lying areas near the seasonal streams (CAZRI, 1966); the drying up of wells or the increased salinity of well water[5]; the replacement of superior perennials by inferior ones or annual grasses including non-edibles in the grasslands (Orakash *et al.*, 1964); the increased population of malformed or stunted trees; and finally the increased human misery reflected by increased seasonal migration and accentuated pauperization through recurrent famine (Jodha, 1975).

Declining crop yields. The impact of extending crop farming on to submarginal land, as encouraged by the tenurial policy, is revealed by Table V. The Table presents the indices of three yearly moving averages of the area and the yield of principal crops of the arid region for the period 1951–2 to 1972–3. The area under all crops shows a rising trend. Only in the case of sorghum (*jowar*) around 1963–4 and sesamum around 1965–6 did the area start to decline. The reasons for this decline will be mentioned shortly. The more striking feature

Table V. Indices of three-yearly moving averages of area and yield of principal crops^a of the arid region: 1951–52—1972–73.^b

Years	Bajra		Kharif pulses		Jowar		Sesamum	
	Area	Yield	Area	Yield	Area	Yield	Area	Yield
1952–53 — 1954–55	115·82	101·09	142·82	109·90	104·92	149·59	117·29	116·05
1953–54 — 1955–56	128·52	100·54	170·31	136·63	118·25	133·33	116·76	104·94
1954–55 — 1956–57	137·29	97·82	190·19	145·45	115·09	102·44	135·61	96·92
1955–56 — 1957–58	149·43	95·65	199·14	154·46	115·80	114·63	138·75	89·51
1956–57 — 1958–59	157·25	96·74	190·43	158·42	116·85	108·13	158·65	79·63
1957–58 — 1959–60	147·83	98·37	183·93	156·44	128·43	103·25	176·45	89·51
1958–59 — 1960–61	159·19	92·39	175·46	151·49	111·59	108·13	183·26	75·31
1959–60 — 1961–62	165·46	87·50	171·29	132·45	115·45	96·75	197·40	58·64
1960–61 — 1962–63	173·71	77·72	171·90	141·58	118·60	77·24	206·82	60·50
1961–62 — 1963–64	170·29	83·69	171·29	148·52	122·46	87·80	201·06	56·16
1962–63 — 1964–65	168·71	85·01	193·16	159·54	117·27	75·49	206·94	46·84
1963–64 — 1965–66	176·49	81·06	198·87	137·72	96·90	42·72	206·22	36·43
1964–65 — 1966–67	181·97	91·96	211·18	126·66	88·50	50·33	227·75	42·14
1965–66 — 1967–68	179·81	86·07	224·49	124·41	72·95	50·50	234·51	45·43
1966–67 — 1968–69	174·10	78·10	223·70	100·81	62·66	41·22	217·73	35·74
1967–68 — 1969–70	164·11	64·82	217·08	112·24	61·53	33·07	180·04	25·96
1968–69 — 1970–71	168·27	100·19	207·13	139·76	66·26	95·28	164·29	9·23
1969–70 — 1971–72	177·01	128·89	212·44	175·98	70·69	124·79	194·61	26·18
1970–71 — 1972–73	191·34	124·13	216·47	144·49	69·85	128·64	191·12	39·37

^a Bajra, kharif pulses, jowar and sesamum covered around 60, 20·5 and 6 per cent of the gross cropped area (average of five years) respectively. The rest of the area was covered by other (nearly 17) minor crops, whose individual share rarely reached even one per cent of the gross cropped area.
^b Calculations are based on districts and the production data of crops from the Directorate of Land Records, Board of Revenue, Government of Rajasthan (for 1951–52 to 1955–56), and Statistical Abstracts of Rajasthan for the rest of the period.

revealed by Table V is that if one excludes the impact of a sequence of extremely good years (1970–1 and 1971–2) the yields in most of the crops show a declining trend. This decline is especially strong in the case of jowar and sesamum, a fact which partly explains the decline in their area in the later period. Once the farmers realized the futility of planting submarginal land to crops like jowar and sesamum (which need relatively good soils). they shifted to kharif pulses, and to some extent to millet (*bajra*). Kharif pulses, being the crops most suited to arid lands, show relatively stable, if not rising, yield trends. Generally, however, an increase in the area under crops has accompanied the decline in the yields of crops. Hence the efforts to raise food production through pushing submarginal land under the plough may prove counter productive in the long run.

V. THE FUTURE COURSE

The main burden of the preceding discussion is that in the past land policies have encouraged a land use and management pattern which is not in keeping with the requirements of arid lands. The reversal of the process will mean the restricting of the use intensity of land to a lower level or the adoption of complementary provisions which permit a high intensity of use without damage to the land resource base. Restriction on the future distribution of submarginal lands for cultivation is the first step. The retirement of lands from cultivation and back to natural vegetation is a difficult task. In fact, in view of the increased human and animal pressure on land and the rigidity of land use encouraged in the past, the lowering of the level of use intensity may not be possible. Hence the alternative lies in the adoption of conservation measures which enhance the permissible level of use intensity of the arid lands.

In this context, it will not be out of place to mention that progress in agriculture productivity is normally associated with the increasing use intensity of the land. The whole process of agricultural growth is sometimes explained in terms of a progessive increase in the intensity of land use.[6] However, the mere intensification of land use unaccompanied by an appropriate technology may prove counterproductive, as seems to have happened in the arid region of Rajasthan (Jodha, 1972). The first step towards raising productivity or maintaining the use intensity of arid lands at a high level is to have the necessary technological complement for both crop lands and grazing lands. In the context of arid lands in the absence of irrigation facility, various conservation measures may offer the required technological complement. Conservation technology covers various measures, some of which are well regulated farm practices, while others involve the creation of physical assets. Rotational grazing and rotations between crops and a long fallow, for instance, fall in the former category. Measures such as the stabilization of sand dunes, the creation of shelter belts and micro-windbreaks, the bunding of croplands and contour furrowing, contour trenching, and many other soil works for range lands fall under the second category. The conservation technology, as evolved and recommended by CAZRI, is yet to reach the farmers (Jodha, 1972). Conservation technology is frustrated by a number of problems, the solution of which is linked to tenurial policies.

One such problem stems from the indivisibility of a number of conservation measures. Measures such as sand dune stabilization, shelterbelt creation or the bunding work or regeneration of range lands through a variety of earthworks, can be adopted effectively only on the basis of the catchment, not on the basis of individual farms. These measures require collective action at the village level and above (Jodha, 1967).

Similarly, some regulatory practices such as the rotational use of past-ures also require collective action. Before land reform when the jagirdar had absolute authority over the village, such collective decisions were not required. In some villages jagirdars used to enforce certain regulatory measures for the utilization of pastures and forests. Rotational grazing around different watering points, known as *toba*, grazing fees per animal, known as *ghasmari*, the zoning of village land for grazing and cutting fodder through declaration known as *chait rakhai* are a few examples. Most of these provisions disappeared with the abolition of the jagirdari system. The new village panchayats which replaced the jagirdars for the purpose of village adminis-tration could not enforce such provisions (Jodha, 1980). In the changed circumstances it is essential to have some form of dual tenure[7] which provides some authority for the group over the lands belonging to individual farmers in a given catchment. This system would not only facilitate the formation of land users' associations but will also help panchayats and revenue authorities to enforce the adoption of conservation technology.

The individual land user's obligations in terms of adherence to conser-vation practices should be specifically incorporated in the land laws. These laws can be enforced with liberal recourse to penalities in the case of default, and the granting of rent remission as reward.

The provisions involving land users' obligations are more important in the case of grazing land. The present unregulated grazing, for instance, results from lack of provision for grazing rights and obligations in terms of grazing fees, taxes and penalties. There is little private cost for resource use. The incorporation of the above provisions in the land laws may greatly help to rationalize the utilization of grazing lands.

Finally, it must be noted that land policies can play only a complementary role in ushering in the era of conservation farming in the arid region. They cannot be a substitute for the conservation oriented development strategy which would encourage production activities appropriate for the arid region.

Endnotes

[1] This chapter is a revised version of N. S. Jodha, 1977, Land Tenure Problems and Policies in the Arid Region of Rajasthan, in Desertification and its Control, New Delhi: ICAR pp. 335–347.

[2] Patwari records are the ultimate source of land statistics and show that a pasture is classified as pasture not because it represents land poorer than the lands classified as crop lands, and is, therefore, suitable for grass only, but because the former Jagirdar or panchayat declared it as such in view of the nearness of the watering points or the village abadi, i.e. the village settlement.

[3] For details, refer to the Rajasthan Protection of Tenants Act, 1949; The Rajasthan Agricultural Rent Control Act, 1954; The Rajasthan Land Reforms and Resumption of Jagir Act, 1952; The Rajasthan Land Reforms and Acquisition of Land Owners Estate Act, 1953, etc.

[4] In recent years the spread of tractors, again encouraged by liberal public loans, has further aggravated the situation in parts of the arid region. Tractors no doubt have a distinct advantage in arid areas where timely planting of crops during the short wet period ensures greater chances of a successful crop. But this has led to decline in both periodic fallowing of crop lands and in permanent fallows, because tractors (available on hire on a harvest-linked deferred payment basis) ensure much more draft power to cover land in the short wet period than was possible in the past with bullocks and camels. According to a case study of tractorization in Nagaur district, the number of tractors increased from 10 in 1964–1965 to 59 in 1973–1974 in a cluster of six villages. Tractor using (as distinct from owning) households constituted 9·8 and 76·8% respectively of total farm households during the two periods. The final consequence was reflected in drastic changes in the land use pattern. The net sown area as percentage of cultivable land increased from 65% in 1964–1965 to 86% in 1973–1974. The current fallow as proportion of cultivable land declined from 13% to 5% during the same period. The corresponding decline in old fallow was from 22% to 9% during the same period. The yield of fodder from *pala* (*Zizyphus nummularia*), which is available even during famine years, declined by 75% during the same period, as tractor use eliminated the roots of this soil binding plant.

[5] In the Jodhpur–Bilara region, the region traditionally famous for its abundance of sweet subsoil water, nearly 22% of 169 selected wells were found to be suffering from the above problems. See Anon (1967).

[6] For instance, the shift from the forest fallow to the bush fallow and hence to the short fallow and then to annual cropping and lastly to multiple cropping represented successive stages towards higher degrees of use intensity, giving rise to large favourable shifts in production possibilities of the land (see Boserup, 1965.)

[7] The term "dual tenure", used in the absence of any more appropriate term, implies that the lands in a given catchment belong to individual farmers. But when it comes to

the catchment-based treatment of the lands, all farms should be treated as belonging to the group having their land parcels in the concerned catchment. The group should have authority to undertake conservation measures. It should also be liable to punishment for the mismanagement of the catchment.

Informal arrangements, similar to the one mentioned above, already exist in predominantly Bisnoi-caste villages in the Jodhpur region where the villagers as a group do not permit an individual farmer to cut trees, etc. in his own field. Nor do they allow the killing of wild animals—deer, rabbits, etc., in the whole of the village territory.

18

Notes on the Experience of Drought Perception, Recollection and Prediction

L. P. Bharara

Central Arid Zone Research Institute
Jodhpur (Raj.), India 342003

This chapter is a brief and preliminary excursion into the meaning of drought. It can only be preliminary because the details of a methodology that would make such an enquiry satisfactory have still not been worked out. The closest examples are the perception studies carried out by geographers in recent years especially under SCOPE and MAB. It is not quantitative, because it deals with ideas, whereas statistics only reflect behaviour. It looks beyond the social dimension of desertification and points the way to an exploration of the cultural dimension. It reflects many of the problems and promises of empirical explorations of meaning, of which the methodological problems are particularly obvious, especially now that the meaning of drought is presumably changing as a function of the increasing rate of social and economic change. The investigation of the cultural dimension of desertification needs eventually to go a step further than either this or the perception studies by not only recording, sampling and analysing variation in environmental perception but by investigating the common sense rationalization of these perceptions and its role in adaptation on the basis of memory.

—Ed.

The incidence of drought in arid and semi-arid regions has received attention from various angles. Generally, it has been studied as a natural calamity creating economic, social and physical disturbance, and the major topics have been its definition, its causes and effects, and the advantages and disadvantages of various relief measures and policies. The present study seeks to add a dimension to these studies, in a form that may be characterized briefly as a contribution to the ethnoscience of droughts. This chapter investigates the memory of drought over a period extending back to the limits of living memory in selected villages in western Rajasthan. An attempt is made to add to our understanding of how memory works in a cultural system which includes the synoptic perception of past droughts as all-encompassing phenomena—a perception on which is based the expectation and prediction of droughts to come. The material takes on additional significance from the changing context of drought, in consequence of population growth, technological development, and economic and social change.

It is easy to forget that the categorization of phenomena as, for example, climatic, biological and social, though convenient and useful, is an arbitrary product of a particular intellectual tradition. In some social situations, such as Rajasthan village communities in western Rajasthan, where these compartmentalizations of knowledge are not so automatically made, drought is not simply the dearth of rain (though everyone knows that rain would remove it)—it is the total quality of life, including, besides weather, animal behaviour and social relations.

The material presented in this chapter[1] shows drought as a pathological condition that recurrently afflicts rural society in western Rajasthan. It represents memory as the *ad hoc* ordering of recollections and their common sense rationalization in the form of sayings. The value of this material should not require explanation: it is a key to understanding the coping mechanisms and potential reactions of a rural population, which (despite certain distinctive aspects of Indian society) must be to some extent representative of a large proportion of the population of the world's drylands, to the droughts which are often harbingers of the severer and longer term process of desertification.

I. RECOLLECTION

The correlation between the farmers' recollection of past harvests and the record of actual rainfall in western Rajasthan is very close. Living memory of the nature of previous years extended as far back as 1899 with substantial agreement. Each year was remembered in terms of *zamana*, a unit of measurement mainly for the kharif harvest. Full zamana is the level of production expected, given sufficient water. The classification of the years from 1899 to 1978 is given in terms of remembered zamana in Tables I(a) and I(b). (Informants actually talked in terms of annas—the old one-sixteenth division of the rupee. Annas have been translated into percentages for convenience.) Out of 80 years 58 were perceived as drought or severe drought (up to 25% zamana); five were mild drought (25–50% zamana); five were average (50–75% zamana); and twelve were good or surplus years (75–100% zamana or more). Each decade witnessed from six to nine severe drought years, from one to three good years, and hardly one average or "normal" year. It is worth noting that the high average rainfall for years of severe drought in Table I(b) show that zamana correlates not with amount of rainfall alone but with a combination of amount, timing and intensity. In this connection it is worth remembering that we have no direct record of one particularly significant indicator of a bad year—the extent of damage from flooding.

Recollection did of course vary. Generally, food producers—farmers and pastoralists—scored better than members of non-food-producing caste groups, though no significant variation was evident in relation to mild drought or "average" years. Much of the variation altogether arose from factors such an uneven spatial distribution of rainfall, land holdings, fallow, and herds.

The basis of the memory of drought years can be related to various factors. Earlier disastrous years—1812–13, 1868–69, 1877–78, 1791–92—were cited by a few old men on the basis of inherited information. Some severe droughts and famines, like 1899, 1901, 1918, 1959, 1960, 1968 and 1969 were remembered with accuracy due to reinforcement or feedback from documentation of migration, mortality, desertion of villages, or scarcity of grain, water, and fodder in sources such as the Rajputana gazeteers and State Administration and Famine Commission Reports. There was also a clear folk memory of good years in 1900, 1908; 1916; 1926; 1933; 1940, 1942; 1950, 1953, 1958; 1961 and 1970. For example, everyone knew that in 1940 the rains had continued till the month of Asoj (September–October)[2] and the moth harvest had excelled all records; that a record bajra crop had been recorded in 1961; and that in 1921 the bajra crop had spoiled as a result of additional late rainfall but moth (*Phaseolus aconitifolius*) production had excelled.

There was an overall tendency to exaggerate the severity of the droughts as they receded into the past, especially on the part of food-producers. Moreover, earlier droughts were associated with not only less rainfall, but more intense storms, more soil erosion, shifting dunes, more crop losses, migration, livestock mortality, desertion of villages, as well as the lack of motorized transport, drinking water, and drought-resistant crop variations. On the other hand, there was a general awareness that population levels were lower, that there was less crop disease, that joint households were more cohesive and resistant to centrifugal forces, that vegetation was denser, and that land was more often fallowed, so that the effects of droughts were not so immediately disastrous, despite their intensity, because of the greater margin between exploitation levels and carrying capacity, and possibly also because of greater social cohesiveness.

Discussion of the causes of drought in Rajasthan ranges over a variety of factors from climatic and biophysical to social and religious. Although most informants related the occurrence of drought primarily to physical factors such as changes in wind direction, cloud cover, frequency and intensity of

Table I. Correlation of rainfall records and recollection of zamana.

(a) by year

| Year | Rainfall Data | |
	Rainfall (mm) at nearest recording station	Zamana (%)
1899	—	0
1900	—	100
1901	205·99	0
1902	193·50	0
1903	461·36	20
1904	331·00	20
1905	79·50	0
1906	274·83	20
1907	443·50	20
1908	644·00	100
1909	504·00	60
1910	339·85	20
1911	230·83	20
1912	307·85	75
1913	131·32	10
1914	243·33	10
1915	322·83	10
1916	462·30	100
1917	806·50	0
1918	62·99	0
1919	236·70	10
1920	172·97	25
1921	191·30	50
1922	313·70	10
1923	399·30	25
1924	198·37	25
1925	177·55	0
1026	304·80	100
1927	323·90	10
1928	468·90	0
1929	426·50	50
1930	327·90	0
1931	246·88	25
1932	180·34	10
1933	403·35	100
1934	377·90	0
1935	229·62	0
1936	304·40	0
1937	334·80	0

Table I—*cont.*

	Rainfall Data	
Year	Rainfall (mm) at nearest recording station	Zamana (%)
1938	165·86	0
1939	111·76	0
1940	454·10	100
1941	150·60	0
1942	514·30	100
1943	278·13	50
1944	382·80	25
1945	345·44	25
1946	209·30	25
1947	343·10	40
1948	86·81	0
1949	336·80	20
1950	429·77	100
1951	140·21	0
1952	135·38	0
1953	813·05	100
1954	276·35	0
1955	382·02	25
1956	382·78	75
1957	285·24	0
1958	522·70	100
1959	324·60	0
1960	175·20	0
1961	297·00	100
1962	342·70	10
1963	585·60	0
1964	586·10	25
1965	235·50	10
1966	147·50	10
1967	352·00	75
1968	285·50	0
1969	182·80	0
1970	571·20	100
1971	466·50	25
1972	331·10	0
1973	661·50	55
1974	295·00	0
1975	695·00	25
1976	438·00	10
1977	436·00	10
1978	367·54	50

Table I—*cont.*
(b) by % zamana

% zamana	Mean annual rainfall (mm) (at nearest station	Mean zamana	Years
0% (severe drought)	248·50	0%	1899; 1901, 05; 1917, 18; 1925, 28; 1930, 34, 35, 36, 37, 38, 39; 1941, 48; 1951, 52, 54, 57, 59; 1960, 63, 68, 69; 1972, 74.
0–25% (drought)	298·66	18%	1902, 03, 04, 06, 07; 1910, 11, 13, 14, 15, 19; 1920, 22, 23, 24, 27; 1931, 32; 1944, 45, 46, 49; 1955; 1962, 64, 64, 65; 1971, 75, 76, 77.
25–50% (mild drought)	353·82	48%	1921, 29; 1943, 47; 1978.
50–75% (average)	441·62	65%	1909; 1912; 1956; 1967; 1973
75–100% (good)	515·20	100%	1900, 08; 1916; 1926; 1933; 1940, 42; 1950, 53, 58; 1961; 1970.
Overall mean	371·56	46%	1899–1978 (80 years)

storms, and the overall quality of cold and hot seasons, many emphasized the role of bio-physical factors such as decline in forest and vegetation cover, increased wind erosion and shifting dunes and silting of village ponds. There was a general consciousness that human activities lay behind these natural factors, particularly indiscriminate cutting for fuel and construction, over-grazing and lopping of trees for fodder, but considerable emphasis was also given to supernatural and moral factors such as luck, apathy, immorality and the revenge of nature.

In discussion of the effects of drought most people first cited economic disorder: failure of crops, loss of livestock, reduction in the value of assets, and forced sale. But there was only slightly less emphasis on social disorder: migration, the failure of institutions, disintegration of social groups and of households.

Drought continues to be seen as a form of instability or disorder in both the natural and the moral world. But the context of instability has changed. Pastoralists may migrate long distances in search of land for grazing in a drought year and, although earlier there was enough vacant territory to absorb them, now farmers complain that they not only exhaust limited water and grass resources, but also destroy the standing crops, disrupt soil conservation measures and create conflict and tension.

The problems caused by the instability have become more complex, not only because of the increased density of the population but because of recent socio-economic changes (see also Malhotra and Mann, Chapter 15). For example, symbiosis of food-producing and occupational caste groups has been disturbed. Members of occupational castes such as potters, leather-workers, oil-pressers, and retailers, have now acquired land and become partially dependent on cultivation, and have thus lost flexibility while individual craftsmen may have gained flexibility through economic diversification, in many cases the crafts have gone into decline. Joint households have broken down into nuclear households causing greater pressure on scarce resources by division and fragmentation of property. Participation in social events has narrowed from the community at large to the smaller extended kinship group. There has been general decrease in family solidarity and increase in conflict, tension and insecurity. During the drought of 1968–1969 people sold land, livestock, ornaments, houses and household articles, and the jajmani or aat system of payment in kind or cash between caste groups for services rendered was disrupted, with the result that the existing trend of socio-economic change was accelerated.

As the social and economic context of drought has changed, so has its meaning. The change in meaning is more difficult to reconstruct. But it is worth noticing that many biophysical effects reported from recent droughts derive from the responses of the weaker segments of the population. For example, particular trees and shrubs (khejri—*Prosopis cineraria*—for fuel, phog—*Calligonum poligonoides*—for charcoal, banwarli for tanning, kumbhat for churns) were cut and sold in larger quantities.

Although the context of the experience and recollection of drought has been changing, the drought lore presented in the next section may not be changing in the same way. It is likely, however, that its use is changing. By analogy with equivalent processes elsewhere it might be expected that this lore will gradually be relegated to the epistemological status of superstition, but (as we also know from elsewhere) such a change in the form of knowledge does not prove a change in the manner of thinking. The study of this type of thinking can be expected to lead to a better understanding of non-scientific thought generally.

II. PREDICTION

Prediction of the nature of the coming year is considered a traditional skill. Its origin is traced to the thirteenth century, the time of the Rajput seer (*pir*)

Harbuji Sankhla, who was Jagirdar of a village in Phalodi tahsil in central Rajasthan. Harbuji Sankhla is said to have observed birds and animals—which implies that they were assumed to be in some way closer to the physical processes of nature—in order to find ways of predicting droughts. Because of his success the skill to predict became the monopoly of the Rajput caste, some of whom cultivated it in each generation. The predictor was considered to be divinely guided. From the day of conception a Rajput mother would make the pious intent to rear her child as a predictor. Throughout pregnancy, each day she would put water in a pitcher by a tree where particular birds could drink from it. On birth a male child's first drink was taken from that pitcher. If the child survived, he was believed to command the language of the local birds and animals and became a *sugni*, an omen seer or predictor.

With the social change of the last few decades the caste monopoly of drought prediction has dissolved. Now the lore concerning drought prediction is a matter of general discussion and effort. It is based on observation of simple qualitative changes in climate and vegetation and behaviour. Arguments from Western science have no apparent impact on it. In the form of a collection of sayings it is widely used as a guide for the coming year. These sayings constitute a type of ethnographic material that lends itself to structural analysis. Here, however, the aim is simply to demonstrate their function in rationalizing a holistically perceived reality. Reality and recollection have been correlated above in relation to precipitation figures. Here the question of reliability is ignored and the focus is directed instead to the synoptic nature of common-sense rationalization.

Winds

(i) *savan men suryo cale, bhadurve purvai*
    asoj men pichvaha cale, bhar bhar gara layi
If the northwestern wind blows in Savan,[2] the eastern wind blows in Bhadun, or the western wind in Asoj, they bring carts full of grain.

(ii) *jad bahe hada hava kun*
    banjara lade lun
When the southwestern wind blows, the Banjaras load salt.

(iii) *nada tankan balad bikavan*
    tu mat cale adha savan
O eastern wind, who cause people to hang up the rope that fastens the yoke and to sell the ox, don't blow up to mid Savan.

(iv) *jeth biti pehli parva, kathak ambar haren*
    asad savan khet sukho, bhadar huve birkha kare
Thunder on the first day after the end of Jeth means two dry months and no rain before Bhadun.

Stars

(v) *diva biti pancmi, som, sukar, guru mul*
    dank kahe he bhadali, nipaje satun tul

On the fifth lunar day after the Divali holiday, if mul (an astrological position of the moon) falls on Monday, Thursday or Friday, then the Brahmin says all the seven grains will grow.

(vi) *sudi asad men budh ko uday bhayo jo pekh*
     sukra ast savan rahe maha kal ava rekh

If Mercury is seen rising in the bright half of the lunar month of Asad, or if Venus is sinking in Savan, a great famine will occur.

(vii) *jeth badi dasami divas je saniscar hoy*
      pani hoy na dharan men birala jiven koy

If Saturday falls on the tenth day of the dark half of Jeth, there will be no water on the earth and only a few people will remain alive.

Clouds

(viii) *titar pankhi badali, bidhava kajal rekh*
       a barase a ghar kare, in men min na mekh

Clouds with wings like a partridge, and a widow with kohl in her eyes, bring rain, without doubt.

(ix) *savan surangi khejri, kati biranga khet*
     savan birangi khejri, kati suranga khet

If the Khejri is colourful in Savan, the fields are colourless in Kati. If the Khejri is colourless in Savan, the fields will be colourful in Kati.

Animal Behaviour

(x) *din mensyal sabad jo kare*
    niscay hi kal halahal pade

When the jackal howls during the day, a great famine is certain.

(xi) *agam sujai sandani daude thalan apar*
     pag patake baise nahin, jad menhavan har

The she-camel knows beforehand and runs to and fro.
She stamps her feet and will not sit when rain comes.

(xii) *cidi nahave dhul men, to pani ave*
      jal men nahave cidakali, to pani jave

When the sparrow bathes in the dust the rain will come,
When the sparrow bathes in the water the rain will go.

Social Relations

(xiii) *akhatij duj ki ren je acanak jace sen*

kacak bic mange nath jay to janije kal subhay
hans kar dey nate nahin koy, mane sahi jamano hoy

The night before the Akhatij festival in March, if one should go suddenly to a friend's house and ask for something and it is refused, there will be famine; if he gives gladly the year will be good.

(xiv) *pag pungal, sar merta, udraj bikaner*
       bhulo cuko jodhpur, thavo jaisalmer

Feet in Pungol, head in Merta, stomach in Bikaner
Famine may move to Jodhpur, stay in Jaisalmer

This small collection of sayings illustrates a concern with rationalization of out-of-the-ordinary events or coincidences such as wind direction, thunder, astrological coincidence, cloud formation, animal and bird behaviour and social relations, as well as the symbolization of drought in the intensification of the salt trade, empty grain carts, idle yokes, the sale of animals, and social monstrosities such as a painted widow, and finally personalization and personification of famine sprawled across the drought-prone region. In discussion the same informants cited many more examples of similar ideas, some of which are well known cultural preoccupations such as astrological coincidence, others—the occupational concern of farmers such as unseasonal natural phenomena. The overall context is defined by an example that gives a perspective beyond the prospect of the immediate year: 7 famines (*kal*), 27 good years (*zamana*), 63 poor years (*kara kaca*), and 3 disastrous years (*ghisan*)—which fits very closely with the table of actual years.

This material is of course the type that is usually classified as superstition, and although every effort was made to collect it from what was generally acceptable across the spectrum of caste, age, and sex differentiation in the communities studied, this sample can realistically be claimed only as characteristic, rather than representative. It is nevertheless important because it demonstrates the ways of thinking which may be assumed to be closely related to the behaviour patterns which programmes of ecological management and economic development set out to change—often unsuccessfully.

   Much of the material—though not all—also falls under the heading of common sense (cf. Geertz, 1975). It appears to derive from *ad hoc* rationalization which generates rules of thumb, and though arbitrary from the point of view of scientific argument, is self-perpetuating. It is important to note that it is also functional, in the sense that it goes towards satisfying the need for order. It serves to order experience and guide expectation. Science can never adequately satisfy that need at the level of everyday thought—there is too much that science does not adequately explain, and many of its explanations are beyond the grasp of people other than professional scientists. Faith in science and in technology, which is spreading and increasing, tends to take over from this type of common sense rationalization, often by discrediting it before replacing it. To the extent that the ordinary man has a limited grasp of scientific explanation he probably suffers some anomie as a result of this reduction or weakening of order caused by the spread of science. But judging from Western experience, faith in science and technology never entirely supplants this common sense or entirely changes this way of thinking. In so far as this way of thinking survives, it deserves more serious and more intensive study.

Endnotes

[1] The data derive from studies carried out in 1977 in selected villages in central Rajasthan, including two pastoral villages, Rohini and Bhojas in Nagaur tahsil (district), and a dry farming village, Rohina in Joyal tahsil, Nagaur district, besides various villages of Jodhpur district, specifically of Shergarh tahsil, which were covered in the process of routine socio-economic surveys. All of these may be considered typical Rajasthan "desert villages". Fieldwork consisted of participant observation, intensive interviews and a field schedule containing open-ended as well as fixed response questions relating to social status, household composition, perception, memory and prediction of drought and its accompaniments. In addition, data on oral tradition, values, and beliefs were collected through informal and often group interviews. All the households of the selected villages were stratified on the basis of caste and from each caste stratum about 10% of the households were selected for intensive interviewing by a method of simple random sampling. Empirical investigations of drought prediction were made in Rohini and then supplemented with data collected from other villages. Here the sample included heads of households belonging to representatives of subcaste communities. The age range of sample heads of households included young (15–34 years), middle aged (35–54) and aged (55 upwards), all of whom had experienced frequent droughts in the areas. In this type of sampling (sometimes referred to as "availability" sampling), which may be the only practical method of obtaining respondents, the investigator attempted to relate as many cases as possible to questions of theoretical interest.

[2] Local and equivalent English months are:

Hindi lunar months (with selected variants)	*English calender*
1. Chetra	March–April
2. Baisakh	April–May
3. Jyestha, Jeth	May–June
4. Asadh	June–July
5. Shravan, Savan	July–August
6. Bhadrapad, Bhadun	August–September
7. Asoj	September–October
8. Kartik, Kati	October–November
9. Mighsar	November–December
10. Pos	December–January
11. Magh	January–February
12. Phalgun	February–March

Bibliography

For the bibliographical convenience of readers with special interest in the regional programmes, references relating to Iran and India have been grouped and are listed separately following the general references.

General

Adam, D. P. and Mehringer, P. J. (1975), Modern Pollen Surface Samples and Analysis of Subsamples, *Journal of Research of U.S. Survey* **3** (6):733–736.

Afghanistan. (1977), Ministry of Agriculture, Forestry Department, Afghanistan Country Statement Prepared for the United Nations Conference on Desertification.

Almagur, U. (1978), Pastoral Partners: Affinity and Bond Partnership Among the Dassanech of South West Ethiopia, Manchester University Press, Manchester.

Altmann, J. (1974), Observational Study of Behavior: Study Methods, *Behavior* **49**:227–287.

Antoun, R. (1972), Arab Village: A Social Structural Study of a Trans-Jordanian Peasant Community, Indiana University Press, Bloomington.

Asad, T. (1970). The Kababish Arabs, C. Hurst, London.

Ashtor, E. (1976), A Social and Economic History of the Near East in the Middle Ages, University of California Press, Berkeley, Ca.

Ault, W. (1972), Open Field Farming in Medieval England: A Study of Village By-Laws, Barnes and Noble, New York.

Baker, R. (1975), "Development" and the Pastoral Peoples of Karamoja, North-Eastern Uganda: An Example of the Treatment of Symptoms, *In* Monod, T., editor (1975), pp. 187–205.

Banfield, E. C. (1958), The Moral Basis of a Backward Society, The Free Press, New York.

Barker, S. and Lange, R. T. (1969), Effects of Moderate Sheep Stocking on Plant Populations of a Black Oak–Bluebush Association, *Australian Journal of Botany* **17**:527–537.

Barry, R. G. and Chorley, R. J. (1976), Atmosphere, Weather and Climate (third edition), Methuen and Co., London.

Barth, F. (1953), Principles of Social Organization in Southern Kurdistan, Oslo, Brødrene Jørgensen A/S-Boktrykkeri (Universitetets Ethografiske Museum Bulletin No. 7).

Barth, F. (1960), Nomadism in the Mountain and Plateau Areas of Southwest Asia, in UNESCO, Problems of the Arid Zone, Paris, UNESCO, pp. 341–355 (Bobbs-Merrill Reprint, No. 9-263).

Bastin, B. (1969), Premiers résultats de l'analyse polinique des loess en Belgique, *Bulletin de l'Association Française pour l'Etude de Quaternaire* **6**(18): 3–11.

Bedoian, W. H. (1978), Economic Alternatives for a Semipastoral Population in Southeast Tunisia, *In* Hyder, D. N., editor Proceedings of the First International Rangeland Congress, Denver (Colorado), Society for Range Management, pp. 71–75.

Behrensmeyer, K. A. and Hill, A. (1979), Fossils in the Making, University of Chicago Press, Chicago.

363

Bennett, J. W. (1976), The Ecological Transition: Cultural Anthropology and Human Adaptation, Pergamon Press, London.

Bennett, J. W. (1978), A Rational-Choice Model of Agricultural Resource Utilization and Conservation, *In* Gonzalez, N. L., editor (1978), pp. 151–186.

Bennett, J. W. (1979), Agricultural Cooperatives in the Development Process: Perspectives from Social Science, Davis (California), University of California, Department of Applied Behavioral Sciences, California Agricultural Policy Seminar Monograph no. 4.

Bernus, E. (1977), Case Study on Desertification: The Eghazer and Azawak Region, Niger, prepared for the United Nations Conference on Desertification. See UNESCO 1980.

Beug, H.-J. (1957), Leitfaden der Pollenbestimmung, Stuttgart, Gustav Fischer.

Beug, H.-J.\(1967), Contributions to the Postglacial Vegetational History of Northern Turkey, *In* Cushing, E. J. and Wright, N. E., editors, Quaternary Palaeoecology, pp. 349–357. Yale University Press, New Haven.

Birks, H. J. B. (1973), Modern Pollen Rain Studies in Some Arctic and Alpine Environments, *In* Birks, H. J. B. and West, R. G., editors, Quaternary Plant Ecology, pp. 143–168. Blackwell Scientific Publications, Oxford.

Boehm, C. (1978), Rational Preselection from Hamadryas to Homo Sapiens: The Place of Decisions in Adaptive Process, *American Anthropologist* **80**:265–296.

Bogdan, A. V. (1958), A Revised List of Kenya Grasses, Government Printer, Nairobi.

Bonnefille, R. (1976), Implications of Pollen Assemblage from the Koobi Flora Formation, East Rudolf, Kenya, *Nature* **264**:403–407.

Boserup, E. (1965), The Conditions of Agricultural Growth: The Economics of Agrarian Change under Population Pressure, George Allen and Unwin, London.

Botschantzev, V. P., (1978), What is *Eremochion pungens* Gilli (Chenopodiaceae), *Botanicheskii Zhurnal* **63** (in Russian).

Box, T. W. and Peterson, D. F. (1978), Carrying Capacity of Renewable Resources Related to Desertification, *In* Reining, P., compiler (1978a), pp. 37–43.

Breckle, S.-W. (1976), Zur Ökologie und zu den Mineralstoffverhaltnissen Absalzender und Nichtabsalzender Xerohalophyten, Vaduz, J. Cramer (*Dissertations Botanicae* **35**).

Breman, H. (1975), Maximum carrying Capacity of Malian Grasslands, in International Livestock Centre for Africa, Evaluation and Mapping of Tropical Rangelands, Addis Ababa, International Livestock Centre for Africa, pp. 249–256.

Brice, W. C., editor (1978), The Environmental History of the Near and Middle East Since the Last Ice Age, Academic Press, London.

Brokensha, D. W., Horowitz, M. and Scudder, T. (1977), The Anthropology of Rural Development in the Sahel, Binghamton (N.Y.), Institute for Development Anthropology.

Brown, L. R. (1978), The Worldwide Loss of Cropland, Washington D.C., Worldwatch Institute, Worldwatch Paper 24.

Caldwell, M. M. (1974), Physiology of Desert Halophytes, *In* Reimold, R. J. and Queen, W. H., editors, Ecology of Halophytes, pp. 3–19. Academic Press, New York and London.

Campbell, D. J. (1978), Coping with Drought in Kenya Maasailand: Pastoralists and Farmers of the Loitokitok Area, Kajiado District, Cyclostyled, IDS Working Paper no. 337, University of Nairobi.

Campbell, J. H. (1964), Honor, Family and Patronage, Oxford University Press, Oxford.

Carter, D. L. (1975), Problems of Salinity in Agriculture, *In* Poljakoff-Mayber, A. and Gale, J., editors, Plants in Saline Environments, pp. 23–35, Springer-Verlag, Berlin and New York.

Casimir, M. J., Winter, R. P., Glatzer, B. (1980), Nomadism and Remote Sensing: Animal Husbandry and the Sagebrush Community in a Nomad Winter Area in Western Afghanistan, *Journal of Arid Environments* **3**:231–254.

Chao, L. L. (1974), Statistics Methods Analysis (second edition), McGraw-Hill Kogakusha Ltd., Tokyo.

Chapman, V. J. (1974), Salt Marshes and Salt Deserts of the World, *In* Reimold, R. J. and Queen, W. H., editors, Ecology of Halophytes, pp. 355–378. Academic Press, New York and London.

Chapman, V. J. (1977), Wet Coastal Systems, *In* Ecosystems of the World Volume 1, Amsterdam, Elsevier Scientific Publications, Amsterdam.

Charney, J., Stone, P. H. and Quirk, W. J. (1975), Drought in the Sahara: A Biogeophysical Feedback Mechanism, *Science* **187**:434–435.

Charney, J., Stone, P. H. and Quirk, W. J. (1977), A Comparative Study of the Effects of Albedo Change on Drought in Semi-Arid Regions, *Journal of Atmospheric Science* **34**:1366–1385.

Clark, J. D. (1971), A Re-examination of the Evidence for Agricultural Origins in the Nile Valley, *Proceedings of the Prehistoric Society* **37**:34–79.

Clark, P. J. and Evans, F. C. (1954), Distance to Nearest Neighbor as a Measure of Spatial Relationships in Populations, *Ecology* **35**:455–453.

Cloudsley-Thompson, J. L. (1974), The Expanding Sahara, *Environmental Conservation* **1**:5–13.

Clyma, W., Ali, A., and Ashraf, N. (1975a), Irrigation Practices and Application Efficiencies in Pakistan, Mona Colony, Bhalwal, Water and Power Development Authority, West Pakistan.

Clyma, W., Ali, A., and Ashraf, M. (1975b), Watercourse Losses, Annual Progress Report, Water Management Research Project, Fort Collins, Water Management Research Project, Colorado State University, Colorado.

Cody, M. L. (1968), On the Methods of Resource Division in Grassland Bird Communities, *American Naturalist* **102**:107–147.

Cody, M. L. (1974), Competition and the Structure of Bird Communities, Princeton University Press, Princeton.

Cohen, M. N. (1977), The Food Crisis in Prehistory, Overpopulation and the Origins of Agriculture, Yale University Press, New Haven.

Conant, F. P. (1965), Korok: A Variable Unit of Physical and Social Space Among the Pokot of East Africa, *American Anthropologist* **67**(2):429–434.

Conant, F. P. (1978), The Use of Landsat Data in Studies of Human Ecology, *Current Anthropology* **19**(2):382–384.

Cook, C. W. and Sims, P. L. (1975), Drought and its Relationship to Dynamics of Primary Productivity and Production of Grazing Animals, *In* International Livestock Centre for Africa, Evaluation and Mapping of Tropical Rangelands, pp. 163–169. Addis Ababa, International Livestock Centre for Africa.

Cooke, R. U. and Warren, A. (1973), Geomorphology in Deserts, London, B. T. Batsford.

Corey, G. L. and Clyma, W. (1975), Improving Farming Water Management in Pakistan, Fort Collins (Colorado), Colorado State University, Water Management Research Project, Water Management Technical Report no. 37 (Pakistan Field Report no. 1).

Crisp, M. D. and Lange, R. T. (1976), Age Structure, Distribution and Survival Under Grazing of the Arid-Zone Shrub *Acacia burkittii*, *Oikos* 27:86–92.

Crowe, B. (1969), The Tragedy of the Commons Revisited, *Science* **166**:1103–1107.

Croxton, F. E. and Cowden, D. J. (1955), Applied General Statistics, London, Sir Isaac Pitman and Sons.

Dahl, G. and Hjort, A. (1976), Having Herds, University of Stockholm, Department of Social Anthropology, Stockholm.

Dahl, G. and Hjort, A. (1979), Pastoral Change and the Role of Drought, Stockholm, Swedish Agency for Research and Cooperation with Developing Countries, SAREC Report R2.

Dale, I. R. and Greenway, P. J. (1961), Kenya Trees and Shrubs, Hatchard's, London.

Darwin, C. (1859), On the Origin of Species by Means of Natural Selection (sixth edition, 1872), Murray, London.

Dasmann, R. (1976), Environmental Conservation (fourth edition), John Wiley and Sons, New York.

Davies, W. and Skidmore, C. L., editors (1966), Tropical Pastures, Faber and Faber, London.

Diarra, L. and Breman, H. (1975), Influence of Rainfall on the Productivity of Grasslands, *In* International Livestock Centre for Africa, Evaluation and Mapping of Tropical Rangelands, pp. 171–174. International Livestock Centre for Africa, Addis Ababa.

Douglas, M. (1980), Purity and Danger Revisited, *Times Literary Supplement*, September 19, 1980, pp. 1045–1047.

Downing, T. E. and Gibson, M., editors (1974), Irrigation's Impact on Society, Tuscon (Arizona), University of Arizona Press, Anthropological Papers of the University of Arizona no. 25.

Draz, O. (1977), Role of Range Management in the Campaign Against Desertification: The Syrian Experience as an Applicable Example for the Arabian Peninsula, A report presented by the UNEP Regional Office for West Asia, Regional Preparatory Meeting for the Mediterranean Area, Algarve, 28 March–1 April 1977, UNCOD/MISC/13.

Dregne, H. E., editor (1970), Arid Lands in Transition, Washington D.C., American Association for the Advancement of Science.

Dregne, H. E. (1977), Desertification of Arid Lands, *Economic Geography* 53(4):322–331.

Dye, W. B. (1956), Chemical Studies on *Halogeton glomeratus*, *Weeds* **4**:55–60.

Dyson-Hudson, N. (1966), Karimojong Politics, Clarendon Press, Oxford.

Dyson-Hudson, R. (1972), Pastoralism: Self Image and Behavior Reality, *Journal of Asian and African Studies* **7**(1–2):30–47.

Earl, D. E. (1975), Forest Energy and Economic Development, Clarendon Press, Oxford.

Eckert, J., Dimick, N., and Clyma, W. (1975), Water Management Alternatives for Pakistan: A Tentative Appraisal, Fort Collins (Colorado), Colorado State University, Water Management Research Project, Water Management Technical Report no. 43.

Eckholm, E. and Brown, L. R. (1977), Spreading Deserts—The Hand of Man, Washington D.C., Worldwatch Institute, Worldwatch Paper 13.

ECOSOC (1977), Popular Participation and Its Practical Implications for Development, Progress Report of the Secretary-General, Geneva, United Nations Economic and Social Council Commission for Social Developments.

Edelberg, L. and Jones, S. (1979), Nuristan, Graz (Austria), Akademische Druck- u. Verlagsanstatt.

Egbert, D. (1979), IMPAC: Image Analysis Package for Microcomputers, Egbert Scientific Software, Greenport (New York).

Elton, C. (1958), The Ecology of Invasions by Animals and Plants, Methuen, London.

Evenari, M., Nessler, U., Rogel, A. and Schenk, O. (1975), Felder und Weiden in Wüsten, Eduard Roether, Darmstadt.

Evenari, M., Shanan, L. and Tadmor, N. (1971), The Negev: The Challenge of a Desert, Harvard University Press, Cambridge (Mass.).

Evenari, M., Shanan, L., Tadmor, N. and Aharoni, Y. (1961), Ancient Agriculture in the Negev, *Science* **133**:979–996.

Faegri, K. and Iversen, J. (1974), Textbook of Pollen Analysis (third edition, revised by K. Faegri), Blackwell Scientific Publications, Oxford.

Farrington, I. S. and Park, C. C. (1978), Hydraulic Engineering and Irrigation Agriculture in the Moche Valley, Peru: c. A.D. 1250–1532, *Journal of Archaeological Science* **5**(3):255–268.

Farvar, M. T. and Milton, J. O., editors (1972), The Careless Technology, The Natural History Press, New York.

Ferguson, W. (1971), Adaptive Behavior of Cattle to Tropical Environments, *Tropical Science* **13**:113–122.

Field, C. (1979), Preliminary Report on Ecology and Management of Camels, Sheep, and Goats in Northern Kenya, Nairobi, Integrated Project on Arid Lands (UNESCO), IPAL Technical Report E1-a.

Flannery, K. (1965), The Ecology of Early Food Production in Mesopotamia, *Science* **147**:1247–1256.

Fleuret, P. C. and Fleuret, A. K. (1978), Fuelwood Use in a Peasant Community: A Tanzanian Case Study, *The Journal of Developing Areas* **12**:315–322.

Fonteyn, P. J. and Mahall, B. E. (1978), Competition Among Desert Perennials, *Nature* **275**:544–545.

Foster, G. M. (1965), Peasant Society and the Image of the Limited Good, *American Anthropologist* **67**: 293–315.

Freeland, W. (1974), Vole Cycles: Another Hypothesis, *American Naturalist* **108**:238–245.

Freeland, W. and Janzen, D. (1974), Strategies in Herbivory by Mammals: The Role of Plant Secondary Compounds, *American Naturalist* **108**: 269–289.

Friedman, J. and Orshan, G. (1975), The Distribution, Emergence and Survival of Seedlings of *Artemisia herba-alba* Asso. in the Negev Desert of Israel in Relation to Distance from Adult Plants, *Journal of Ecology* **63**:627–632.

Fukuda, S. (1976), Effect of Risk and Uncertainty on Decisions Regarding Herd Size and Composition in Arid Areas, unpublished M.A. dissertation, University of Sussex.

Garcia-Moya, E., and McKell, C. M. (1970), Contribution of Shrubs to the Nitrogen Economy of a Desert-Wash Plant Community, *Ecology* **51**:81–88.

Geertz, C. (1975), Common Sense as a Cultural System, *The Antioch Review* **33**:5–26.

George, S. (1977), How the Other Half Dies: The Real Reason for World Hunger, Osman and Co., Montclair (New Jersey) Allanheld.

Gibson, M. (1974), Violation of Fallow and Engineered Disaster in Mesopotamian Civilization, *In* Downing, T. E. and Gibson, M. (editors), 1974, pp. 7–19.

Gillespie, R., Horton, D. R., Ladd, P., Macumber, P. G., Rich, T. H., Thorne, R. and Wright, R. V. S. (1978), Lancefield Swamp and the Extinct Australian Megafauna, *Science* **200**:1044–1048.

Glantz, M. H., editor (1976), The Politics of Natural Disaster, The Case of the Sahel Drought, Praeger, New York.

Glantz, M. H., editor (1977), Desertification: Environmental Degradation in and Around Arid Lands, Westview Press, Boulder (Colorado).

Goldschmidt, W. (1975), A National Livestock Bank: An Institutional Device for Rationalising the Economy of Tribal Pastoralists, *International Development Review* **17**(2):2–7.

Gonzalez, N. L., editor (1978), Social and Technological Management in Dry Lands: Past and Present, Indigenous and Imposed, Westview Press, Boulder (Colorado).

Goodwin, James R. (1979), Pelagic Shark Fishing in Rural Mexico: A Context for Co-operative Action, *Ethnology* **18**(4):325–337.

Gould, R. A. (1971), Uses and Effects of Fire Among the Western Desert Aborigines of Australia, *Mankind* **8**:14–24.

Grayson, D. K. (1977), Pleistocene Avifauna and the Overkill Hypothesis, *Science* **195**:691–693.

Greig-Smith, P. (1964), Quantitative Plant Ecology (second edition), Butterworths, London.

Griffiths, J. F. (1958), An Initial Investigation of the Annual Rainfall in East Africa, Nairobi, East African Meteorological Department Memoirs, Volume III, no. 5.

Gronhaug, R. (1978), Scale as a Variable in Analysis: Fields in Social Organization in Herat, Northwest Afghanistan, *In* Barth, F., editor, Scale and Social Organization, pp. 78–121. Universitetsforlaget, Oslo.

Grove, A. T. (1973), Desertification in the African Environment, *In* Dalby, D. and Harrison, R., editors, "Drought in Africa" (Report of the 1973 Symposium), pp. 33–45. University of London Centre for African Studies, London.

Gulliver, P. (1955), The Family Herds, Routledge and Kegan Paul, London.

Hall, M. (1976), Dendroclimatology, Rainfall and Human Adaptation in the Later Iron Age of Natal and Zululand, *Annals of Natal Museum* **22**(3):693–703.

Hamilton, W. J. and Watt, K. E. F. (1970), Refuging, *Annual Review of Ecology and Systematics* **1**:263–286.

Hardin, G. (1968), The Tragedy of the Commons, *Science* **162**:1234–1248 (also reprinted in Hardin, G. and Baden, J., editors, 1977, pp. 16–30).

Hardin, G. (1977), Ethical Implications for Carrying Capacity, *In* Hardin G. and Baden, J., editors, (1977), pp. 112–125.

Hardin, G. and Baden, J., editors (1977), Managing the Commons, W. H. Freeman and Company, San Francisco.

Hare, F. K., Kates, R. and Warren, A. (1977), The Making of Deserts: Climate, Ecology, and Society, *Economic Geography* **53**(4):332–345.

Harper, J. L. (1977), Population Biology of Plants, Academic Press, New York and London.

Harrington, G. N. (1980), Grazing Arid and Semi-Arid Pastures, *In* Morley, F. H. W., editor, Grazing Animals, pp. 181–202. Elsevier Scientific Publishing Company, Amsterdam.

Harshvardhan, and Cess, R. D. (1978), Effects of Troposphere Aerosols Upon Atmospheric Cooling Rates, *Journal of Quantitative Spectroscopy and Radiative Transfer* **19**:621–633.

Hasan, S. B. (1976), Case Study in Irrigated Areas with Problems of Salinization and Waterlogging. Planning and Co-ordination Cell, Irrigation Drainage and Flood Control Research Council, Lahore (Pakistan), Gulberg-III. See UNESCO 1980.

Hastings, J. R. (1965), On Some Uses of Non-Normal Coefficients of Variation, *Journal of Applied Meteorology* **4**:475–478.

Heathcote, R. L. (1969), The Pastoral Ethic, *In* McGinnies, W. G. and Goldman, B. J., editors, Arid Lands in Perspective, Washington D.C., American Association for the Advancement of Science and Tucson (Arizona), University of Arizona Press, pp. 311–324.

Henkel, P. A. and Shakhov, A. A. (1945), The Ecological Significance of the Water Regime of Certain Halophytes, *Botanicheskii Zhurnal* **30**:154–166.

Henning, I. and Henning, D. (1976), Die Klimatologische Trockengrenze, *Meteorologische Rundschau* **29**:142–151.

Herring, R. J. and Kennedy, C. Jr. (1979), The Political Economy of Farm Mechanization Policy: Tractors in Pakistan, *In* Hopkins, R. F., Puchala, D. J. and Talbot, R. B., editors, Food, Politics, and Agricultural Development, Westview Press, Boulder (Colorado).

Hevly, R. (1968), Studies of the Modern Pollen Rain of Northern Arizona, *Journal of the Arizona Academy of Science* **5**:116–135.

Hevly, R., Mehringer, P. J., and Yokum, H. (1965), Modern Pollen Rain in the Sonoran Desert, *Journal of the Arizona Academy of Science* **3**:123–135.

Horowitz, A. and Zak, I. (1968), Preliminary Palynological Analysis of an Evaporitic Sequence from Mount Sedom, Israel, *Reviews of Palaeobotany and Palynology* **7**:25–30.

Houérou, H. N. Le (1976), Ecological Management of Arid Grazing Land Ecosystems, *In* Glantz, M., editor, (1976), pp. 267–281.

Houérou, H. N. Le (1977a), Biological Recovery vs. Desertization, *Economic Geography* **53**(4):413–420.

Houérou, H. N. Le (1977b), Nature and Causes of Desertification, *In* Glantz, M., editor, (1977), pp. 17–38.

Houérou, H. N. Le and Hoste, C. H. (1977), Rangeland Production and Annual Rainfall Relations in the Mediterranean Basin and in the African Sahelo-Sudanian Zone, *Journal of Range Management* **30**:181–189.

Hudson, N. W. (1969), Field engineering for Agricultural Development, Oxford, Clarendon Press.

Hunt, R. C. and Hunt, E. (1976), Canal Irrigation and Social Organization, *Current Anthropology* **17**:389–411.

Huntingford, G. W. B. (1953a), The Northern Nilo-Hamites (East Central Africa Part VI), International African Institute, Ethnographic Survey of Africa, London.

Huntingford, G. W. B. (1953b), The Southern Nilo-Hamites (East Central Africa Part VIII), International African Institute, Ethnographic Survey of Africa, London.

Hutchings, S. S. and Stewart, G. S. (1953), Increasing Forage Yields and Sheep Production on Intermountain Winter Ranges, Washington, D.C., U.S. Department of Agriculture Circular 925.

Hutchinson, Sir J. (1969), Erosion and Land Use: The Influence of Agriculture on the Epirus Region of Greece, *Agricultural History Review* **17**:85–90.

Imbrie, J. and Imbrie, K. P. (1979), Ice Ages, MacMillan Press, London.

Institute de Investigaciones Agropecurias (INIA), (1977), Case Study on Desertification, Region of Combarbala, Chile, prepared for the United Nations Conference on Desertification. See UNESCO 1980.

Irons, W. and Chagnon, N., editors, (1979), Evolutionary Biology and Human Social Behaviour, Duxbury Press, North Scituate (Mass.).

Jackson, I. J. (1970), Annual Rainfall Probability and the Binomial Distribution, *East African Agriculture and Forestry Journal* **35**:265–272.

Jacobsen, T. and Adams, R. M. (1958), Salt and Silt in Ancient Mesopotanian Agriculture, *Science* **128**(3334):1251–1258.

Jahania, Ch. M. H. (1973), The Canal and Drainage Act, 1873 (with Rules), (third edition), Mansoor Book House, Lahore.

Janzen, D. H. (1976), Why Bamboos Wait So Long to Flower, *Annual Review of Ecology and Systematics* **7**:347–391.

Janzen, D. H. (1977), The Interaction of Seed Predators and Seed Chemistry, *In* Labeyrai, V., editor, Comportement des Insects et Milieu Trophique, Paris, Centre National de la Recherche Scientifique, Colloques Internationaux du C.N.R.S. no. 265, pp. 415–428.

Janzen, D. H. (1978), Complications in Interpeting the Chemical Defenses of Trees Against Tropical Arboreal Plant-Eating Vertebrates, *In* Montgomery, G., editor, (1978), pp. 73–84.

Janzen, D. H. (1979), New Horizons in the Biology of Plant Defenses, *In* Rosenthal, G. A. and Janzen, D. H., editors, (1979), pp. 331–350.

Jeffreys, M. D. W. (1975), Pre-Columbian Maize in the Old World: An Examination of Portuguese Sources, *In* Arnott, M. L., editor, Gastronomy: The Anthropology of Food and Food Habits, pp. 23–66. Mouton, The Hague and New York.

Johnson, D. L. (1979), Management Strategies for Dry Lands: Available Options and Unanswered Questions, in Mabbutt, J. A., editor, Proceedings of the Khartoum Workshop on Arid Lands Management, The University of Khartoum—The United Nations University, 22–26 October, 1978, Tokyo, The United Nations University.

Johnson, D. III, Early, A. C. and Lowdermilk, M. K. (1977), Water Problems in the Indus Food Machine, *Water Resources Bulletin* **13**:1253–1268.

Jones, R., editor (1970), The Biology of *Atriplex* (based on a symposium held in Deniliquin, N.S.W., 14–15 October, 1969), Canberra, Division of Plant Industry, Commonwealth Scientific and Industrial Research Organization (Studies of the Australian Arid Zone).

Jones, R. (1977), Australia Felix—The Discovery of a Pleistocene Prehistory, *Journal of Human Evolution* **6**:353–361.

Jones, R. (1979), The Fifth Continent: Problems Concerning the Human Colonisation of Australia, *Annual Review of Anthropology* **8**:445–466.

Jowsey, P. C. (1966), An Improved Peat Sampler, *New Phytologist* **65**:245–248.

Kassas, M. (1970), Desertification Versus Potential for Recovery in Circum-Sahara Territories, *In* Dregne, H. E., editor (1970), pp. 123–142.

Keen, B. A. (1946), The Agricultural Development of the Middle East, His Majesty's Stationery Office, London.

Kenya (1969), Central Bureau of Statistics, Population Census, Nairobi Government Printer.

Kershaw, A. P. (1976), A Late Pleistocene and Holocene Pollen Diagram from Lynch's Crater, North-Eastern Queensland, Australia, *New Phytologist* **77**: 469–498.

Kershaw, K. A. (1974), Quantitative and Dynamic Plant Ecology (second edition), Edward Arnold, London.

King, T. J. and Woodell, S. R. J. (1973), The Causes of Regular Pattern in Desert Perennials, *Journal of Ecology* **61**:761–765.

Kingsbury, J. M. (1978), Ecology of Poisoning, *In* Keeler, R. F. *et al.*, editors, Effects of Poisonous Plants on Livestock, pp. 81–91, Academic Press, New York and London.

Kirkby, M. (1972), The Physical Environment of the Nochixtlan Valley, Oaxaca, Vanderbilt University Publications in Anthropology No. 2, Nashville, Tenn.

Koster, H. (1977), The Ecology of Pastoralism in Relation to Changing Patterns of Land Use in the Northeast Peloponnese, Philadelphia, University of Pennsylvania, Ph.D. Thesis, Department of Anthropology.

Kowal, J. M. and Kassam, A. H. (1978), "Agricultural Economy of Savanna. A Study of West Africa," Clarendon Press, Oxford.

Kreeb, K. H. (1964), "Ökologische Grundlagen der Bewässerungskulturen in den Subtropen," Gustav Fischer, Stuttgart.

Kreeb, K. H. (1965), Die Ökologische Bedeutung der Bodenversalzung, *Angewandte Botanik* **39**:1–15.

Lambrecht, F. L. (1972), The Tsetse Fly: A Blessing or Curse?, *In* Farvar, M. T. and Milton, J. P., editors, (1972), pp. 726–741.

Lamprey, H. (1978), The Integrated Project on Arid Lands (IPAL), *Nature and Resources* **14**(4):2–10.

Langdale-Brown, I., Osmaston, H. A., and Wilson, J. G. (1964), The Vegetation of Uganda and Its Bearing on Land Use, Entebbe, Government of Uganda Printer.

Lange, R. T. (1969), The Piosphere: Sheep Trace and Dung Patterns, *Journal of Range Management* **22**: 396–400.

Larcher, W. (1978), Ökologie der Pflanzen (third edition), Ulmer, Stuttgart.

Lees, F. A. and Brooks, H. C. (1977), The Economic and Political Development of the Sudan, Westview Press, Boulder (Colorado).

Lees, S. H. (1974), The State's Use of Irrigation in Changing Peasant Society, *In* Downing, T. E. and Gibson, M., editors, (1974), pp. 123–128.

Legesse, A. (1973), Gada: Three Approaches to the Study of African Society, Academic Press, New York and London.

Leroi-Gourhan, A. (1974), Etudes Palynologique des derniers 11,000 ans en Syrie semi-desertique, *Paleorient* **2**:442–451.

Lewis, O. (1959), Five Families—Mexican Case Studies in the Culture of Poverty, New American Library, New York.

Lhote, H. (1973), The Search for the Tassili Frescoes, Hutchinson, London.

Lichti-Federovich, S. and Richie, J. C. (1968), Recent Pollen Assemblages from the Western Interior of Canada, *Review of Palaeobotany and Palynology* **7**:297–344.

Lieftinck, P., Sadove, A. R., Creyke, T. C. (1969), Water and Power Resources of West Pakistan: A Study in Sector Planning (three volumes), Johns Hopkins Press, for the World Bank, Baltimore.

Litchfield, W. H. (1975), Palynological Studies in Brigalow (*Acacia harpophylla* F. Muell.) communities: Pollen Analysis in Gilgai Soil Under the Primary Forest, *Australian Journal of Botany* **23**:355–371.

Low, W. A. (1977), Behavior of Herbivores (Except Sheep) Influencing Rangelands in Australia, *In* Western Australian Rangeland Society, The Impact of Herbivores on Arid and Semi-Arid Rangelands, pp. 149–162. Western Australian Rangeland Society, Perth.

Lowdermilk, M. K., Freeman, D. M. and Early, A. C. (1978), Farm Irrigation Constraints and Farmers' Responses: Comprehensive Field Survey in Pakistan. 6 volumes. Joint Contribution of Colorado State University and Directorate of Watercourse Chak Farming Survey and Review Division, Publication No. 2. Lahore: Pakistan, Water and Power Development Authority.

Lowe, J. W. (1977), The IFC and the Agribusiness Sector, *Finance and Development* **14**(1):26–28.

Lundholm, B. (1976), Domestic Animals in Arid Ecosystems, *In* Rapp, A., Le Houérou, H. N., and Lundholm, B., editors, (1976), pp. 29–42.

McArthur, I. D. and Harrington, G. N. (1978), A Grazing Ecosystem in Western Afghanistan, *In* Hyder, D. N., editor, Proceedings of the First International Rangelend Congress, Denver (Colorado), Society for Range Management, pp. 596–599.

MacArthur, R. and Wilson, E. O. (1967), The Theory of Island Biogeography, Princeton University Press, Princeton.

McClure, H. A. (1976), Radiocarbon Chronology of Late Quarternary Lakes in the Arabian Desert, *Nature* **263**:755–756.

McKell, C. M. (1975), Shrubs—A Neglected Resource of Arid Lands, *Science* **187**:803–809.

McKey, D. (1978), Soils, Vegetation, and Seed-Eating by Black Colobus Monkeys, *In* Montgomery, G., editor, (1978), pp. 423–437.

McNaughton, S. J. (1968), Autotoxic Feedback in the Regulation of *Typing* Populations, *Ecology* **49**:367–369.

Machin, J. (1971), Plant Microfossils from Tertiary Deposits of The Isle of Wight, *New Phytologist* **70**:851–872.

Maher, L. J. (1964), *Ephedra* Pollen in the Sediments of the Great Lakes Region, *Ecology* **45**:391–395.

Maley, J. (1977), Palaeoclimates of Central Sahara During the Early Holocene, *Nature* **269**:573–577.

Malik, B. A. (1978), Some Aspects of Concept and Practice of Land Reclamation, *The Pakistan Times* (Lahore), October 24, 1978.

Mann, C. K. (1972), Formulating a Consistent Strategy Toward On-Farm Land Development in Turkey, Ankara, Agency for International Development, Discussion Paper 8.

Martin, H. A. (1973), Palynology and Historical Ecology of some Cave Excavations in the Australian Nullabor, *Australian Journal of Botany* **21**:283–316.

Martin, P. S., Sabels, B. E., and Shutler, D. (1961), Rampart Cave Coprolite and Ecology of the Shasta Ground Sloth, *American Journal of Science* **259**:102–127.

Marx, E. (1977), The Tribe as a Subsistence Unit: Nomadic Pastoralism in the Middle East, *American Anthropologist* **79**:343–363.

Mellars, P. M. (1975), Ungulate Populations, Economic Patterns, and the Mesolithic Landscape, *In* Evans, J. G., Limbrey, S. and Cleere, H., editors, The Effect of Man on the Landscape: the Highland Zone, pp. 49–56. Council for British Archaeology, Research Report No. 11, London.

Merrey, D. J. (1978), Reorganizing Local Level Water Management in Pakistan, A Case Study Prepared for Symposium on Desertification and Anthropology, Jodhpur, Central Arid Zone Research Institute.

Merrey, D. J. (1979), Irrigation and Honor: Cultural Impediments to the Improvement of Local Level Water Management in Punjab, Pakistan, Fort Collins (Colorado), Colorado State University, Water Management Research Project, Water Management Technical Report no. 53 (Pakistan Field Report no. 9).

Merrey, D. J. (1980), Problems of Farmer Organization in an "Appropriate Technology" Project: Lessons from the Water Course Improvement Project in Pakistan, *In* International Association for the Advancement of Appropriate Technology in Developing Countries, New Dimensions of Appropriate Technology, Selected Proceedings of the 1979 Symposium by the International Association for the Advancement of Appropriate Technology in Developing Countries, Ann Arbor, the University of Michigan.

Merrey, D. J. (in preparation), Irrigation and Social Change in Punjab, Pakistan,

Philadelphia, University of Pennsylvania, Ph.D. Thesis, Department of Anthropology.

Merrey, K. L. (in preparation), The Punjabi Wedding as an Ethnosociological Mode, Philadelphia, University of Pennsulvania, Ph.D. Thesis, Department of Anthropology.

Michel, A. A. (1959), The Kabul, Kunduz, and Helmand Valleys and the National Economy of Afghanistan, Washington D.C., National Academy of Sciences, National Research Council, Foreign Field Research Program sponsored by Office of Naval Research Report no. 5.

Michel, A. A. (1967), The Indus Rivers: A Study of the Effects of Partition, Yale University Press, New Haven.

Michel, A. A. (1972), The Impact of Modern Irrigation Technology in the Indus and Helmand Basins of Southwest Asia, *In* Farvar, M. T. and Milton, L. P., editors, (1972), pp. 257–275.

Mirazai, N. A. and Breckle, S.-W. (1978), Untersuchungen an Afghanischen Halophyten I. Salzverhältnisse in Chenopodiaceen Nord-Afghanistans, *Botanische Jahrbuecher fuer Systematik der Pflanzengeschichte und Pflanzengeographie* **99**: 565–578.

Mirza, A. H. (1975), A Study of Village Organizational Factors Affecting Water Management Decision Making in Pakistan, Fort Collins (Colorado), Colorado State University, Water Management Technical Report no. 34.

Mirza, A. H. and Merrey, D. L. (1979), Organizational Problems and their Consequences on Improved Watercourses in Punjab, Fort Collins (Colorado), Colorado State University, Water Management Research Project, Water Management Technical Report no. 55.

Monod, T., editor (1975), Pastoralism in Tropical Africa, Oxford University Press, London.

Montgomery, G., editor, (1978), The Ecology of Arboreal Folivores, Smithsonian Institute Press, Washington D.C.

Moore, P. D. (1973), The Influence of Prehistoric Cultures Upon the Initiation and Spread of Blanket Bog in Upland Wales, *Nature* **241**:350–353.

Moore, P. D. (1975), Origin of Blanket Mires, *Nature* **256**:267–269.

Moore, P. D. and Webb, J. A. (1978), An Illustrated Guide to Pollen Analysis, Hodder and Stoughton, London.

Moore, R. T., Breckle, S.-W., and Caldwell, M. M. (1972), Mineral Ion Composition and Osmotic Relations in *Atriplex confertifolia* and *Eurotia lanata, Oecologia* **11**:67–78.

Mosimann, J. E. and Martin, P. S. (1975), Simulating Overkill by Paleoindians, *American Scientist* **45**:304–313.

Mueller-Dombois, D. and Spatz, G. (1975), The Influence of Feral Goats on the Lowland Vegetation in Hawaii Volcanoes National Park, *Phytoenologia* **3**:1–29.

Mulvaney, D. J. (1975), The Prehistory of Australia, Penguin, London.

Nagy, J. G., Steinhof, H. W., and Ward, F. M. (1964), Effects of Essential Oils of Sagebrush on Deer Rumen Microbial Function, *Journal of Wildlife Management* **28**:785–790.

Netting, R. (1976), What Alpine Peasants Have in Common: Observations of Communal Tenure in a Swiss Village, *Human Ecology* **4**(2):135–146.

Nicholson, S. E. (1978), Drought vs. Desertification, *In* Gonzalez, N. L., editor, (1978), pp. 187–194.

Niklewski, J. and van Zeist, W. (1970), A Late-Quaternary Pollen Diagram from North-Western Syria, *Acta Botanica Neerlandica* **19**:737–754.

Novikoff, G. F., Wagner, M. H. *et al.* (1973–1977), Systems Analysis of the PreSaharan

Ecosystem of Southern Tunisia (Progress Reports 1 to 6 for the years 1972–1977—6 volumes), Logan (Utah), Utah State University, UNESCO, Man and the Biosphere Project 3 (Grazing Lands) and US IBP Desert Biome.

Noy-Meir, I. (1975), Stability of Grazing Systems: An Application of Predator–Prey Graphs, *Journal of Ecology* **63**:459–581.

Nulty, L. (1972), The Green Revolution in West Pakistan: Implications of Technological Change, Praeger Publishers, New York.

Olson, M. (1965), The Logic of Collective Action—Public Goods and the Theory of Groups, Harvard University Press, Cambridge (Mass.).

Orians, G. and Solbrig, O. (1977), Convergent Evolution in Warm Deserts, Dowden, Hutchinson and Ross, Stroudsberg (Pa.).

Ormerod, W. E. (1976), Ecological Effect of Control of African Trypanosomiasis, *Science* **191**;815–821.

Ostrom, E. (1969), Collective Action and the Tragedy of the Commons, *In* Hardin, G. and Baden, J., editors, (1977), pp. 173–181 (originally written for the Workshop in Political Theory and Policy Analysis, Bloomington, Indiana University, 1969).

Otterman, J. (1974), Baring High Albedo Soils by Overgrazing: A Hypothesized Desertification Mechanism, *Science* **86**:531–533.

Otterman, J. (1977), Anthropogenic Impact on the Albedo of the Earth, *Climatic Change* **1**:137–155.

Paine, R. (1972), The Herd Management of Lapp Reindeer Pastoralists, *In* Irons, W. and Dyson-Hudson, N., editors, Perspectives on Nomadism, Leiden: E. J. Brill, pp. 76–87.

Pakistan, (1978), Planning Commission, The Report of the Indus Basin Research Assessment Group, Islamabad, Planning Commission.

Passmore, J. (1974), Man's Responsibility for Nature, Charles Scribner and Son, New York.

Pearse, C. K. (1971), Grazing in the Middle East—Past, Present and Future, *Journal of Range Management* **24**(1):13–16.

Peristiany, J. G., editor (1966), Honor and Shame, University of Chicago Press, Chicago.

Pianka, E. R. (1970), On *r*- and *K*-Selection, *American Naturalist* **104**:592–597.

Pielou, E. C. (1962), The Use of Plant-to-Neighbor Distances for the Detection of Competition, *Journal of Ecology* **50**:357–367.

Porter, P. W. (1965), Environmental Potentials and Economic Opportunities—A Background for Cultural Adaptation, *American Anthropologist* **67**(2):409–420.

Pratt, D. J. and Gwynne, M. D. (1977), Rangeland Management and Ecology in East Africa, Hodder and Stoughton, London.

Prentice, I. C. (1978), Modern Pollen Spectra from Lake Sediments in Finland and Finmark, North Norway, *Boreas* **7**:131–153.

Price Williams, D. (1973), "Environmental Archaeology in the Western Negev," *Nature* **242**:501–503.

Radosevich, G. E. (1975), Water User Organizations for Improving Irrigated Agriculture: Applicability to Pakistan, Fort Collins (Colorado), Colorado State University, Water Management Research Project, Water Management Technical Report no. 44.

Radosevich, G. E. and Kirkwood, C. (1975), Organizational Alternatives to Improve On-Farm Water Management in Pakistan, Fort Collins (Colorado), Colorado State University, Water Management Technical Report no. 36.

Raikes, R. (1965), The Ancient Gabarbands of Baluchistan, in *East and West*, **15**:26–35.

Raikes, R. (1967), Water, Weather and Prehistory, Hutchinson, London.

Raikes, R. (1978), The African Aqualithic: *Comments Antiquity* 52:233–235.

Raikes, R. L. and Dyson, R. H. Jr., (1961), The Prehistoric Climate of Baluchistan and the Indus Valley, *American Anthropologist* 63:265–281.

Rainy, M. (1976), Preliminary Analysis of a Continuous 71 Year Record of a Single Rendille Subclan Settlement Group-Discussion, *In* Man and the Environment in Marsabit District, Proceedings of a meeting held at the UNESCO/UNEP Arid Lands Project Field Station, Gatab, Mt. Kulal, Kenya, (mimeo).

Rapp, A. (1974), A Review of Desertification in Africa: Water, Vegetation, and Man, Secretariat for International Ecology, Stockholm.

Rapp, A., Le Houérou, H. N., and Lundholm, B., editors (1976), Can Desert Encroachment Be Stopped?, Swedish Natural Science Research Council, Stockholm.

Rasul, S. H. (1977), Desertification in the ESCAP Region, Nairobi, United Nations Conference on Desertification.

Reining, P., compiler (1978a), Desertification Papers, Washington D.C., American Association for the Advancement of Science.

Reining, P., compiler (1978b), Handbook on Desertification Indicators, Washington D.C., American Association for the Advancement of Science.

Reuss, J. O. and Kemper, W. D. (1978), Water Management as Affected by Size of Holding and Cropping Systems, *In* Water Management Research Project, Improving Irrigation Water Management on Farms, Annual Technical Report, Water Management Research Project, Fort Collins (Colorado), Colorado State University, Water Management Research Project.

Reuss, J. O., Skogerboe, G. V., and Merrey, D. J. (1979), Water Management Programs to Improve Agricultural Productivity in Pakistan: An Appraisal, Lahore (Pakistan), Colorado State University, Water Management Research Project.

Rittenberger, V. (1978), The International Order and United Nations Conference Politics, New York, UNITAR, Science and Technology Working Papers Series.

Rodin, L. E. and Basilevich, N. I. (1967), Production and Mineral Cycling in Terrestrial Vegetation (English edition), Oliver and Boyd, Edinburgh and London.

Rosenthal, G. A. and Janzen, D. H., editors (1979), Herbivores: Their Interactions with Plant Secondary Compounds, Academic Press, New York and London.

Sagan, C., Toon, O. B., and Pollack, J. B. (1979), Anthropogenic Albedo Changes and the Earth's Climate, *Science* 206:1363–1368.

Sandford, S. (1976), Pastoralism Under Pressure, Overseas Development Institute Review no. 2:45–68.

Sandford, S. (1977a), Dealing with Drought and Livestock in Botswana, unpublished report for the government of Botswana.

Schmidt-Nielsen, K. (1964), Desert Animals—Physiological Problems of Heat and Water, Oxford University Press, London.

Schmithüsen, J. (1976), Atlas zum Biogeographie, B-I-Hochschulatlanten.

Schoenwetter, J. (1973), Archaeological Pollen Analysis of Sediment Samples from Asto Village Sites, *In* Lavalée, D. and Julien, M., editors, Les Etablissements Asto à l'époque préhispanique, Lima (Peru), Institute Français d'Etudes Andines, Travaux de l'Institut Français d'Etudes Andines 15, pp. 101–111.

Schoenwetter, J. (1974a), Pollen Analysis of Human Paleofeces from Upper Salts Cave, *In* Watson, P. J., editor, Archaeology of the Mammoth Cave Area, pp. 49–58, Academic Press, New York and London.

Schoenwetter, J. (1974b), Pollen Records of Guila Naquitz Cave, *American Antiquity* **39**:292–303.

Schoenwetter, J. and Doerschlag, L. (1971), Surficial Pollen Records from Central Arizona. 1. Sonoran Desert Scrub, *Journal of the Arizona Academy of Science* **6**:216–221.

Schrire, C. (1980), An Inquiry into the Evolutionary Status and Apparent Identity of San Hunter-Gatherers, *Human Ecology* **8**(1):9–32.

Shackleton, N. J. (1980), Deep Drilling in African Lakes, *Nature* **288**:211–212.

Shackleton, N. J. and Opdyke, N. D. (1976), Oxygen-Isotope and Palaeomagnetic Stratigraphy of Pacific Core V28-239, Late Pliocene to Latest Pleistocene, *Memoirs of the Geological Society of America* **145**:449–464.

Shahrani, M. N. (1978), Ethnic Relations and Access to Resources in Northeast Badakhshan, *In* Anderson, J. W. and Strand, R. F., editors, Ethnic Processes and Intergroup Relations in Contemporary Afghanistan, New York, Afghanistan Council of the Asia Society, Occasional Paper no. 15, pp. 15–25.

Shahrani, M. N. M. (1979), The Kirghiz and Wakhi of Afghanistan, Adaptation to Closed Frontiers, University of Washington Press, Seattle.

Sharma, M. L. and Tongway, D. J. (1973), Plant Induced Soil Salinity Patterns in Two Saltbush (*Atriplex spp.*) Communities, *Journal of Range Management* **26**:121–125.

Shaw, B. D. (1976), Climate, Environment and Prehistory in the Sahara, *World Archaeology* **8**(2):133–149.

Shihabi, M. (1935), al-Zira'a al-'Ilmiyya al-Haditha (Modern Practical Agriculture), Moderation Press, Damascus.

Silverman, S. (1968), Agricultural Organization, Social Structure, and Values in Italy: Amoral Familism Reconsidered, *American Anthropologist* **70**(1):1–20.

Simon, J. L. (1980), Resources, Population, Environment: An Oversupply of False Bad News, *Science* **208**:1431–1437 (27 June 1980).

Sobania, N. (n.d.), Background History of the Mt. Kulal Region of Kenya, Nairobi, Integrated Project on Arid Lands (UNESCO), IPAL Consultancy Report.

Soloviev, V. A. (1967), Pathways of Regulation of the Content of Excess Absorbed Ionsin Plant Tissues (with sodium ions as an example), *Soviet Plant Physiology* **14**:915–923.

Spencer, P. (1973), Nomads in Alliance, Oxford University Press, London.

Spooner, B. (1973), The Cultural Ecology of Pastoral Nomads, Addison-Wesley Module in Anthropology No. 45.

Spooner, B. (1979), Environmental Problems and the Organization of Development in Arid Lands of South-West Asia, Regional Seminar on Environment and Development. United Nations Economic and Social Commission for Asia and the Pacific and United Nations Environment Programme, Bangkok.

Spooner, B. and Mann, H. S. (1979), The Organization of Deserts, Report of an international symposium entitled "Anthropology and Desertification" held at C.A.Z.R.I., 21–23 December, 1978, Jodhpur, C.A.Z.R.I.

Stager, L. E. (1976), Farming in the Judaean desert during the Iron Age, *Bulletin of the American School of Oriental Research*, **221**:145–158.

Steiner, M. (1934), Zur Ökologie der Salzmarschen der Nordöstilichen Vereinigten Staaten von Nordamerika, *Jahrbuch für Wissenschaftliche Botanik* **81**:94–202.

Stoddart, L. A., Smith, A. D., and Box, T. W. (1975), Range Management (third edition), McGraw-Hill Book Co., New York.

Swaminathan, M. S. and Sinha, S. K. editors (in press), Global Aspects of Food Production, Academic Press, London.

Sweet, L. (1965), Camel Raiding of North Arabian Bedouin: A Mechanism of Ecological Adaptation, *American Anthropologist* **67**: 1132–1150.

Tadros, H. R. (1976), Problems Involved in the Human Aspects of Rural Settlement Schemes in Egypt, *In* Rapoport, A., editor, The Mutual Interaction of People and Their Built Environment, A Cross-Cultural Perspective, pp. 453–483, Mouton, The Hague.

Tadros, H. R. (1978), Rural Settlement in Egypt, Cairo Papers in Social Sciences, Volume One, Monograph Four.

Tadros, H. R. (1979), The Human Aspects of Rural Settlement Schemes in Egypt, *In* Bernichewsky, B., editor, Anthropology and Social Change in Rural Areas, The Impact of Agrarian Reform Upon Local Communities, pp. 121–148, Mouton, The Hague.

Thalen, D. C. P. (1979), Ecology and Utilization of Desert Shrub Rangelands in Iraq. Dr. W. Junk, The Hague.

Thimm, H.-U. (1978), Socio-Economic Aspects of Development Projects—Democratic Republic of Sudan, A Report to the United Nations University, July 1978.

Thomas, A. S. (1966), Ecology and Human Influence, *In* Davies, W. and Skidmore, C. L., editors, (1966), pp. 46–58.

Tucker, W. (1977), Environmentalism and the Leisure Class—Protecting Birds, Fishes—and Above All Social Privilege, *Harpers* **255**(1531): 49.

Turner, C. (1970), The Middle Pleistocene Deposits at Marks Tey, Essex, *Philosophical Transactions of the Royal Society of London* **B257**: 373–440.

Turner, J. (1964), The Anthropogenic Factor in Vegetational History. I. Treharon and Whixall Mosses, *New Phytologist* **63**: 73–90.

Tyldesley, J. B. (1973), Long Range Transmission of Tree Pollen to Shetland. I. Sampling and Trajectories, *New Phytologist* **72**: 175–181.

UNCOD (1977a), Desertification: Its Causes and Consequences, compiled and edited by the Secretariat of the United Nations Conference on Desertification, Nairobi, Pergamon Press, Oxford.

UNCOD (1977b), United Nations Conference on Desertification. A Report. 29 August–9 September 1977, Nairobi.

UNEP (United Nations Environment Program), (1978), Additional Measures and Means of Financing for the Implementation of the Plan of Action to Combat Desertification, Note by the Executive Director for the Sixth Session of the Governing Council, Nairobi, May 9–25, 1978, UNEP/GC. 619/Add. 2.

UNESCO (1974), Task Force on: the Contribution of the Social Sciences to the MAB Programme, Final Report, Paris, UNESCO, MAB Report no. 17.

UNESCO (1977), Development of Arid and Semi-Arid Lands: Obstacles and Prospects, Paris, UNESCO, MAB Technical Notes 6.

UNESCO (1979), Trends in Research and the Application of Science and Technology for Arid Zone Development, Paris, UNESCO, MAB Technical Notes 10.

UNESCO (1980), Case Studies on Desertification, Paris, UNESCO.

United Nations Centre for Economic and Social Information (1980), The World Conservation Strategy, *Development Forum* **8**(2).

Vinje, J. M. and Vinje, M. M. (1955), Preliminary Aerial Survey of Microbiota in the Vicinity of Davenport, Iowa, *American Midland Naturalist* **54**: 418–432.

Vita-Finzi, C. (1969), The Mediterranean Valleys: Geological Changes in Historical Time, Cambridge University Press, Cambridge.

Wadia, D. N. (1960), The Post-Glacial Desiccation of Central Asia, *Monograph of the National Institute of Science, India* **10**: 1–25.

Waisal, Y. (1972), Biology of Halophytes, Academic Press, New York and London.

Wallace, A., F. C. (1971), Administrative Forms of Social Organization, Addison-Wesley Modular Publications no. 9.

Walter, H. (1974), Die Vegetation Osteuropas, Nord- und Zentralasiens, Stuttgart, Gustav Fischer, Vegetationsmonographien der Einselnen Grossräume VII.

Walter, H. and Kreeb, K. H. (1970), Die Hydratation und Hydratur des Protoplasmus der Pflanzen und ihre Ökophysiologische Bedeutung, Vienna and New York, Springer-Verlag, Protoplasmatiologia II C6.

Walton, K. (1969), The Arid Zones, Chicago, Aldine.

Warren, A. and Maizels, J. K. (1977), Ecological Change and Desertification, *In* UNCOD 1977a, pp. 169–260.

Water Management Research Project (1976), Institutional Framework for Improved On-Farm Water Management in Pakistan. A Special Technical Report, Fort Collins (Colorado), Colorado State University, Water Management Research Project.

Watkins, N. D., Keany, J., Ledbetter, M. T., and Huang, T. C. (1974), Baring High-Albedo Soils by Over-Grazing: A Hypothesized Desertification Mechanism, *Science* **186**:531–536.

Weiss, T. G. and Jordan, R. S. (1976), Bureaucratic Politics and the World Food Conference: The International Policy Process, *World Politics* **28**(3):422–439.

Welkie, G. W. and Caldwell, M. M. (1970), Leaf Anatomy of Species in Some Dicotyledon Families as Related to the C3 and C4 Pathways of Carbon Fixation, *Canadian Journal of Botany* **48**:2135–2146.

Wendorf, F., Schild, R., El Hadidi, N., Close, A. E., Kubusiewicz, M., Wieckowska, H., Issawi, B., and Haas, H. (1979), Use of Barley in the Egyptian Late Palaeolithic, *Science* **205**:1341–1347

Wendorf, F., Schild, R., Said, R., Haynes, C. V., Gautier, A., and Kobusiewicz, M. (1976), The Prehistory of the Egyptian Sahara, *Science* **193**:103–114.

Western, D. and Dunn, T. (1979), Environmental Aspcts of Settlement Site Decisions Among Pastoral Maasai, *Human Ecology* **7**:75–98.

Welten, M. (1957), Über das Glaziale und Spätglazial Vorkommen von *Ephredra* an Nord Westlichen Alpenrand, *Berichte der Schweizen Botanischen Gesellschaft* **67**:33–54.

Westoby, M. (1974), An Analysis of Diet Selection by Large Generalist Herbivores, *American Naturalist* **108**:290–304.

White House, (1964), Report on Land and Water Development in the Indus Plain, Washington D.C., United States, Department of Interior Panel on Waterlogging and Salinity in West Pakistan.

Whitson, R. E. (1975), Range Decision-Making Under Uncertainty—An Illustration, *Journal of Range Management* **28**(4):267–270.

Whyte, A. V. T. (1977), Guidelines for Field Studies in Environmental Perception, MAB Technical Note no. 5, UNESCO, Paris.

Wiebe, H. H. and Walter, H. (1972), Mineral Ion Composition of Halophytic Species from Northern Utah, *American Midland Naturalist* **87**:241–245.

Williams, D. P. (1973), Environmental Archaeology in the Western Negev, *Nature* **242**:501–503.

Wilson, E. O. and Bossert, W. H. (1971), A Primer of Population Biology, Sinaver Associates, Stanford (Conn.).

Winstanley, D. (1973), Rainfall Patterns and General Atmospheric Circulation, *Nature* **245**:190–194.

Wittfogel, K. Z. (1957), Oriental Despotism: A Comparative Study of Total Power, Yale University Press, New Haven.

Woodell, S. R. J., Mooney, H. A. and Hill, A. J. (1969), The Behavior of *Larrea divaricata* (Creosote Bush) in Response to Rainfall in California, *Journal of Ecology* **57**:37–44.

Worthington, E. B. (1958), Science in the Development of Africa, London: Commission for Technical Cooperation in Africa South of the Sahara.

Wright, H. E. (1967), The Use of Surface Samples in Quaternary Pollen Analysis, *Reviews of Paleobotany and Palynology* **2**:321–330.

Young, S. B. and Schofield, E. K. (1973), Pollen Evidence for Late Quaternary Climate Changes on Kergulen Islands, *Nature* **245**:311–312.

Zeist, W. van, Timmers, R. W., and Bottema, S. (1968), Studies of Modern and Holocene Pollen Precipitation in Southeastern Turkey, *Palaeohistoria* **14**:19–39.

Zeist, W. van, Woldring, H. and Stapert, D. (1975), Late-Quaternary Vegetation and Climate of Southwestern Turkey, *Palaeohistoria* **17**:55–143.

Zohary, M. (1973), Geobotanical Foundations of the Middle East (2 vols.), Gustav Fischer Verlag, Stuttgart.

Zummer-Linder, M. (1976), Botswana, *In* Rapp, A., Le Houérou, and Lundholm, B., editors, (1976).

Iran

Adams, R. M. (1962), Agriculture and Urban Life in Early Southwestern Iran, *Science* **136**:109–122.

Afshar, A. (1978), Camels at Persepolis, *Antiquity* **52**:228–231.

Aresvik, O. (1976), The New Agricultural Development of Iran, Praeger, New York.

Barth, F. (1961), Nomads of South Persia: The Basseri Tribe of the Khamseh Confederacy, Little, Brown, and Co., Boston.

Barth, F. (1964), Capital, Investment and Social Structure of a Nomadic Group in South Persia, *In* Capital Saving and Credit in Peasant Societies, Raymond Firth and B. S. Yamey, editors, Aldine, Chicago.

Beck, L. (1980), Herd Owners and Hired Shepherds: The Qashqa'i of Iran, *Ethnology* **19**(3):327–352.

Bharier, J. (1972), The Growth of Towns and Villages in Iran, 1900–1966, *Middle Eastern Studies* **8**:51–61.

Bobek, H. (1959), Features and Formation of the Great Kavir and Masileh Arid Zone Centre, University of Tehran Publication no. 2:143–168.

Bowen-Jones, H. (1968), Agriculture, *In* Fisher, W. B., editor, The Cambridge History of Iran I: The Land of Iran, pp. 565–598, Cambridge University Press, Cambridge.

Bradburd, D. (1980), Never Give a Shepherd an Even Break: Class and Labor Among the Komachi, *American Ethnologist* **7**(4):603–620.

Breckle, S.-W. (1979), Cool Deserts and Shrub Semideserts in Afghanistan and Iran, *In* Ecosystems of the World, Volume 5, Amsterdam, Elsevier Scientific.

Breckle, S.-W. (1981), The Time-Scale of Salt-Accumulation in a Desert Hydrotope in North-Central Iran, in Tübinger Atlas des Vorderen Orients.

Centlivres-Demont, M. (1971), Une Communauté de Potiers en Iran, Weisbaden, Dr. Ludwig Reichert Verlag.

English, P. W. (1966), City and Village in Iran—Settlement and Economy in the Kirman Basin, University of Wisconsin Press, Madison (Wisconsin).

Enloe, C. H. (1977), Iran: The Politics of Pollution and Energy Development, *In* Kelley, D., editor, The Energy Crisis and the Environment: An International Perspective, pp. 218–232, Praeger, New York.

Firouz, E. and Harrington, F. A. Jr. (1976), Iran: Concepts of Biotic Community Reservation, in IUCN, Ecological Guidelines for the use of Natural Resources in the Middle East and South West Asia, Morges (Switzerland), International Union for Conservation of Nature and Natural Resources, IUCN Publication New Series—No. 34, pp. 147–169.

Flower, D. J. (1968), Water Use in North-East Iran, *In* Fisher, W. B., editor, The Cambridge History of Iran I: The Land of Iran, pp. 599–610, Cambridge University Press, Cambridge.

Freitag, H. (1977), Notes to the Provisional Vegetation Map of the Touran Protected Area in Iran 1977.

Freitag, H. (in press), Human Impact on the Vegetation of Central Iran, Exemplified by the Northern Fringe of the Great Kavir, Tübinger Atlas Des Vorderen Orients.

Gabriel, A. (1935), Durch Persiens Wüsten, Strecker Schroder, Stuttgart.

Horne, L. (1980a), Dryland Settlement Location: Social and Natural Factors in the Distribution of Settlements in Turan, *Expedition* **22**(4);11–17.

Horne, L., (1980b), Settlement and Environment in Tauran, *In* Spooner *et al.* (1980), pp. 26–64.

Horne, L. (1980c), Village Morphology: The Distribution of Structures and Activities in Turan Villages, *Expedition* **22**(4):18–23.

Horne, L. (in press), Recycling in an Iranian Village, Archaeology.

Horne, L. (in preparation), The Spatial Organization of Rural Settlement in Tauran, Iran, Philadelphia, University of Pennsylvanis, Ph.D. Thesis, Department of Anthropology.

Iran. (1977), Department of the Environment, Case Study on Desertification, Iran: Turan, Department of Environment. See UNESCO 1980.

Iran. (1961), Ministry of the Interior, National and Province Statistics of the First Census of Iran: November 1956, vol. 1, Tehran, Statistical Center of Iran.

Iran. (1976), Plan and Budget Organization, Center for National Spatial Planning (1976), National Spatial Strategy Plan—Working Document Number 6—Population and Employment, Tehran, Center for National and Spatial Planning.

Iran. (1969), Plan Organization, Village Gazetteer, vol. 7, Tehran, Statistical Center of Iran.

Iran. (1973), Plan Organization, Agricultural Census, Tehran, Statistical Center of Iran.

Krinsley, D. B. (1970), A Geomorphological and Palaeoclimatological Study of the Playas of Iran, Washington D.C., U.S. Government Printing Office.

Lambton, A. K. S. (1969), The Persian Land Reform 1962–1966, Clarendon Press, Oxford.

Middle East Research and Information Project (1976), Land Reform and Agribusiness in Iran, MERIP Report no. 43.

Mojtahedi, Naser (1955), *Zughāl-e Chub va Farāvarda-hā-ye Shimiāi-ye Ān*, Tehran.

Moore, P. D. and Bhadresa, R. (1978), Population Structure, Biomass, and Pattern in a Semi-Desert Shrub, *Zygophyllum eurypterum*, in the Turan Biosphere Reserve of North-Eastern Iran, *Journal of Applied Ecology* **15**:837–845.

Nyerges, A. E. (1977), Traditional Pastoralism in the Middle East: The Ecology of Domesticated Sheep and Goats in the Turan Biosphere Reserve, Iran, Philadelphia, University of Pennsylvania, M.A. Thesis, Department of Anthropology (unpublished).

Nyerges, A. E. (1979), The Ecology of Domesticated Animals Under Traditional Pastoral Management: Preliminary Results from the Turan Biosphere Reserve, Iran, London, Overseas Development Institute, Pastoral Network Paper 9d.

Nyerges, A. E. (1980), The Impact of Settlement in Vegetation with Special Reference to the Role of Pastoralism, *In* Spooner, B. *et al.* (1980), pp. 161–211.

O'Donnell, T. (1980), Garden of the Brave in War—Recollections of Iran, New York, Ticknor and Fields.

Okazaki, Shoko (1968), The Development of Large-Scale Farming, *In* Iran: The Case of the Province of Gorgan," The Institute of Asian Economic Affairs, Tokyo (Japan)

Prickett, M. E. (1979), Settlement and the Development of Agriculture in the Rud-i Gushk Drainage, Southeastern Iran, in German Archaeological Institute, Tehran Chapter, editor, Archaeologisch Mitteilungen aus Iran, vol. 6 (7th International Congress for Iranian Art and Archaeology), Berlin, Dietrich Reimen, pp. 44–56.

Rechinger, K.-H. (1963–), Flora Iranica, Akademische Druk-u. Verlagsanstalt, Graz.

Rechinger, K.-H. (1977), Plants of the Touran Protected Area, Iran, *The Iranian Journal of Botany* **1**:155–180.

Rechinger, K.-H. and Wendelbo, P. (1976), Plants of the Kavir Protected Region, Iran, *Iranian Journal of Botany* **1**(1):23–56.

Ruttner, A. W. and Ruttner-Kolisko, A. E. (1972), Some Data on the Hydrology of the Tabas-Shirgesht-Ozbak-Kuh Area (East Iran), *Jahrb. Geol. B.-A.* **115**:1–48.

Ruttner-Kolisko, A. (1964), Kleingewässer am Ostrand der Persischen Salzwüste—Ein Beitrag zur Limnologie Arider Gebiete, *Verh. Internat. Vereinig. Limnol.* **15**:201–208.

Ruttner-Kolisko, A. (1966), The Influence of Climatic and Edaphic Factors on Small Astatic Waters in the East Persian Salt Desert, *Verh. Internat. Vereinig. Limnol.* **16**:524–531.

Sabeti, H. (1976), Forests, Trees, and Shrubs of Iran, Tehran, Ministry of Agriculture and Natural Resources (in Persian).

Sandford, S. (1977b), Pastoralism and Development in Iran, London, Overseas Development Institute, Pastoral Network Paper 3C.

Schulz, A. (1979), Food in Iran: The Politics of Insuffiency, *In* Hopkins, R. F., Puchala, D. J., and Talbot, R. B., editors, Food, Politics, and Agricultural Development, Westview Press, Boulder (Colorado).

Spooner, B. (1976), Flexibility and Interdependence in Traditional Pastoral Land Use Systems: A Case for the Human Component in Ecological Studies for Development (Touran), in IUCN, Ecological Guidelines for the Use of Natural Resources in the Middle East and South West Asia, Morges (Switzerland), International Union for Conservation of Nature and Natural Resources, IUCN Publication New Series no. 34, pp. 86–93.

Spooner, B. (1982), Irrigation, Encyclopaedia Iranica.

Spooner, B. and Horne, L., editors (1980), Cultural and Ecological Perspectives for the Turan Program, Iran, *Expedition* **22**(4).

Spooner, B., Horne, L., Martin, M., Nyerges, A. E., and Hamlin, C., (1980), The Ecology of Rural Settlements in Arid Lands: A Pilot Study. The Environmental Impact of Settlement in Tauran, A report prepared for UNESCO's Programme on Man and the Biosphere Project no. 11, Paris.

Uhart, E. (1952), Rapport au governement de l'Iran sur la carbonisation, FAO Rapport no. 44, FAO/52/10/6281.

UNDP/FAO (1972), Contribution of National Rangelands to the Economy of Iran, Rome, UNDP/FAO, (AGP:SF/Iran Technical Report 2).

Wasylikova, K. (1967), Late Quaternary Plant Macrofossils from Lake Zeribar, Western Iran, *Review of Palaeobotany and Palynology* **2**:313–318.

Wendelbo, P. (1976), An Annotated Check List of the Ferns of Iran, *Iranian Journal of Botany* **1**(1):11–17.

Woollam, V. J. (1977), Iranian Village Study: Feasibility of Using Solar Energy (Revised version, 1980), Report prepared for MERC, Arya-Mehr University of Technology, Tehran.

Wright, H. E., McAndrews, J. H. and van Zeist, W. (1967), Modern Pollen Rain in Western Iran, and Its Relation to Plant Geography and Quaternary Vegetational History, *Journal of Ecology* **55**:415–443.

Wulff, H. E. (1966), The Traditional Crafts of Persia, MIT Press, Cambridge (Mass.).

Zeist, W. van (1967), Late Quaternary Vegetation History of Western Iran, *Review of Palaeobotany and Palynology* **2**:301–311.

Zeist, W. van and Bottema, S. (1977), Palynological Investigations in Western Iran, *Palaeohistoria* **19**:19–95.

Zeist, W. van and Wright, H. E. (1963), Preliminary Pollen Studies at Lake Zeribar, Zagros Mountains, South-West Iran, *Science* **140**:65–67.

India

Agrawal, D. P. and Pande, B. M., editors, (1977), Ecology and Archaeology of Western India, Delhi, Concept Publishing Company.

Alavi, H. (1972), Kinship in West Punjab Villages, *Contributions to Indian Sociology* n.s. **6**:1–27.

Allchin, B., Goudie, A., and Hegde, K. (1978), The Prehistory and Palaeogeography of the Great Indian Desert, Academic Press, London and New York.

Allchin, B., Hagde, K. T. M., and Goudie, A. (1972), Prehistory and Environmental Changes in Western India. A Note on the Budha Pushkar Basin, Rajasthan, *Man and Environment* **74**:542–564.

Anon. (1961), Report of the State Land Utilisation Committee, Government of Rajasthan, Government Press, Jodhpur.

Anon. (1963), The Pattern of Rural Development in Rajasthan, Evaluation Organisation, Government of Rajasthan.

Anon. (1965), Accumulation of Desert Sand, *Agricultural Research*, **5**(3), ICAR, New Delhi.

Anon. (1967), Quarterly Report on Preliminary Geohydrological Survey of Jodhpur-Nagaur Sector, Report S,& I.2 Rajasthan Groundwater Board, Jodhpur.

The Arid Zone Research Association of India (1977), Annals of Arid Zone 16(3)—Special Issue on Desertification.

Bhadani, B. L. (1980), Economic Conditions in Pargana Merta (Rajasthan), in Devra, G. L., editor, Some Aspects of Socio-Economic History of Rajasthan, Jodhpur, Rajasthan Sahitya Mandir, pp. 113–129.

Bhati, N. S., editor (1969), Marwar ra parganan ri Vigat, Parts I, II, and III, Jodhpur, Rajasthan Oriental Research Institute.

Bhatt, V. V. (1978), Decision Making in the Public Sector: Case Study of Swaraj Tractor, *Economic and Political Weekly* (Bombay) **13**(12):30 and 45.

Bryson, A. A. and Barreis, D. A. (1967), Possibilities of Major Climatic Modifications and their Implications: Northwest India, A Case for Study, *Bulletin of the American Meteorological Society* **48**(3):136–142.

CAZRI (1966), Scientific Progress Report 1964–65, Jodhpur, Central Arid Zone, Research Institute (mimeo).

CAZRI (1974), Basic Resources of Bikaner District (Rajasthan), Jodhpur, Central Arid Zone Research Institute.

CAZRI (1977), Case Study on Desertification, Luni Development Block, India, prepared by the Central Arid Zone Research Institute (Jodhpur, India) for the United Nations Conference on Desertification. See UNESCO 1980.

CAZRI (1978), Arid Zone Research in India (1952–1977), Souvenir, Jodhpur, Central Arid Zone Research Institute.

Datta, R. K. and George, C. J. (1964), Some Climatic and Synoptic Features of the Arid Zone of West Rajasthan, *In* Proceedings of the Symposium on Problems of the Indian Arid Zone, Jodhpur.

Dhir, R. P. (1977a), Palaeo-Climatic Inferences from Quaternary Pedogenetic Processes in Arid Zones, *In* Agrawal, D. P. and Pandey, B. M., editors, Ecology and Archaeology of Western India, New Delhi, Concept Publishing Co.

Dhir, R. P. (1977b), Saline Waters—Their Potentiality as a Source of Irrigation, *In* Jaiswal, P. L., editor, (1977), pp. 130–148.

Dhir, R. P. (1977c), Soil Degradation Due to Over-Exploitative Human Effort, *Annals of Arid Zone* **16**:321–330.

Dhir, R. P. (1977d), Western Rajasthan Soils: Their Characteristics and Properties, *In* Jaiswal, P. L., editor, (1977), pp. 102–115.

Dhir, R. P. and Kolarkar, A. S. (1977), Observations on Genesis and Evolution of Arid Zone Soils, Journal of the Indian Society of Soil, *Science* **25**:260–264.

Dhir, R. P., Singh, N. and Sharma, B. K. (1977), Some Observations on the Behaviour of Soils and Sediments Under Late Quaternary Environment and Human Settlement in Western Rajasthan, *Man and Environment* **1**:21–24.

Doherty, V. S. and Jodha, N. S. (1977), Conditions for Group Action Among Farmers, Hyderabad (India), International Crops Research Institute for the Semi-Arid Tropics, Economics Program Occasional Paper no. 19.

Geertz, C. (1975), Common Sense as a Cultural System, *The Antioch Review* **33**:5–26.

Ghose, B., Kar, A., and Hussain, Z. (1980), Comparative Role of the Aravalli and the Himalayan River Systems in the Fluvial Sedimentation of the Rajasthan Desert, *Man and Environment* **4**:8–12.

Ghose, B., Pandey, S., and Singh, S. (1968), Processes and Extent of Erosion and Its Effect on Land Use in the Central Luni Basin, Western Rajasthan, *Annals of Arid Zone* **7**(1):15–30.

Ghose, B., Pandey, S., Singh, S., and Lal, G. (1966), Geomorphology of the Central Luni Basin, Western Rajasthan, *Annals of Arid Zone* **5**:1–25.

Ghose, B., Singh, S., and Kar, A. (1977), Geomorphology of the Rajasthan Desert, *In* Jaiswal, P. L., editor (1977), pp. 69–76.

Ghosh, A. (1952), The Rajputana Desert—Its Archaeological Aspect, Bulletin of the National Institute of Science, India (*Proceedings of the Symposium on Rajputana Desert*) **1**:37–42.

Goudie, A. S., Allchin, B., and Hegde, K. T. M. (1973), The Former Extensions of the Great Indian Sand Desert, *Geographical Journal* **139**:243–257.

Gupta, R. K. and Saxena, S. K. (1971), Ecological Studies on the Over-Grazed Rangelands in the Arid Zone of West Rajasthan, *Journal of the Indian Botanical Society* **50**:289–300.

Gupta, R. K. and Saxena, S. K. (1972), Potential Grassland Types and Their Ecological Succession in Rajasthan Desert, *Annals of Arid Zone* **11**: 198–218.

Gustafson, W. E. and Reidinger, R. B. (1971), Delivery of Canal Water in North India and West Pakistan, *Economic and Political Weekly* (special issue: Review of Agriculture) **6**(52):A-157–62.

ICAR (1965), Accumulation of Desert Sand, New Delhi, Indian Council of Agricultural Research, *ICAR Agricultural Research* **5**(3).

ICAR (1979), 50 Years of Agricultural Research and Education, ICAR Golden Jubilee 1929–1979, New Delhi, ICAR.

Jain, K. C. (1972), Ancient Cities and Towns of Rajasthan, Delhi, Moti Lal Banarsi Das.

Jaiswal, P. L., editor (1977), Desertification and Its Control, New Delhi, Indian Council of Agricultural Research.

Jodha, N. S. (1967), Capital Formation in Arid Agriculture: A Study of Resource Conservation and Reclamation Measures Applied to Arid Agriculture, Jodhpur, University of Jodhpur, Ph.D. Thesis (unpublished).

Jodha, N. S. (1970), Land Policies of Rajasthan: Some Neglected Aspects, *Economic and Political Weekly* (Quarterly Review Agriculture) **5**(26):A81–A84.

Jodha, N. S. (1972), A Strategy for Dry Land Agriculture, *Economic and Political Weekly* (Quarterly Review of Agriculture) **7**(13):A7–A12.

Jodha, N. S. (1974), A Case of the Process of Tractorisation, *Economic and Political Weekly* (Quarterly Review of Agriculture) **9**(52):A111–A118.

Jodha, N. S. (1975), Famine and Famine Policies: Some Empirical Evidence, *Economic and Political Weekly* **10**(40):1609–1623.

Jodha, N. S. (1980), The Process of Desertification and the Choice of Interventions, *Economic and Political Weekly* **15**(32):1351–1356.

Jodha, N. S. and Purohit, S. D. (1971), Weather and Crop Instability in the Dry Region of Rajasthan, *Indian Journal of Agricultural Economics* **26**(4):286–295.

Jodha, N. S. and Vyas, V. S. (1969), Conditions of Stability and Growth in Arid Agriculture, Vallabh Vidyanagar, Gujarat, Agro-Economic Research Centre.

Kalla, J. C. (1977), Statistical Evaluation of Fuel Yield and Morphological Variates for Some Promising Energy Plantation Tree Species in Western Rajasthan, *Annals of Arid Zone* **16**:117–126.

Kar, A., Singh, S., and Ghose, B. (1977), Geomorphological Characteristics and Palaeo-Climatic Significance of Rhyolite Pediments in Jodhpur District, Western Rajasthan, *Man and Environment* **1**:16–20.

Kaul, R. N. and Misra, D. K. (1961), Land Utilisation Problem in Arid Zone and Its Significance in Soil Conservation, Proceedings of the National Academy of Sciences, India, Section B31 P III.

Krishnan, A. (1968), Distribution of Arid Areas in India, in 21st International Geographical Congress, Proceedings of the Symposium on the Arid Zone, Jodhpur, pp. 11–19.

Krishnan, A. K. and Thanvi, K. P. (1969), Water Budget in the Arid Zone of Rajasthan During 1941–1960, *Annals of Arid Zones* **8**(2):291–299.

Krishnan, A. K. and Thanvi, K. P. (1971), In Rajasthan. Irrigation Department, Souvenir, Jaipur, Rajasthan Irrigation Dept.

Krishnan, M. S. (1952), Geological History of Rajasthan in its Relation to Present Day Conditions, *Bulletin of the National Institute of Science, India* (Proceedings of the Symposium on Rajputuna Desert) **1**:19–31.

Malhotra, S. P. (n.d.), Socio-Economic Structure of Population in Arid Rajasthan, Jodhpur, Central Arid Zone Research Institute.

Malhotra, S. P. (1977), Socio-Demographic Factors and Nomadism in the Arid Zone, *In* Jaiswal, P. L., editor, 1977, pp. 310–323.

Mann, H. S., editor (1977), Desert Ecosystem and Its Improvement, Central Arid Zone Research Institute, Jodhpur.

Mann, H. S. (1980), Arid Zone Research and Development, Scientific Publishers, Jodhpur.

Mann, H. S., Malhotra, S. P., and Shankarnarayan, K. A. (1977), Land and Resource Utilization in the Arid Zone, *In* Jaiswal, P. L., editor, (1977), pp. 89–101.

Meher-Homji, V. M. (1973), In the Sind-Rajasthan Desert the Result of Recent Climatic Changes, *Geoforum* **15**:47–57.

Mehra, T. N. and Sen, A. K. (1977), Groundwater Resources of the Arid Zone of India, *In* Jaiswal, P. L., editor, (1977), pp. 156–159.

Misra, V. N. (1962), Palaeolithic Culture of Western Rajputana, *Bulletin of the Deccan College Research Institute* **21**:86–156.

Misra, V. N. (1977), Evolution of the Pattern of Human Settlement in the Arid and Semi-Arid Regions, *In* Jaiswal, P. L., editor, (1977), pp. 10–24.

Misra, V. N. and Rajaguru, S. N. (1975), Some Aspects of Quaternary Environment and Early Man in India, Workshop, Problems of the Deserts in India, Geological Survey of India, Jaipur.

Misra, V. N., Rajaguru, S. N., Agrawal, D. P., Thomas, P. K., Husain, Z., and Dutta, P. S. (1980), Prehistory and Palaeoenvironment of Jayal Western Rajasthan, *Man and Environment* 4:19–31.

Mohapatra, G. C., Bhatia, S. B., and Sahu, B. K. (1963), The Discovery of a Stone Age Site in the Indian Desert, *Research Bulletin of the Punjab University, Science Section* (n.s.) 14 (3 and 4):205–223.

NCAER (1965), Land and Livestock in Rajasthan, New Delhi, National Council of Applied Economic Research.

Piggot, S. (1950), Pre-Historic India, Penguin Books, Harmondsworth.

Prakash, M. and Ahuja, L. D. (1964), Studies on Different Range Condition Class Grasslands in Western Rajasthan, *Annals of Arid Zone* 3(1–2):91–98.

Raheja, P. C. and Sen, A. K. (1964), Resources in Perspective, Recent Developments in Rajasthan (Souvenir volume), Central Arid Zone Research Institute, Jodhpur.

Rajasthan, India. Evaluation Organization (1963), The Pattern of Rural Development in Rajasthan, Jaipur, Government of Rajasthan.

Rajasthan, India. Groundwater Board (1967), Quarterly Report on Preliminary Geohydrological Survey of Jodhpur-Nagour Sector, Jodhpur, Rajasthan Groundwater Board, Report S and I.2.

Rajasthan, India. State Land Utilisation Committee (1961), Report, Rajasthan Government Press, Jodhpur.

Rao, A. V. and Mehta, D. J. (1977), Desalination Techniques for Conversion of Brackish Water into Potable Water for Small Communities, *In* Jaiswal, P. L., editor, (1977), pp. 149–155.

Reidinger, Richard B. (1974), Institutional Rationing of Canal Water in Northern India: Conflict Between Traditional Patterns and Modern Needs, *Economic Development and Cultural Change* 23:79–104.

Roy, B. B. and Pandey, S. (1971), Expansion or Contraction of the Great Indian Desert, *Proceedings of the Indian National Science Academy* 36B(6):331–343.

Sandford, S. (1978), Some Aspects of Livestock Development in India, London, Overseas Development Institute, Pastoral Network Paper 5C.

Saxena, S. K. (1977), Desertification Due to Ecological Changes in the Vegetation of the Indian Desert, *Annals of Arid Zone* 16:367–373.

Seth, S. P. and Mehta, N. K. (1963), Fertility Survey and Soil Test: Summaries of Some Districts of Arid Regions of Rajasthan, *Annals of Arid Zone* 2(1):61–68.

Sharma, D. (1966), Rajasthan Through the Ages. Rajasthan State Archives, Bikaner.

Sharma, R. C. (1972), Settlement Geography of the Indian Desert, New Delhi, Kumar Bros.

Singh, G. (1971), The Indus Valley Culture Seen in the Context of Post-Glacial Climatic and Ecological Studies in North-West India, *Archaeology and Physical Anthropology in Oceania* 6(2):177–189.

Singh, G. (1977), Climatic Changes in the Indian Desert, *In* Jaiswal, P. L., editor, (1977), pp. 25–30.

Singh, G., Chopra, S. K., and Singh, A. B. (1973), Pollen Rain from the Vegetation of North-West India, *New Phytologist* 72:191–206.

Singh, G., Joshi, R. D., Chopra, S. K., and Singh, A. B. (1974), Late Quaternary History of Vegetation and Climate of the Rajasthan Desert, India, *Philosophical Transactions of the Royal Society of London* B267:467–501.

Singh, S. (1977), Sand Dunes and Palaeo-Climate in Jodhpur District, Western Rajasthan, *Man and Environment* **1**:7–15.

Singh, S., Pandey, S., and Ghose, B. (1971), Geomorphology of Middle Luni Basin, *Annals of Arid Zone* **10**:1–14.

Singh, G., Kershaw, A. P. and Clark, R. (1981). Quaternary Vegetation and Fire History in Australia. *In* Fire and the Australian Biota. Ed. (Gill, A. M., Groves, R. H. and Nobel, I. R.), pp. 23–54. Aust. Academy of Science, Canberra, Australia.

Stein, Sir A. (1942), A Survey of Ancient Sites, Along the Lost Saraswati River, *Geographical Journal* **9**:173–182.

Tod, Col. J. (1957, first published 1829–32), Annals and Antiquities of Rajasthan I and II (2 vols.), Routledge and Kegan Paul, London.

Vishnu Mittre (1974), Plant Remains and Climate from the Late Harappan and Other Chalcolithic Cultures of India—A Study in Interrelationships, *Geophytology* **4**:46–53.

Wadia, D. N. (1960), The post-glacial desiccation of Central Asia, New Delhi, Monographs of the National Institute of Science.

Index